THE EVOLUTIONARY PROCESS

A Critical Review of Evolutionary Theory

Verne Grant

Columbia University Press
New York   1985

Library of Congress Cataloging in Publication Data

Grant, Verne.
The evolutionary process.

Bibliography: p.
Includes indexes.
1. Evolution. I. Title.
QH366.2.G67 1985   575   85-7733
ISBN 0-231-05752-0
ISBN 0-231-05753-9 (pbk.)

Columbia University Press
New York   Guildford, Surrey
Copyright © 1985 Columbia University Press
All rights reserved
Printed in the United States of America

This book is Smyth-sewn and printed on permanent and durable acid-free paper.

To Karen

Other Books by the Same Author

Natural History of the Phlox Family   (1959)
The Origin of Adaptations   (1963)
The Architecture of the Germplasm   (1964)
Flower Pollination in the Phlox Family.
 With Karen A. Grant   (1965)
Hummingbirds and Their Flowers.
 With Karen A. Grant   (1968)
Genetics of Flowering Plants   (1975)
Organismic Evolution   (1977)
Plant Speciation   (1971, 1981)

Contents

Preface

The primary objective of this book is to describe and discuss the processes that bring about evolutionary change and the factors that affect these processes. Evolutionary processes are considered at various levels from local populations to major groups, and on different time scales from generation time to geological time. The focus throughout is on evolution in whole organisms, especially animals and plants, rather than on molecular changes or mathematical models.

Numerous controversies are simmering and bubbling in evolutionary biology at the present time. A second objective of this book is to highlight these controversies and analyze the issues involved.

A third objective was to keep the treatment concise while covering a broad range of subjects. This objective could only be achieved by emphasizing general principles and limiting the number of examples. The corollary is that many good examples and unessential details had to be omitted.

Literature sources for statements made in the text are given in the usual form, that is, as parenthetical author/year citations in the text. In addition I have used footnotes to guide the reader to further readings on various subjects.

The book was not written for any one class of readers exclusively. It was written around the subject. It should prove readable to a spectrum of readers ranging from advanced biology students to professional workers in evolutionary biology and bordering fields.

This is my third generalized treatment of organic evolution, the preceding ones being *The Origin of Adaptations* (1963) and *Organismic Evolution* (1977). It scarcely needs to be said that many changes have occurred in the subject itself, in my perception of it, and in my way of dealing with it during the intervals between volumes. On the other hand, some subjects have required little modification through successive treatments. *The Evolutionary Process* is the successor to *Organismic Evolution* (Freeman), and is similar to it in many places, but differs in level, policy, and range of topics, as well as in the inclusion of the more recent literature.

My secretary and aide, Kathleen Feeley, typed the manuscript with care and accuracy, and hunted up special items in the library for me. Karen A. Grant read chapters 31–39 in manuscript. Joan McQuary shepherded the manuscript through the press with skill and understanding. Several book publishers granted permission to use previously published illustrations as noted in the respective captions. To all these parties my sincere thanks.

University of Texas   V.G.
Austin, Texas
July 1984

PART I
Introduction

CHAPTER 1
The Problem

Organic Diversity
Adaptation
The Adaptive Landscape

The world of living organisms exhibits several general features which have always aroused feelings of wonder in mankind. The first of these general phenomena is the great structural complexity of organisms. The second is the apparently purposive or adaptive nature of many of the characteristics of these organisms. The third striking general feature is the existence of a tremendous diversity of forms of life. The problem of biological complexity and adaptation is thus compounded by the fact that there are many diverse kinds of organisms with these properties in the world.

The questions evoked by these phenomena are obvious. How have complex organisms come into being? What forces have molded their adaptive characteristics? How has organic diversity originated and how is it maintained? To which can be added the special but relevant questions: What is the place of mankind in the organic world, and what is the ancestry of man?

In all ages man has sought intellectually satisfying answers to these questions. In pre-scientific societies the explanations have taken the form of myths. Some of these have been carried over into the world religions. The scientific explanations are embodied in the theory of evolution. Before entering into our discussion of evolutionary theory, however, let us outline the problems that this theory has to explain in somewhat greater detail.

Organic Diversity

There are some 3,700 species of mammals and 8,600 species of birds in the Recent fauna. About 20,000 species of Recent fishes have been described. The living vertebrates as a collective whole comprise about 42,000 known species.

The number of known Recent species rises to greater heights in some other dominant major groups: to about 107,000 in the mollusks, 215,000 in the flowering plants, and 750,000 in the insects. Estimates of the species diversity in the various major groups of organisms are summarized in tables 1.1 and 1.2. The tables show that the total number of known species of organisms in the modern world is approximately 1.4 million.

The Herculean task of taxonomic exploration and description is well advanced in the birds and mammals, but is far from finished in most other groups. Substantial numbers of marine invertebrates, flowering plants, and other groups remain to be described taxonomically. Ichthyologists estimate that the number of Recent fish species described is about 20,000, but that the total number of Recent fishes, described and undescribed, is close to

Table 1.1. Estimated number of described Recent species in the animal kingdom (Data from Mayr 1969)

Chordates		43,025
Mammals	3,700	
Birds	8,600	
Reptiles	6,300	
Amphibians	2,500	
Bony fishes	20,000	
Sharks and rays	550	
Jawless fishes	50	
Tunicates and lancelets	1,325	
Arthropods		838,000
Insects	750,000	
Crustaceans	20,000	
Spiders	57,000	
Mollusks		107,250
Gastropods	80,000	
Bivalves	25,000	
Echinoderms		6,000
Segmented worms (Annelida)		8,500
Flatworms (Platyhelminthes)		12,750
Nematodes and relatives		12,500
Coelenterates		5,300
Bryozoans and relatives		3,750
Sponges		4,800
Miscellaneous small groups		ca. 2,050
Total		ca. 1,043,925

Table 1.2. Estimated numbers of described Recent species in four kingdoms

Kingdom and Major Group	Approximate No. Species		Source
Plant kingdom			
Flowering plants	215,000		Cronquist 1981
Gymnosperms	640		Jones 1941
Ferns and fern allies	10,000		"
Mosses and liverwort	23,000		"
Green algae	7,500		Bold et al. 1980
Brown, red, and golden-brown algae	11,400		"
Total		267,540	
Fungus kingdom			
True fungi	40,000		Ainsworth and Bisby 1954
Slime molds	4,000		"
Total		44,000	
Protistan kingdom			
Protozoans	28,350		Mayr 1969
Plant flagellates (Euglenoids)	800		Bold et al. 1980
Dinoflagellates	ca. 1,000		Bold, pers. comm.
Total		30,150	
Moneran kingdom			
Blue-green algae	1,500		Bold et al. 1980
Bacteria	1,700		"
Total		3,200	
Grand Total for 4 kingdoms		1,388,815	

40,000. The unfinished business in insect taxonomy bulks larger still. Entomologists suggest that the described species of insects (ca. 750,000) represent only a small fraction, perhaps one-fifth or one-tenth, of the insect species that actually exist in the modern world.

By making some conservative assumptions regarding the proportion of known to unknown species in the various major groups, it is possible to arrive at rough estimates of the total existing species diversity. A fair estimate is that there are at least 4.5 million species of modern organisms (Grant 1963: 81–82).

Two independent estimates have been made of the number of species, both living and extinct, that have existed on earth throughout geological time (Simpson 1952; Grant 1963). The ranges of the two estimates overlap broadly. Using the zone of overlap, one can conclude tentatively that the total organic diversity throughout the history of life is in the neighborhood of one to several billion species.

Adaptation

Many of the hereditary characters of organisms conform to some feature of their normal environment in such a way as to benefit the organism. Such characters are adaptive. The eighteenth-century naturalist Buffon described many adaptive characters in birds, fish, and other animals in the *Histoire Naturelle*. A typical statement is quoted here to illustrate the historical usage (Buffon 1770, 1808):

> As to the external structure of Birds, it is peculiarly adapted for swiftness of motion; it is . . . designed to rise in the air . . . Wide adaptation of means to ends [appears] in the configuration of the feathered race . . .

Darwin's stock example of adaptation was the woodpecker. One statement in *The Origin of Species* will suffice (Darwin 1859: ch. 6):

> Can a more striking instance of adaptation be given than that of a woodpecker for climbing trees and for seizing insects in the chinks of the bark?

One can think of the whole adaptive character combination of woodpeckers in this connection: the chisel bill, the strengthened head bones and head muscles, the extensile tongue with barbed tip, the feet with two pairs of sharp-pointed toes for clinging to a vertical surface, and the stout tail feathers for propping up the body while clinging in a vertical position.

It is useful to distinguish between general adaptations and special adaptations (Simpson 1953a: ch. 6). The former fit the organism for life in some broad zone of the environment; the latter are specializations for some particular way of life. Thus the wing of birds is a general adaptation, while the chisel bill and clinging foot of woodpeckers are special adaptations. General adaptations are among the distinguishing characteristics of major groups of organisms.

General and special adaptations undoubtedly exist and are undoubtedly common and widespread in the living world. Some large classes of phenomena, such as concealing and mimetic coloration in animals and ecotypes in plants, have been shown experimentally to have adaptive value.

But is adaptation universal? Or, do non-adaptive characters occur with substantial frequency? The pendulum of biological opinion has swung back and forth on this question for a century. A tendency to jump to conclusions regarding adaptations occurred in some turn-of-the-century natural history writings and provoked the inevitable counter-reaction. In recent decades many morphologists and taxonomists have questioned the adaptive significance of the so-called "trivial" morphological characters used in diagnostic taxonomic keys. Currently certain schools of molecular biologists and biomathematicians are suggesting that some protein variations are non-adaptive (see King and Jukes 1969; Kimura and Ohta 1971a, 1971b; Kimura 1979, 1981). Recently, a general attack on the alleged overemphasis on adaptation in modern evolutionary theory has been launched by Gould and associates (Gould and Lewontin 1979; Gould 1980, 1982). This attack is poorly founded and poorly documented as pointed out in a reply (Grant 1983).

Where do we stand? It is possible, indeed probable, that some non-adaptive characters exist. There are evolutionary mechanisms for establishing such characters. But identifying them as such is a procedure filled with uncertainties. Some human observers have a tendency to label as non-adaptive any character whose adaptive properties are not obvious to them. Such anthropocentric interpretations of nature can be misleading.

It would not be obvious at first glance that the difference between colored onions and white onions is adaptive. However, red and yellow onions are resistant to a fungus disease, a smudge (*Colletotrichum cincinans*), whereas white onions are susceptible to the same disease organism. The disease resistance is due to the presence in the onion bulbs of catechol and protocatechuic acid, which are toxic to smudge, and these compounds are associated with red or yellow pigments in the bulbs (Jones et al. 1946; Walker and Stahmann 1955; Levin 1971).

The color of onions is not an isolated instance. Time and again, a character which has been considered to be of no adaptive significance has later been found to serve a definite useful function in the life of the organism.

Furthermore, the scope for non-adaptive characters appears to be limited to relatively simple characters. Complex characters are determined by many genes. The assembling of the gene combinations underlying such complex characters requires selection in one form or another. And selection is the way to adaptation. It is a fair presumption, therefore, that complex characters in general are either adaptive in themselves or are correlates or byproducts of adaptive characters.

The Adaptive Landscape

The environment in all its varied aspects can be visualized as an adaptive landscape consisting of hills and valleys (Wright 1932; Dobzhansky 1970). The topography symbolizes the distribution of adaptive fields. The hilltops are adaptive peaks, and the low-lying areas between them constitute a no-man's-land of adaptive valleys. Organisms occupy the various adaptive peaks by virtue of their adaptive character combinations and the underlying gene combinations. Some adaptive peaks are narrow, and others broad, corresponding to the relative breadth of specialization. Some adaptive peaks are low, and relatively easy to climb; others are high and difficult to climb.

Each adaptive peak is occupied by a different species of organism. The low ground between the peaks represents the absence or rarity of interspecific hybrids with inadaptive gene combinations. Each species is more or less isolated on its own adaptive peak. An adaptive landscape containing numerous peaks symbolizes the diversity of environments in the world and the corresponding diversity of organisms.

The metaphor can be extended to embrace species groups and genera. The peaks on the adaptive landscape are not randomly distributed, but are clustered in series of separate ranges. The peaks in one range are likely to be occupied by related species belonging to one genus, while a second range provides the habitat for the members of another genus.

The environment can also be portrayed as a series of adaptive zones grouped into adaptive grids (Simpson 1944, 1953). Thus tree-tops form an adaptive zone for one set of bird species, while the ground is the adaptive zone of another and distinct set of ground-feeding birds, and the tree-tops, trunks, and ground together form an adaptive grid. The adaptive zone may be narrow or broad. Typically the species of a genus occupy different ecological facies of a relatively narrow adaptive zone. By comparison, the members of a family, or most members, generally live in different subdivisions of a broad adaptive zone.

CHAPTER 2

The Solution

Creation or Evolution

For centuries in Western civilization the standard explanation of the phenomena of life highlighted in the preceding chapter was the biblical story of creation. The Book of Genesis (ch. 1 and 2) related that God created plants on the third day (or stage) of creation, animals on the fifth day, and man on the sixth day of creation. Adam and Eve were the parents of all mankind. An analysis of biblical records placed the date of creation at about 4000 B.C.

This account was the official truth, resting on Authority, and it dominated thought to the exclusion of any other viewpoint in Western societies. The monolithic hold of creationism was brought to an end, however, by the publication of Darwin's *Origin of Species* in 1859. After 1859 there was an alternative explanation of life, a scientific theory of evolution based, not on religious authority, but on verifiable evidence.

The general idea of organic evolution was not new in 1859. That idea had been put forward earlier by Lamarck in the *Philosophie Zoologique* (1809), by Chambers in *Vestiges of Creation* (1844), and by others. Lamarck's work was unconvincing as to both factual evidence and theory, and so were the treatments of the other early nineteenth-century authors. These treatments did, however, prepare the way for the reception of Darwin's *Origin of Species*.[1]

1. For good recent accounts of the history of the evolution theory see Harris (1981), Mayr (1982a), and Ruse (1979, 1982).

In *The Origin of Species*, Darwin marshaled an overwhelming body of evidence, derived from several independent lines of investigation, pointing to the past and continuing occurrence of organic evolution. And he developed the theory of natural selection to provide a motive force for the evolutionary changes. Darwin's treatment of the subject, unlike those of previous authors, was convincing to zoologists, botanists, and geologists. The years and decades after 1859 saw a widespread acceptance of the evolution theory throughout the scientific world. There was, to be sure, much controversy after 1859, but the controversy in that period took place mainly in lay circles and involved theologians and religious groups rather than scientists.

The Evidence for Evolution

Let us give a brief summary here of the various lines of evidence that point to the occurrence of evolution. Most of the lines of evidence listed below were brought to bear on the question by Darwin in *The Origin of Species*. The amount of evidence in each line is, of course, much greater today than it was in Darwin's time. A sample of the evidence in each category will be presented in later chapters of this book.

1. Direct observation. Evolutionary changes within populations and some modes of species formation are established by direct observational and/or experimental evidence.

2. Extrapolation to larger groups. There is a continuous series of levels from the population and geographical race through the species to the species group, subgenus, and genus. To anyone who has studied living nature and has become imbued with a sense of the unity of nature, it is inconceivable that organic diversity should be produced by evolution at the lower levels but by some other means at supraspecific levels. Separate hypotheses to account for the origin of low-level and high-level organic groups are unwarranted.

3. Fossil record. Many groups of organisms with good fossil records show a succession of forms through geological time. In some cases transitional forms between two different major groups are preserved as fossils.

4. Taxonomic pattern of relationships among living species. Species are clustered naturally into genera, genera into families, families into orders, and so on. This "natural subordination of organic beings in groups under groups," as Darwin put it, is a result of a branching phylogeny. A natural genus Y is composed of species with a common ancestry; a related genus Z contains

another set of species; the genera Y and Z are two branches of an older branch, the family; and so on through the taxonomic hierarchy. The hierarchical nature of taxonomic relationships was not invented by evolutionary biologists; in fact it was discovered by pre-evolutionary taxonomists who accepted creationism, but it was later correctly interpreted by evolutionists. Living species would not be expected to cluster in groups within groups if they were products of separate acts of creation.

5. Geographical distribution. Many genera, families, and other groups of medium taxonomic rank are confined to one geographical region—a particular archipelago, subcontinent, or continent, etc. The group has its center of distribution in this region. Meanwhile another isolated geographical region harbors a distinct organic group. Thus the hummingbirds are an American family of birds and the Hawaiian honeycreepers a Hawaiian family. The logical inference in such cases is that the species belonging to each group, or at least many of them, evolved in the region of their present abundance and diversity. The doctrine of creation provides no explanation for the observed patterns of geographical distribution of supraspecific groups.

6. Homology. When the members of a major group are compared, they are found to possess a common general plan of structural organiation, but to differ with respect to certain homologous parts in the body. One can think of the different forms taken by the homologous forelimbs in different orders of mammals, of the homologous hindlimbs in different mammalian orders, of the limbs in different classes of terrestrial vertebrates, of the corolla parts in the flowers of different angiosperm families and orders, and so on.

Let Darwin discuss the significance of homologous organs in his own words (1872: ch. 14):

> What can be more curious than that the hand of a man, formed for grasping, that of a mole for digging, the leg of the horse, the paddle of the porpoise, and the wing of the bat, should all be constructed on the same pattern, and should include similar bones, in the same relative positions? . . .
>
> Nothing can be more hopeless than to attempt to explain this similarity of pattern in members of the same class, by utility or by the doctrine of final causes. The hopelessness of the attempt has been expressly admitted by Owen in his most interesting work on the 'Nature of Limbs.' On the ordinary view of the independent creation of each being, we can only say that so it is;—that it has pleased the Creator to construct all the animals and plants in each great class on a uniform · plan; but this is not a scientific explanation.
>
> The explanation is to a large extent simple on the theory of the selection of successive slight modifications,—each modification being

profitable in some way to the modified form, but often affecting by correlation other parts of the organisation. In changes of this nature, there will be little or no tendency to alter the original pattern, or to transpose the parts. The bones of a limb might be shortened and flattened to any extent, becoming at the same time enveloped in thick membrane, so as to serve as a fin; or a webbed hand might have all its bones, or certain bones, lengthened to any extent, with the membrane connecting them increased, so as to serve as a wing; yet all these modifications would not tend to alter the framework of the bones or the relative connection of the parts. If we suppose that an early progenitor—the archetype as it may be called—of all mammals, birds, and reptiles, had its limbs constructed on the existing general pattern, for whatever purpose they served, we can at once perceive the plain signification of the homologous construction of the limbs throughout the class.

7. Vestigial organs. Some members of a major group often possess an organ that is atrophied and non-functional. This rudimentary organ is homologous with a well-developed functional organ in other members of the same group. Thus flightless birds have rudimentary wings and some snakes, including the python, had rudimentary hindlimbs. These structures are interpreted as reduced vestiges of their well-developed homologues in other members of the same major group. The subgroup possessing the rudimentary organ entered into a habitat or way of life in which the formerly functional organ was no longer useful, and it was greatly reduced by selection, but vestiges of it persist as phylogenetic remnants. There is no good explanation for the existence of useless rudimentary organs in the doctrine of creationism.

8. Biochemical similarities. A modern line of evidence, not available in Darwin's time, is the close similarity in the biochemical composition and molecular structure of homologous proteins in members of different related families or orders. Good examples are the homologous forms of hemoglobin and of cytochrome c in humans, apes, and monkeys.

It should be noted, finally, that the case for evolution rests, not on one or two lines of evidence alone, but on the concurrent testimony of several independent classes of facts, as Darwin showed in *The Origin of Species*. Concurrence between separate lines of evidence, or consilience, is a powerful argument for the truth of the conclusions, as noted by Ruse (1979, 1982).

The Creationism Controversy

The publication of Darwin's *Origin of Species* sparked a bitter public controversy between creationism and evolutionism in England, Europe, and North America during the late nineteenth century.[2] The ranks of creationists have dwindled since then, as laymen have generally come to accept the conclusions of science.

In the United States, however, certain fundamentalist religious sects have continued to insist on a literal interpretation of the biblical account of creation. These sects have continuously and militantly opposed the teaching of evolution in the public schools.[3] A state law in Tennessee in the 1920s, making it a crime to teach evolutionary ideas, led to the famous Scopes trial, in which a high-school teacher was tried and found guilty (and given a light fine).

The fundamentalists have changed their tactics but not their aims in the modern era. The focus is still on the teaching of biology in public schools. Fundamentalists now want to see an even-handed presentation of the biblical creation story and Darwinian evolution in biology textbooks and classrooms. The former is dressed up as "scientific creationism" for the occasion. Furthermore, evolution should be played down in the schools and presented as "a theory rather than a fact." The fundamentalist groups have influenced state school boards along the above lines, and again the question has been tried in the courts. In the "Scopes Trial II" in Arkansas in 1981–1982 a state act requiring "balanced treatment for creation-science and evolution-science" was found unconstitutional.[4] At this writing the fundamentalists are waging their pro-creationist campaign at the level of local school boards.

Since this controversy is certain to continue in the future, anachronistic as it is, some comments on the merits of late twentieth-century creationism may be in order.

The demand for equal time in textbook and classroom presentations of creationism and evolution is based on the premise that the two subjects are commensurate—that they warrant equal consideration—and this premise is false. Creationism and evolutionism are not in the same league. Creationism is not a scientific theory that can be weighed against the evolution the-

2. For readings on this period see F. Darwin (1958), Eiseley (1958), Fothergill (1952), Glick (1974), and Ruse (1979).
3. Creationism in the United States is reviewed by Numbers (1982), and Ruse (1982).
4. The incisive and courageous opinion of Judge Overton in this case has been reprinted in several scholarly journals; see Overton (1982a, 1982b).

ory. It is a religious dogma. There is no independent evidence to support its account of the origin of plant, animal, and human life. The acceptance of the creationist story depends on faith rather than reason.

Let us consider the point, stated so often by creationists, that evolution is a theory, not a fact. We will accept for the moment the naive assumption that a scientific subject can be categorized as either theory or fact. And we will refrain from pointing out that creationism is neither a theory nor a fact.

Evolution was indeed a theory at one time, but a bit of water has flowed over the dam since 1859. Much of evolution is now in the category of verified fact. Evolutionary changes at the levels of microevolution and speciation have been observed by competent biologists and can now be regarded as proven facts.

Evolutionary changes at the level of macroevolution are in a somewhat different position, in that they are historical phenomena and consequently were not observed directly by any human observer. Lack of eye-witness testimony is an inherent problem in any attempt to reconstruct past events. The biblical account of creation, which was written long after the event, suffers from this same difficulty.

The simple dichotomy of theory vs. fact does not cover the situation in macroevolution. Evolutionary history cannot be observed directly, but can be inferred from the fossil record. The fossils themselves are facts. Macro-evolution is a necessary inference from these facts.

Evolutionary Explanations

The salient phenomena of organic complexity, organic diversity, and adaptiveness are explained scientifically as products of the process of evolution. Two types of evolutionary explanation can be discerned: the historical and the causal. The first traces the phylogenetic sequence leading up to the observed end result, whereas the second investigates the causal mechanisms involved. Both approaches are valid and both are necessary. This book is concerned primarily with the causes of evolution, but phylogenetic evidence is frequently and necessarily brought into the picture.

We have to deal with evolutionary phenomena at different levels. Three broad levels are recognized in this book. These are: evolutionary changes within populations (microevolution), evolution of races and species (speciation), and evolution of major groups (macroevolution). The three levels call for different methods of research, which in turn produce different types of evidence.

Most treatments recognize two levels microevolution and macroevolution, and deal with speciation under the former. But the species level is quite different from the local population level. The evolutionary processes are not entirely the same at the two levels and therefore should be discussed separately. A further advantage of the three-level treatment becomes apparent when we treat macroevolution. In the past, explanations of macroevolution have often been sought in population genetics. To a large extent they should. But some crucial elements are likely to be missed in so highly reductionist an approach to the problem. There is a growing realization at present that speciation, as well as microevolution, provides a key to the understanding of macroevolution.

The Synthetic Theory

Mid-nineteenth-century Darwinism emphasized natural selection as the main driving force of evolution. A supplementary role was assigned to the inheritance of acquired characters. "Disuse, aided sometimes by natural selection, will often have reduced organs when rendered useless under changed habits or conditions of life; and we can understand on this view the meaning of rudimentary organs." (Darwin 1872: ch. 15). Late nineteenth-century Weismannism or neo-Darwinism ruled out the inheritance of acquired characters and consequently assigned a much more exclusive role to natural selection (Weismann 1889–1892, 1893).

The weak point in the aforementioned causal theories, as well as in early nineteenth-century Lamarckism, was lack of knowledge concerning variation, its nature and transmission. This deficiency was remedied by early twentieth-century Mendelism. Although early Mendelism was not directly concerned with evolution, it did clear the ground for the next successful causal theory.

The early Mendelian geneticists considered the behavior of genes in family pedigrees. But the P, F_1, F_2, and F_3 of formal genetics are an artificial abstraction. In nature a mutant allele does not simply become united with the normal allele in an F_1 individual and become passed on to one-half of its gametes. Under natural conditions, outside the genetics laboratory or breeding plot, the mutation that develops in an individual organism does not remain within a family pedigree, but enters into the gene pool of a population. And there its fate, if it is either deleterious or beneficial, will be to change in frequency. It follows that the genetics of populations cannot be studied properly without taking cognizance of the effects of natural selection on hereditary variations. The originally separate lines of thought of Mendelism and Darwinism must be merged.

This merger was made in the 1930s and led directly to the modern synthetic theory of evolution, which began as a population-genetical approach to evolution. Key works around 1930 which considered the behavior of genes in populations, as affected by natural selection were Chetverikov (1926), Fisher (1930), Wright (1931), and Haldane (1932).

These were essentially theoretical treatments. They were followed by Dobzhansky's *Genetics and the Origin of Species* (1st ed., 1937; later editions 1941, 1951a). In this extremely important book, Dobzhansky puts experimental population genetics in the center of the stage. The theoretical results of the previous authors are taken up and related to experimental evidence concerning variation and selection. The inquiry extends to race formation and speciation. Dobzhansky laid down the precept that our understanding of macroevolution should be sought through a comprehension of microevolutionary processes that can be observed in experiments or in nature during a human lifetime (1937a:12).

The next stage saw the production of a remarkable series of books which showed that the data of other fields of biology are consistent with the population genetical theory of evolution. The population genetical approach was extended to animal systematics (Mayr 1942; Huxley 1942), general biology (Huxley 1942), paleontology (Simpson 1944; Rensch 1947), animal cytology (White 1945), animal macroevolution (Rensch 1947), plant systematics (Stebbins 1950), and plant cytogenetics (Stebbins 1950). Simpson's demonstration in *Tempo and Mode in Evolution* (1944) that the paleontological evidence is congruent with population genetical mechanisms was a particularly difficult and crucial achievement.

The synthetic theory can be characterized as the population genetical approach to microevolution and its extensions to other evolutionary levels and to other biological fields. In its core it represents a combination of the population geneticist's approach, which provides theoretical precision, with the naturalist's approach to living populations and species, which brings in the touch with reality. In its entirety it encompasses a much larger range of fields. Thus considered, it is not a special theory, which can be verified or falsified, but a general theory, a paradigm, which can absorb changes and modifications within wide limits, and has done so over the years since its inception.

Some current authors refer to the synthetic theory of evolution as neo-Darwinism (e.g., Gould 1980, 1982). This terminology is misleading for three reasons. The term neo-Darwinism is already preempted historically as a synonym of Weismannism. Second, there are significant differences between the natural selection of the synthetic theory and that of Darwinism. And finally, the synthetic theory, while it does have roots in Darwinism, also has roots in Mendelism and classical population genetics.

According to Mayr (Mayr 1982a; Mayr and Provine 1980), the synthetic theory was built in a short period from approximately 1936 to 1947. This was indeed an active period in evolutionary theory. Nevertheless, the dates given seem unduly restrictive since they exclude the immediate forerunners around 1930 and important contributions coming after 1947. Broader perspectives on the formation of the synthetic theory are given by Dobzhansky (1965) and Simpson (1966). It is perhaps best to think of the synthetic theory as developing in a series of stages, as we have presented it above.

Other Theories

Several other theories of evolution have figured prominently in the past and still have a few adherents in the modern period. These theories and some of their historical and modern advocates warrant mention.[5]

1. Orthogenesis. Some inner force impels organisms to change in a given direction. This force is envisioned differently by different authors. It may be a built-in tendency toward greater complexity of organization (Lamarck 1809, 1815–1822), a mystical force (Bergson 1911; Teilhard de Chardin 1959), or a postulated process of oriented mutation (Osborn 1934; Werth 1956).

2. Inheritance of acquired characters. Changes in individual organisms become hereditary, and lead to permanent changes in their descendants. There are various versions of this doctrine. The effects of the use and disuse of organs (Lamarck 1809). Induction of permanent change by the external environment (Geoffroy Saint-Hilaire 1830). Neo-Lamarckism (Lysenko 1948; Koestler 1972).

3. Saltationism. Large mutations produce new species suddenly: de Vries (1901–1903), Goldschmidt (1940), Schindewolf (1950).

5. For an in-depth treatment see Mayr (1982a, part 2).

In addition, a number of modern students are more or less strongly opposed to the synthetic theory. They do not necessarily espouse any alternative theory but may lean towards one or the other of the above. Examples are Grassé (1977), Koestler (1972), Koestler and Smythies (1969:1 and 427), and Gould (1980). Grassé seems to lean toward theory 1, Koestler (1972) definitely leans toward theory 2, and Gould (1977) sometimes leans toward theory 3.

Theories 1 and 2 make postulations about variation that are blatantly at odds with our experimental knowledge of the subject. Theory 3 in the extreme forms envisioned by its original authors is not supported by genetics. The nature of variation is the crucial question on which all three of these theories have foundered.

PART II
Microevolution

main present but the cross-linkages among individuals due to mating bonds are greatly reduced. We hesitate to say that mating bonds are absent entirely in asexual organisms, since many normally asexual organisms have some alternative, parasexual method of reproduction or have occasional reversions to sexuality. The population is usually an interbreeding group, whether the interbreeding is regular or occasional, and in all cases is a reproductive entity.

The population is also an ecological unit. The individuals composing it are genotypically similar in their ecological tolerances, occupy a definite area in some particular ecological niche or habitat, and make similar demands on their environment.

Actual populations appear in a variety of sizes and shapes. Population size, reckoned as the number of adult breeding individuals in any generation, can vary from one or a few to many millions of individuals. Population structure has four main components: population size, spatial configuration, breeding system, and dispersal rate. The last of these will be discussed in chapter 9.

Population size, the number of adult breeding individuals in any generation (N), can range from a few to many millions. As regards the spatial characteristics of populations, three extreme conditions and their various intermediate conditions can be recognized. The extremes are: (a) large continuous populations, (b) small isolated colonial populations (or populations conforming to the island model), and (c) linear populations.

Large continuous populations are exemplified by plains grasses, covering areas scores or hundreds of miles wide. Organisms with a colonial population system exist in a series of scattered, disjunct, and often small populations. Examples are furnished by terrestrial organisms in island archipelagoes, freshwater forms in series of lakes, mountaintop inhabitants in mountainous country, and organisms restricted to a particular soil or rock substrate that has a spotty distribution. Linear populations develop along rivers, coastlines, and similar habitats that are long and more or less continuous in one dimension but short and restricted in the other.

The various intermediate conditions are common. A large population may be continuous in some parts of its area but interrupted and semi-continuous in other parts. Likewise, the colonies belonging to an island system may be only semi-isolated rather than completely isolated. In the next section we will describe a concrete example of population structure in the giant sequoia (*Sequoiadendron giganteum*) which exhibits a variety of conditions ranging from isolated colonies in the north to an interrupted forest belt in the south.

As regards the breeding system, the extremes are wide outcrossing and self-fertilization. Common intermediate conditions are: outcrossing between close neighbors; inbreeding by systems other than self-fertilization, such as brother-sister matings in animals; and mixtures of outcrossing and selfing, as in hermaphroditic but self-compatible flowering plants.

The spatial configurations and breeding systems are combined in numerous ways to give a great diversity of actual population structures. Thus a large continuous population may be composed of wide outcrossers, as in many wind-pollinated plains grasses, but it may also be composed of narrow outcrossers or inbreeders. Similarly, a small isolated colony may consist of either outcrossers or inbreeders. The type of population structure affects the variation pattern of the population, as will be brought out later.

Populations of Giant Sequoia

The giant sequoia *(Sequoiadendron gianteum)*, a wind-pollinated, outcrossing, seedling-reproducing tree, occurs in pine-fir forest at middle elevations (5,000–8,000 feet) on the west slope of the Sierra Nevada in California. Its distribution area forms a narrow band about 250 miles long (figure 3.1A). Within this area, the giant sequoia occurs in a series of disconnected populations (figure 3.1B).

There are some 32 to 75 local populations depending on whether particular dumbbell-shaped populations are counted as one or more than one. The number of groves recognized by different botanists is 32 (Jepson 1909), 71 (Fry and White 1938), and 75 (Rundel 1972a).[1] We are following Jepson in order to make use of his old demographic data.

At the turn of the century the 32 local populations recognized here had the sizes indicated in table 3.1. The smaller populations are known as groves and the larger ones as forests, with a few exceptions. The forests had been extensively lumbered by the time of the surveys summarized in table 3.1, but the stumps of felled trees were still standing. The census figures given in the table thus refer to the size of the previous, undisturbed natural populations. The table reveals a wide range in population size, from groves of a few trees to forests consisting of thousands of individuals. A later census (Fry and White 1938) shows a smaller size in most populations.

The northern and southern parts of the species area differ in population

1. For a dot map of *Sequoiadendron giganteum* see Griffin and Critchfield 1976.

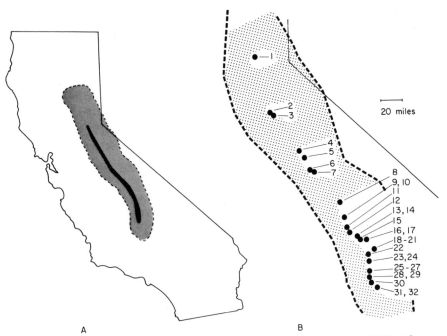

Figure 3.1. Geographical distribution of *Sequoiadendron giganteum*. (A) Distribution area of *Sequoiadendron* (black) in Sierra Nevada (stippled area). (B) Central and southern Sierra, enlarged, with populations of *Sequoiadendron* indicated by dots and numbered as in table 3.1.

structure. In the northern part of its range, the giant sequoia usually exists in small, widely separated groves. Gaps of 10 to 50 miles between groves are not uncommon. In the southern area large forests are, or were, more common and the gaps are bridged by scattered trees so that the populations are linked into a semi-continuous belt.

The wide gaps between the northern groves correspond to areas occupied by glaciers during the last glacial period. Presumably the giant sequoia formed larger and more continuous populations in the central Sierra Nevada prior to the last glaciation, was decimated and fragmented in that area by the glacial climate, and has been unable to regain its lost territory since. The populations in the southern Sierra were less severely affected by glaciation (Sudworth 1908; Axelrod 1959).

Local population boundaries are determined today by soil moisture in at least some cases that have been investigated. The groves occur in sites which have ample soil moisture during the dry summer months. The source of the soil moisture is summer rains in the high sierras. Some sites at middle ele-

Table 3.1. Populations of giant sequoia *(Sequoiadendron giganteum)* in Sierra Nevada of California (Data from Jepson 1909)

Grove or Forest	Size of Area (acres)	Number of Trees
NORTHERN AREA		
1. North Grove	—	6
2. Calaveras Grove	51	101
3. Stanislaus Grove	1,000	1,380
4. Tuolumne Grove	10	40
5. Merced Grove	20	33
6. Mariposa Grove	125	547
7. Fresno Grove	2,500	1,500
SOUTHERN AREA		
8. Dinky Grove	50	170
9. Converse Basin Forest	5,000	12,000
10. Boulder Creek Forest	3,200	6,450
11. General Grant Grove	2,500	250
12. Redwood Canyon Forest	3,000	15,000
13. North Kaweah Forest	500	800
14. Swanee River Grove	20	129
15. Giant Forest	8,000	20,000
16. Redwood Meadow Grove	50	200
17. Harmon Meadow Grove	10	80
18. Atwell Forest	1,500	3,000
19. Lake Canyon Grove	20	80
20. Mule Gulch Grove	25	70
21. Homer's Peak Forest	5,500	1,500
22. South Kaweah Forest	160	300
23. Dillon Forest	3,600	3,500
24. Tule River Forest	15,000	5,000
25. Pixley Grove	850	500
26. Fleitz Forest	4,000	1,500
27. Putnam Mill Forest	4,000	900
28. Kessing Groves	2,800	700
29. Indian Reservation Grove	1,500	350
30. Deer Creek Grove	300	100
31. Freeman Valley Forest	1,000	400
32. Kern River Groves	700	200

vations collect ground water from these rains, while others do not, and the giant sequoias are restricted to the former (Rundel 1972b).

Polymorphism

Polymorphism is defined as the coexistence of two or more discontinuous segregating forms in a population, where the frequency of the rare type is not due to mutation alone (Ford 1964, 1965). In other words, polymorphism is the variation in a local breeding population that exhibits distinct or sharp Mendelian segregation.

The term polymorphism as defined above excludes certain types of variation. It excludes purely phenotypic variation (because this is nongenetic); it excludes geographical variation (which does not exist in one population); it excludes polygenic variation (which does not segregate into discontinuous classes); and finally it excludes genetic variation due to new or recurrent mutation.

Polymorphic variation can be classified in several ways. A useful distinction is drawn between genetic polymorphism and chromosomal variation. The former is segregational variation in respect to homologous alleles of the same gene locus; the latter is polymorphism for chromosome types, such as sex chromosomes, or rearrangements, such as inversions.

Another distinction is that between transient and balanced polymorphism. In the case of transient polymorphism the diversity is temporary; it occurs while one form is in the process of replacing another under the controlling influence of natural selection. In balanced polymorphism the polymorphic types are more or less permanent components of the population, being preserved by selection in favor of diversity (Ford 1964, 1965).

The various forms of polymorphism—genetic, chromosomal, transient, and balanced—are all common and widespread in the living world. Polymorphism is a virtually universal feature of populations in sexual organisms. An example involving the blood types in man will be described in the next section.

The phenomenon of polymorphism leads us to the concept of the gene pool, which in turn gives us another way of looking at the local breeding population. Consider a population that is polymorphic for a gene A and contains the alleles A_1, A_2, and A_3. This population will form the diploid genotypes A_1A_1, A_1A_2, A_2A_2, etc., and these genotypes will be the observed units in any sampling of individuals, but clearly the underlying genetic polymorphism is a fundamental feature of the population in question. The population can be said to have a gene pool consisting of A_1, A_2, and A_3 alleles. Furthermore, these alleles will occur at characteristic frequencies in the gene pool; let us assume that their frequencies are 60%, 30%, and 10%, respectively. Thus the population can be described quantitatively in terms of the types and frequencies of genes in its gene pool.

It should be noted here that the gene pool concept is broader than the polymorphism concept. The gene pool of a population is made up of all its genes. Thus the gene pool of our hypothetical population might be polymorphic for gene A, harbor a rare mutant allele of another gene B, and be monomorphic for other genes C and D.

The local breeding population can now be defined, or at least characterized, as a group of individuals sharing in a common gene pool (Dobzhansky 1950). The individuals constituting the population in any given generation are the various genotypic products of the gametes drawn from its gene pool in the preceding generation.

Polymorphism in Human Blood Groups

Human individuals vary in their reactions to blood transfusions. Some transfusions result in agglutination or clumping of the red blood cells, whereas other transfusions do not. The agglutination reaction depends on the immunological relationships between the antigens in the red cells and the antibodies in the serum of the two bloods.

On the basis of the type of antigen present, four blood groups are recognized (A, B, AB, and O). Individual humans fall into one or the other of these four phenotypic classes. A person belonging to blood group A can donate blood to another person of type A without causing agglutination. Similarly, type B blood can be transfused into type B blood without agglutination. But transfusion of A into B, or B into A, results in heavy agglutination. It is not necessary to describe the reactions resulting from all combinations of blood types here (see Stern 1960; Race and Sanger 1962). In general, agglutination results from transfusions between members of different blood groups.

The blood types are determined by a series of three alleles: I^A, I^B, and I^O. The allele I^O is recessive to I^A and I^B (and is sometimes written i). The alleles I^A and I^B are codominant. The six diploid genotypes of the three alleles give rise to the four phenotypic blood groups of the ABO series, as indicated in table 3.2. Actually there are different but immunologically similar isoal-

Table 3.2. Genetics of ABO Blood Groups (Stern 1960)

Genotype	Phenotype (blood group)
$I^A I^A$ and $I^A I^O$	Type A
$I^B I^B$ and $I^B I^O$	Type B
$I^A I^B$	Type AB
$I^O I^O$	Type O

leles of I^A (I^{A1}, I^{A2}, etc.), and therefore more than six possible genotypes, but we can ignore these fine differences in I^A for our present discussion.

Human populations are generally polymorphic for the ABO blood groups. The frequencies of the blood-group phenotypes and of the underlying gene alleles are known for hundreds of local populations throughout the world. The allele frequencies in three populations are listed below (data from Mourant 1954; Mourant et al. 1976):

Uppsala, Sweden	$0.319\ I^A + 0.079\ I^B + 0.603\ I^O$
Punjab, India	$0.181\ I^A + 0.259\ I^B + 0.560\ I^O$
Navajo Indians, New Mexico	$0.133\ I^A + 0.000\ I^B + 0.867\ I^O$

These examples illustrate the point that human populations are alike in being polymorphic for the ABO series, but differ in their allele frequencies. Each local population has a gene pool with its own characteristic composition of I alleles.

The local populations are parts of larger, regional, racial groupings. Related local populations in the same region tend to have slightly different gene pools. Thus the frequency of I^A is 31.9% in the Uppsala, Sweden, population, as noted above, and is 28.4% in Falun, Sweden. Conversely, consistent differences in allele frequencies appear between geographical races.

Almost all native human populations in Western Europe have a high frequency of I^A and a low frequency (under 10%) of I^B. Central Asia shows a high frequency (20–30%) of I^B. Among American Indians, I^O is high in frequency, whereas I^B is either low or absent (Mourant 1954; Mourant et al. 1976). The polymorphic balance shifts in passing from one geographical area to another. We will return to the geographical distribution of the ABO blood groups in a later chapter.

It is very interesting that parallel polymorphic variation in the ABO groups is found in the great apes. The orangutan and gibbon have A, B, and AB types (Mourant 1954; Wiener and Moor-Jankowski 1971). The ABO polymorphism is thus more ancient than the human species itself, and is shared with man's closest relatives in the primate order.

There are several other systems of blood groups in man: the Rh system, MN system, and others. Human populations are polymorphic for these systems too (Race and Sanger 1962; Mourant et al. 1976). The polymorphic variation in the Rh and other systems seems to be independent of that in the ABO series.

Enzyme Polymorphism

The technique of gel electrophoresis makes it possible to detect polymorphisms in enzymes and some other proteins that could not be detected by ordinary genetic methods. A tissue extract is put in a gel and exposed to an electric potential to set up an electric field. The different types of enzymes in the gel, with their characteristic mobilities in the electric field, then become physically separated, and, when the gel is stained, show up as distinct bands. Allelic differences with respect to one enzyme system, and genic differences between enzymes, can be detected in this way.

Application of the electrophoretic technique to samples of natural populations has produced surprising results. Unexpectedly high levels of polymorphism have been found in populations of various species of organisms. Table 3.3 presents some examples from various major groups of organisms (taken from a more extensive tabulation of Nevo 1978). It will be noted that the percentages of polymorphic enzyme loci range up to 40% or more in some species.

The average frequency of enzyme polymorphism for a sample of species belonging to a given large group is as follows (Nevo 1978):

Mammals	14.7%
Birds	15.0
Reptiles	21.9
Amphibians	26.9
Bony fishes	15.2
Drosophila	43.1
Other insects	32.9
Plants	25.9

Enzyme polymorphism has turned out to be widespread in the organic world.[2]

The Population Concept

The concept of genetically variable populations as reproductive units is by no means obvious. This concept did not exist in biology during the eigh-

2. For reviews see Gottlieb (1971), Hamrick (1979), Lewontin (1973, 1974), and Nevo (1978). Two significant early papers on the subject are Hubby and Lewontin (1966), and Lewontin and Hubby (1966).

Table 3.3. Frequency of enzyme polymorphisms in various organisms
(Data from Nevo 1978)

Organism	No. Loci Assayed	Percent Loci Polymorphic
PRIMATES		
Homo sapiens, European	71	28
Macaca fuscata	29	10
RODENTS		
Mus musculus, Denmark	41	26
Peromyscus polionotus	32	23
BIRDS		
Passer domesticus	15	33
Pipilo erythrophthalmus	18	17
LIZARDS		
Sceloporus grammicus	19	21
Uta stansburiana, mainland	17–18	14
AMPHIBIANS		
Bufo americanus	14	26
Rana ridibunda	27	34
BONY FISHES		
Salmo gairdneri	23	15
Mugil cephalus	30	20
DROSOPHILA		
D. melanogaster	19	42
D. pseudoobscura	18, 24	30, 42
D. willistoni	31	46
OTHER INSECTS		
Solenobia triquetrella	16	47
Periplaneta americana	20	40
GRASSES		
Hordeum spontaneum	28	30
Avena barbata	16, 19	21, 31
OTHER FLOWERING PLANTS		
Phlox drummondii	16	19
Stephanomeria exigua	14	34
FERN ALLEY		
Lycopodium lucidulum	18	10

teenth and early nineteenth centuries, nor does it exist in some branches of
biology even today. According to Mayr (1972a, 1982) it was introduced into
biology by Darwin in 1859. The population concept was one of the elements in the Darwinian revolution in scientific thought.

The population concept stands in contrast to essentialism. Essentialism is the view that the observed things in the world are the expressions of an underlying essence. The things appear in varying forms, but the essence is immutable. The members of a class of objects, including the individuals in a population, are the variable expressions of the same essence.

Essentialism in one version or another was the traditional philosophy in Europe. Platonic philosophy, Christian theology, and philosophical idealism represented different versions of essentialism. Essentialism naturally dominated the thinking in the early history of biology. Here it took the form that Mayr (1957a, 1957b, 1972a, 1982) has called typological thinking. This is the view that individual organisms are the imperfect and hence variable manifestations of the archetype of the species to which they belong.

Typological thinking is an obstacle to understanding evolution, which requires population thinking instead, since evolution is a change in the genetic composition of populations. A great but subtle accomplishment of the Darwinian revolution, according to Mayr (1972a, 1982), was the replacement of typological thinking by population thinking in biology. The introduction of the population concept removed an old and strong obstacle to gaining an understanding of natural selection in particular and evolution in general.

CHAPTER 4
The Statics of Populations

The Hardy-Weinberg Law

The gene pool of a local population normally contains various polymorphic genes in addition to the monomorphic ones. Furthermore, the polymorphic genes will exhibit certain definite allele frequencies in any given generation. Thus for a gene A present in two allelic forms, A and *a*, the gene pool in one generation might consist of 70% A alleles and 30% *a* alleles. The question arises: What are the expected allele frequencies in the next generation?

In a diploid population these alleles will be carried in the homozygous and heterozygous genotypes AA, *aa*, and A*a*. These genotypes will also be found in certain proportions in any given generation. They are the parents of the genotypes in the next generation. The related question then arises: What are the expected proportions of genotypes in the second and later generations?

The expected allele frequencies and genotype frequencies are given by the Hardy-Weinberg law. This law is operative under the following conditions. The population is assumed to be large, so that errors of sampling do not affect the frequencies significantly from generation to generation. The population is isolated; there is no immigration. The individuals in the population are assumed to contribute equal numbers of functioning gametes; in other words, the various genotypes are equally successful in reproduction. Finally, random mating is assumed to prevail in the population.

Random mating, or panmixia, can be defined either in terms of individuals or in terms of gametes, with equal results. Considered from the standpoint of individuals, random mating occurs when individuals with different genetic constutions mate without regard to their genotype. For example, an AA female might mate with AA, A*a*, or *aa* males without any preference for one type of male.

The condition can be defined somewhat more precisely in terms of the array of gametes in the gamete pool. Random mating in this sense means that any given female gamete has a chance of being fertilized by any type of male gamete, and that this chance is directly proportional to the frequency of that type of male gamete in the gamete pool. In short, gametes carrying different alleles combine in pairs in proportion to their respective frequencies in the gamete pool. The individuals constituting the population in any generation then represent the products of different pairs of gametes drawn at random from the gamete pool of the preceding generation.

The Hardy-Weinberg law makes two statements about a population that conforms to the foregoing conditions: (a) the allele frequencies will tend to remain constant from generation to generation; (b) the genotypes will reach an equilibrium frequency in one generation of random mating and will remain at that frequency thereafter. These two generalizations will be illustrated in the following sections.

Allele Frequencies

We will illustrate the principle of constancy in allele frequencies by means of a numerical example. Assume that a diploid population polymorphic for A contains the following proportions of genotypes in the starting generation: 60% AA, 20% A*a*, and 20% *aa*. We will trace the A alleles step by step through two generations.

1. Allele frequencies in first generation. Since the genotype frequencies are given as

$$0.60 \text{ AA} + 0.20 \text{ A}a + 0.20 \text{ } aa$$

the allele frequencies (*q*) in the same generation must be

$$q \text{ A} = \frac{0.60 + 0.60 + 0.20}{2} = 0.70$$

$$q \text{ } a = \frac{0.20 + 0.20 + 0.20}{2} = 0.30$$

2. Gamete pool of first generation. The individuals are assumed to be equally fecund. Therefore the diploid individuals will produce haploid gametes in the proportions 70% A and 30% a. The allele frequencies in the gamete pool are the same as those in the original gene pool.
3. Random mating. The gametes are drawn at random to form the zygotes of the second generation. The paired combinations are

♀ gametes		♂ gametes
0.70 A	×	0.70 A
0.70 A	×	0.30 a
0.30 a	×	0.70 A
0.30 a	×	0.30 a

4. Zygotic frequencies in second generation. The products of the above system of random mating are

$$0.49 \ AA$$
$$0.21 + 0.21 = 0.42 \ Aa$$
$$0.09 \ aa$$

The zygotes are assumed to be equally viable. Therefore these figures give the expected equilibrium frequencies of genotypes in the second generation.

It may be noted that the population was not in equilibrium as regards genotype frequencies in the first generation, but reached that equilibrium condition in just one generation of random mating.
5. Allele frequencies in second generation. The gene pool of the second generation can be seen to contain the two alleles in the following frequencies:

$$q \ A = \frac{0.49 + 0.49 + 0.42}{2} = 0.70$$

$$q \ a = \frac{0.42 + 0.09 + 0.09}{2} = 0.30$$

Thus the allele frequencies are the same now as they were in the first generation.

The Hardy-Weinberg Formula

The Hardy-Weinberg formula provides a short, direct method of calculating the expected frequencies of genotypes in a random-mating population from the known allele frequencies in the gene pool.

Consider again a gene pool containing two alleles of a gene A. Let p represent the frequency of allele A and q the frequency of allele a (where $p+q=1$). The random combinations of A and a gametes will then produce zygotes in the proportions given by the expansion of the binomial square $(p+q)^2$. The equilibrium frequency of genotypes, in other words, will be

$$p^2 \text{ AA}$$
$$2\ pq \text{ A}a$$
$$q^2\ a\text{A}$$

We can apply this formula to our previous example (where q is used in a slightly different sense), and arrive at the same results by a simple operation. That is, if $p = 0.70$, the equilibrium frequency of AA is $p^2 = 0.49$.

Where the polymorphism involves a series of multiple alleles, the Hardy-Weinberg formula takes the form of a polynomial square. For three alleles $(A_1, A_2,$ and $A_3)$, whose frequencies are p, q, and r, so that $p + q + r = 1$, the equilibrium frequency of genotypes is given by the trinomial square $(p + q + r)^2$.

Effects of Inbreeding

Genotype frequencies quickly reach and then remain at an equilibrium condition under random mating in a large population. Departures from random mating bring about deviations from the Hardy-Weinberg equilibrium of genotypes. Let us consider another simple hypothetical example.

Assume that a population in generation 0 consists entirely of Aa heterozygotes. Reproduction is by self-fertilization. In generation 1 the population will contain 25% AA, 50% Aa, and 25% aa. The heterozygous class continues to decay at a regular rate in subsequent generations. After seven generations of selfing (i.e., in inbred generation 7), the population will contain almost 50% AA and almost 50% aa.

The formula giving the proportion of homozygotes derived from a heterozygous ancestor, for a single gene (A), after m generations of self-fertilization, is

$$\frac{2^m - 1}{2^m}$$

As applied to the case at hand, with seven generations of selfing, this formula gives the following expected genotype frequencies in inbred generation 7:

127/128 (AA + aa)
1/128 Aa

Inbreeding thus alters the genotype frequencies in the direction of a preponderance of homozygotes. But another aspect should be noted: the allele frequencies are not affecting by the inbreeding. In our hypothetical example the allele frequencies were 50% A and 50% a in generation 0; and they remain the same in inbred generation 1 and again in generation 7.

Conclusions

In a large, polymorphic, panmictic population composed of equally viable and equally fertile individuals, the various homozygous and heterozygous genotypes will quickly reach certain equilibrium frequencies, which are dependent upon the existing allele frequencies. The genotype frequencies, having reached the Hardy-Weinberg equilibrium, will then remain constant during all succeeding generations of random mating.

Allele frequencies tend to remain constant from generation to generation in a large polymorphic population composed of equally viable and equally fertile individuals. This constancy is not dependent upon random mating. The aspect of the Hardy-Weinberg law dealing with allele frequencies is therefore more general than that dealing with genotype frequencies, and it is also more basic for evolutionary theory.

It is useful to examine the special conditions required for the operation of the Hardy-Weinberg law. These conditions exclude any factors affecting gene frequency other than the process of gene reproduction itself. Insofar as the level of variation in the gene pool is determined by gene reproduction per se, that level remains steady and unchanging through successive generations.

This is the statics of populations. Change in the level of variation of populations—the dynamic side of the question—requires the action of specific factors or forces. These are the forces of microevolution, to be considered next.

CHAPTER 5
The Dynamics of Populations

Microevolution Defined
Microevolutionary Changes in *Drosophila pseudoobscura*
The Primary Evolutionary Forces
Interactions Between Evolutionary Forces
Conclusions

Microevolution Defined

The frequencies of alleles in a large gene pool do not change by themselves, as noted in the preceding chapter. Yet it is a fact of observation that local populations do undergo changes in allele frequencies during successive generations. The gene pool does change with time.

We can define microevolution as a systematic change in the frequencies of homologous alleles, chromosome segments, or chromosomes in a local population. In other words, microevolution is any increase or decrease in the frequency of a variant form in the gene pool that continues to occur over a sequence of generations.

Examples of microevolutionary changes in natural populations are legion. It will suffice for our present purpose to describe one good example in *Drosophila pseudoobscura*.

Microevolutionary Changes in *Drosophila pseudoobscura*

Drosophila pseudoobscura is characterized by variation in the gene arrangement in a segment of chromosome III. The segment in question differs by inversions in different strains of the fly. The various inversion types are designated by names and letter symbols: Standard *(ST)*, Arrowhead *(AR)*, Chiricahua *(CH)*, Tree Line *(TL)*, Pikes Peak *(PP)*, etc. There are sixteen such inversion types in the species. These can be identified by their characteristic banding patterns in the salivary-gland chromosomes of the larvae.

Most natural populations of *Drosophila pseudoobscura* in western North America are polymorphic with respect to chromosome III. The polymorphic populations contain two or more inversion types, and consist of their various homozygous and heterozygous combinations (e.g., *ST/ST, CH/CH, ST/CH*). In any given population the inversion types tend to occur in certain average annual frequencies.

The common inversion types in *D. pseudoobscura* in the Sierra Nevada of California are *ST, AR,* and *CH.* Four other inversions occur at low frequencies in this area *(TL, OL, SC, PP).* Our story centers on one of these rare types *(PP)* in a particular local population (Mather). The Mather population, in the Yosemite region of California, was studied intensively by Dobzhansky and his coworkers over a period of many years (see Dobzhansky 1956, 1958, 1971).

The Pikes Peak inversion type was present but exceedingly rare in Sierran populations of *Drosophila* in and before 1945, and was unknown in the Mather population at that time. In 1946 *PP* appeared in the Mather population at the very low frequency of 0.3%. During the course of the next decade (1947–1957) *PP* rose rapidly in frequency to highs of 10 and 12%, as shown in table 5.1. It fluctuated at somewhat lower but moderate frequencies during the following years (1959–1965). Since then it has dropped to a fairly low level (table 5.1).

The rise in frequency of *PP* in the Mather population was accompanied by a decline of the formerly common *CH* to unprecedented lows of 2–6%. In recent years *CH* has come back up to its earlier level (table 5.1). Corresponding adjustments took place in the proportions of *ST* and *AR* during the same period (Dobzhansky 1971).

The spectacular increase in frequency of the *PP* inversion type was not confined to the Mather population alone. Parallel increases in *PP* were observed in other separate populations of *D. pseudoobscura* in scattered localities throughout California and Arizona. Thus in the San Jacinto Mountains in southern California, *PP* appeared for the first time in 1952, rose to peaks of 10 and 11% in the next few years, and declined again in the 1960s (Dobzhansky 1971). The trends involving *PP* were general over a wide geographical area.

The causes of the observed microevolutionary changes are unknown. Enough is known to rule out some possible factors. Polymorphic laboratory populations of *Drosophila pseudoobscura* reared under favorable conditions

Table 5.1. Frequency of two inversion types (*PP* and *CH*) in a population of *Drosophila pseudoobscura* at Mather, California, over a 28-year period (Dobzhansky 1971; Anderson et al. 1975)

Year	Frequency of *PP*, %	Frequency of *CH*, %	Number of Chromosomes Assayed
1945	0.0	17	308
1946	0.3	17	336
1947	0.7	20	425
1950	3	17	812
1951	5	11	856
1954	12	13	1312
1957	10	4	316
1959	4	11	298
1961	6	3	350
1962	9	2	450
1963	7	6	446
1965	6	11	534
1969	2	3	312
1971	3	12	390
1972	6	17	576

conform to Hardy-Weinberg expectations as regards the proportions of the inversion types and the corresponding diploid genotypes. Experimental evidence confirms theory in ruling out any internally directed changes in this case. The possible effects of chance are ruled out by the large size of the populations themselves and of the population samples studied. Mutation pressure is ruled out by the fact that parallel changes have taken place in widely separated local populations. Natural selection is almost certainly responsible for the observed changes. However, the selective factor involved remains unidentified despite extensive efforts to determine it experimentally (Dobzhansky 1956, 1958, 1971).

The Primary Evolutionary Forces

The factors that bring about changes in gene frequencies or chromosome frequencies can be designated as the primary evolutionary forces. Four such forces are known. They are: mutation, gene flow, natural selection, and genetic drift.

Assume that a population gene pool consists predominantly of A_1 alleles and secondarily of A_2 alleles. The original gene frequencies can be altered by each one of the foregoing evolutionary forces, as follows.

The A_1 allele may mutate to A_2, occasionally or repeatedly, thereby increasing the frequency of the latter. Individuals or gametes carrying A_2 may migrate into the population from some other population in which A_2 is more common. Such migration or gene flow will also alter the pre-existing gene frequencies in the recipient population.

The carriers of A_1 and of A_2 may differ in phenotypic characteristics affecting their ability to survive and reproduce and contribute offspring to the next generation. If individuals carrying A_2 are superior to those carrying A_1 in these respects, A_2 will gradually increase in frequency. This is the force of natural selection, which is the most important of the factors controlling gene frequencies in natural populations.

Finally, the frequencies of A_1 and A_2 may shift significantly by chance alone in small populations. Genetic drift refers to the random component in gene frequency changes.

Each of the evolutionary forces varies in the intensity of action. The intensity can be quantified. Mutation rate (u) can vary from 0 to 1, where 0 is complete gene stability and 1 is complete instability of the gene. Similarly, the rate of gene flow can range from $m = 0$ (no migration) to $m = 1$ (complete swamping).

The selection coefficient (s) measures the average increase per generation of one allele relative to other, competing alleles. This coefficient can vary from 0 to 1, where $s = 0$ means no selection and $s = 1$ means complete gene replacement in a single generation.

The possibility of the action of genetic drift is expressed by a relationship between population size (N) and the other variables. Drift can be effective in controlling gene frequencies when N is low relative to s, m, and u.

Interactions Between Evolutionary Forces

The first two forces mentioned above, mutation and gene flow, produce variability. The second two forces, selection and drift, sort out this variability. The variation-producing forces start the microevolutionary process, and the variation-sorting forces go on to establish the variant types in new frequencies. Evolutionary change within populations can be viewed as a resultant of the opposing forces that produce and sort out genetic variations.

One of the old theories of evolution, the theory of orthogenesis, which still has some adherents today, postulates that evolutionary changes are directed mainly by the mutation process. There are two fatal arguments against this view.

First, mutational changes occur more or less at random with respect to the adaptive requirements of the organism, and could not alone bring about the adaptive characteristics actually seen in organisms. The additional factor of natural selection is necessary to take the process "from the chemical level of mutation to the biological level of adaptation" (Darlington 1939:127).

In the second place, selection happens to come into action after mutation in the time sequence of microevolution. No matter how strong or how oriented the mutation pressure may be in any population, its effects for good or bad are always subject to censure by natural selection. Selection always has the last word.

It is equally true that selection could not act without the mutation process to create a supply of new genetic variations. Microevolution is due not to the operation of any single force, but to the interaction between two or three or four forces.

The quantification of the primary evolutionary forces opens the way for a quantification of their interactions. The latter is a relatively simple matter in theory, and is sometimes feasible in synthetic experimental populations, but is extremely difficult to carry out in natural populations because of the numerous uncontrolled factors involved. Nevertheless, some approaches to the problem have been made in real populations.

Interaction between mutation and selection is exemplified by hemophilia in man. The bleeder disease, hemophilia, involving failure of clotting of blood, is usually fatal at an early age. The most common type of hemophilia, hemophilia A, is due to a recessive allele *(hh)* of the sex-linked or X-linked gene *Hh*. Heterozygous female carriers *(Hh/hh)* produce some sons of the constitution *h/0* who exhibit the disease (Stern 1973).

The hemophiliac males mostly, though not invariably, die before reaching reproductive age, and therefore the *hh* allele has a low selective value in the semilethal range. The *hh* alleles are continuously being eliminated from human populations by selection. Yet they persist in the population at a low frequency. Their persistence is attributed to recurrent mutation from *Hh* to *hh* (Stern 1973), the estimated mutation rate being 1.3×10^{-5} (Cavalli-Sforza and Bodmer 1971). The persistent low frequency of *hh* in human populations thus represents an equilibrium between recurrent mutation and strong counterselection.

The production of variability by a combination of mutation and gene flow is illustrated by the land snail, *Cepaea nemoralis*, in Western Europe. The snail colonies are generally polymorphic for the presence or absence of brown bands on the shells. The phenotypic differences are controlled by a gene *B* present in two allelic forms; the dominant allele *B* determines bandless shells, and the recessive allele b, banded shells. The B gene mutates from bandless to banded and vice versa at an estimated rate of $u = 0.0001-0.0005$ (Lamotte 1951).

The population structure of the European land snail differs in different areas. These differences affect the rate of gene flow. In the district of Aquitaine in France the colonies lie fairly close together and are frequently bridged by migrant individuals. Gene flow here is estimated to occur at the rate of $m = 0.003-0.004$. Gene flow is a more important source of population variability than gene mutation in Auqitaine. In the province of Brittany in France, by contrast, the colonies are widely spaced and only rarely connected by migrant individuals. Mutation is relatively more important here than in Aquitaine (Lamotte 1951).

This is by no means the whole story. The banding pattern in the snails is also controlled by selection and probably by drift (see Part III).

The action of genetic drift depends entirely on the interrelationship between the factors, N, s, u, and m. Thus, drift can be an effective force controlling gene frequencies when $N < \frac{1}{2}s$, if u and m are negligible, but moderate mutation rates and migration rates can nullify the action of drift. A most important interaction in nature is that involving the combination of drift and selection. These relationships will be described in more detail in a later chapter.

Conclusions

The definition stated at the beginning of this chapter is usually stated as a definition of evolution in general: "Evolution is a change in gene frequencies in populations." Some evolutionists regard this as an inadequate definition of evolution (Mayr 1982:400). I agree and that is why I introduced it as a definition of microevolution in particular.

Thus restricted to its proper sphere, however, the gene-frequency-change definition is valuable and should be retained, since it is the expression of a useful fundamental concept about the evolutionary process. The genetic composition of populations does not change systematically as a result of reproduction per se. Yet such changes do occur. And the causes of change can be sought in certain known forces. The extent of the change is determined by the interaction and balance among these forces. The whole population genetical approach to evolution is subsumed in the gene-frequency-change concept of microevolution.

CHAPTER 6
Mutation

A mutation is a sudden hereditary change in a phenotypic character caused by an abrupt structural and functional alteration in the genetic material. The genetic material is organized in a hierarchy of structural-functional units ranging from molecular sites within a gene to whole chromosomes and genomes. Correspondingly, there are different types of mutations from gene mutations to genome mutations. This chapter is concerned primarily with gene mutations.

Sudden hereditary changes in phenotype can be caused by genetic processes other than structural changes in the genes or chromosomes. There are spurious as well as true mutations. The phenotypic changes do not in themselves reveal the genetic process involved. It is very difficult to distinguish between the various types of true and spurious mutations on the basis of direct observational evidence alone. Conversely, there is a type of sudden change in the genetic material that does not produce a phenotypic effect, as we shall see.

Gene Mutations

A gene or point mutation is an alteration that occurs in the nucleotide sequence within the limits of a gene and changes the mode of gene action.

It is, in general, and with one exception to be described in the next section, a molecular change in the gene that brings about an altered phenotypic effect. Suppose that a gene contains a codon or triplet CTT at some point on one strand, specifying the amino acid glutamic acid in a polypeptide chain. The codon CTT could change to GTT by the substitution of a single nucleotide. The altered codon now specifies glutamine instead of glutamic acid in the polypeptide product of gene action. The original and mutant protein molecules differ themselves and may well determine other second-order phenotypic differences.

The stability of genes through successive cell generations and individual generations, and hence the conservative aspect of heredity, is due to the exactness of the copying process during gene replication. But the copying process is not perfect. Copying errors occur occasionally. Gene mutations can be regarded as such errors of copying.

The new mutant allele then replicates itself faithfully until the next mutational change occurs. The net result of gene mutation is thus the existence of a pair or series of homologous alleles. Conversely, the presence of allelic variation in any gene goes back ultimately to the process of gene mutation.

Any gene in the genotype is presumably subject to mutation. At any rate, mutations do observably occur in genes controlling a wide diversity of characters. In *Drosophila melanogaster*, for example, there are wing mutants with slightly shriveled wings, greatly reduced wings, or no wings; eye-color mutants with white or vermilion eyes; bristle mutants of various sorts, and so on. An array of leaf mutants with different leaf forms is known in the wild tomato, *Lycopersicon pimpinellifolium*. Biochemical mutations affecting steps in metabolic pathways are well known in microorganisms and are present though less easily studied in higher organisms.[1]

In terms of the magnitude of their phenotypic effects, gene mutations form a spectrum ranging from those with slight effects to those producing some major change in phenotype. The extreme types are called minor mutations and macromutations, respectively. Conspicuous but non-drastic mutations are typical of the mid-region of the spectrum. Minor mutations are exemplified by mutant types in *Drosophila melanogaster* with statistically slight deviations from normal viability or normal bristle number. An example of a macromutation is the mutant *tetraptera* in *D. melanogaster* with four in-

1. For catalogs of the mutants in *Drosophila melanogaster* and *Zea mays* see Braver (1956), Lindsley and Grell (1968), and Neuffer et al. (1968). A comprehensive review of the evolutionary genetics of gene mutations in plants is given by Gottschalk (1971).

stead of two wings. It represents a departure from the two-winged condition characteristic of the family Drosophilidae and order Diptera.

In diploid animals and plants, a high proportion of new mutations are recessive, the wild-type genes being dominant. An important consequence of the recessive condition of many mutant alleles is that they are not exposed immediately and directly to the action of selection, but may be kept in storage for generations in the diploid population.

Types of Point Mutations

Several types of point mutation can be recognized, depending on the mode of molecular change within the gene. Four types will be briefly described here (following Wallace 1981).

1. Missense mutation. This is what we described in the preceding section. A base substitution occurs within a triplet (e.g., CTT→GTT), and the altered triplet codes for an amino acid which is different from that specified by the ancestral triplet.

2. Frameshift mutation. A new base or base pair is inserted at the beginning of a DNA sequence. All the triplets are then changed downstream from the insertion, with a resulting change in the polypeptide product. For example, let the ancestral sequence be ATT—TAG—CGA; and let T be inserted at the beginning; then the new sequence becomes TAT—TTA—GCG—A . . .

3. Nonsense mutation. A base substitution gives a new triplet which is a chain-terminating codon. There are three of these in the genetic code. Now the resulting polypeptide chain stops at a new point and consequently has different properties than the ancestral polypeptide product.

4. Synonymous missense mutation. The genetic code contains a considerable amount of redundancy, with two or several triplets coding for the same amino acid. It can be expected, therefore, that base substitutions will sometimes occur so as to change one triplet into another synonymous one that specifies the same amino acid. In this case, as a result of the redundancy in the code, we have a molecular change within a gene that does not produce a phenotypic effect. Such synonymous mutations are probably fairly common.

Mutation Rates

Some representative rates of spontaneous mutation in the usual sense of the term, that is, mutations with phenotypic effects, are given in table 6.1. It will be noted, first, that the mutation rates are generally low; and second, that different genes in the same species often exhibit a fairly wide range of mutability. In corn, for example, the plant color gene is quite mutable while the waxy starch gene is highly stable.

A third point to note is that the mutation rates in bacteria are lower by one to several orders of magnitude than those in multicellular organisms. It can be stated that the few bacterial genes listed in the table are representative

Table 6.1. Rates of spontaneous mutation from normal to mutant type at specific loci in several organisms (Compiled from Stadler 1942; Cavalli-Sforza and Bodmer 1971; Merrell 1975; and Suzuki et al. 1981)

Organism and Gene	Frequency of Mutation to Mutant Allele
I. Multicellular organisms	Frequency per gamete
Homo sapiens	
Albinism	3×10^{-5}
Hemophilia A	1.3×10^{-5}
Hemophilia B	5.5×10^{-7}
Huntington's chorea	2×10^{-6}
Drosophila melanogaster	
Yellow body	12×10^{-5}
Eye color	4×10^{-5}
Ebony body	2×10^{-5}
Zea mays	
Plant color	4.9×10^{-4}
Color inhibitor	11×10^{-5}
Sugary endosperm	2×10^{-6}
Shrunken seeds	1×10^{-6}
Waxy starch	0 in 1.5 million gametes tested
II. Bacteria	Frequency per cell per division
Escherichia coli	
Lactose fermentation	2×10^{-7}
Histidine requirement	4×10^{-8}
Phage T1 resistance	2×10^{-8}
Streptomycin sensitivity	$10^{-8}-10^{-9}$

of a larger body of data. Bacterial genes seem to be more stable, in general, than genes in eukaryotic organisms.

There are reasons for believing that at least some of the estimates of mutation rates in higher organisms are too high. One source of error is the difficulty of distinguishing between true intra-genic mutations and rare recombinations of very closely linked genic units. The two events can lead to the same observable result, namely, an abrupt true-breeding change in phenotype. Any large collection of mutations in a diploid organism, as observed at the phenotypic level, is likely to contain a fraction of undetected rare recombinations in addition to the true gene mutations, thus distorting the estimated mutation rate upward (see chapter 8).

Undetected selective advantage of the heterozygous type over the corresponding homozygotes in diploid populations is another possible source of overestimates of the mutation rate.

Even if the true gene mutation rates in higher organisms are lower by, say, one order of magnitude than present estimates indicate, there would still be an adequate supply of mutational variation in populations. A moderate-sized population with an output of 10 million gametes would produce at least some new mutations in an average gene in each generation of gametes.

Genotypic Control

A gene is known in *Drosophila melanogaster* that produces high mutation rates in other genes in the complement. This mutator gene is designated *Hi*. Flies homozygous for *Hi* have mutation rates 10 times higher than the normal rates; heterozygotes for *Hi* have 2–7 times the normal mutation rates. The gene *Hi* induces both visible and lethal mutations in many genes. It also causes inversions, a type of chromosome mutation (Ives 1950; Hinton, Ives, and Evans 1952).

We have described gene mutations previously as chance errors of copying during gene reproduction, and undoubtedly this is a correct view as far as it goes, but the evidence of mutator genes indicates that there is another aspect. The production of new mutational variation, which is important for the longterm success of the species in evolution, may not be left to chance entirely, but may be promoted by mutator genes. The mutation rate of a species may be, in part, a genotypically controlled component of its overall genetic system.

The closely related, tropical American species, *Drosophila willistoni* and *D. prosaltans*, have different mutation rates. The rates were measured for the class of lethal mutations on chromosome II and chromosome III. The mutation rates are as follows:

	Chromosome II	Chromosome III
D. willistoni	2.2×10^{-5}	3.0×10^{-5}
D. prosaltans	1.1×10^{-5}	2.1×10^{-5}

Drosophila willistoni, with the higher mutation rate, is common and widespread, occurring in many ecological niches, whereas *D. prosaltans* is rare and ecologically restricted. It is plausibly suggested that the high mutation rate of *D. willistoni*, by producing a supply of new variations, is a factor in its ecological versatility and hence in its abundance (Dobzhansky, Spassky, and Spassky, 1952).

Adaptive Value

Most new mutant types have a lower viability than the normal or wild type. The reduction in viability ranges from a slight subvital condition to semilethal and lethal. A series of mutations in the X chromosome of *Drosophila melanogaster* were scored for viability. Ninety percent of the X-chromosome mutants had lower viability than normal flies, and 10% of the mutants were above normal, or supervital. The 90% fraction ranged from slightly subvital (45%) through intermediate degrees of reduced viability to semilethal (6%) and lethal (14%) (Timofeeff-Ressovsky 1940).

More generally, the adaptive value of new mutants is usually reduced. Adaptive value includes fertility and the functional usefulness of morphological characters as well as physiological viability. Many mutant types are infertile whether or not they are normally viable. Morphological macromutations usually have impaired functional efficiency. Nearly 99% of a large sample of induced mutations in barley *(Hordeum vulgare)* had a lowered adaptive value (Gustafsson 1951).

These observations can readily be explained. The genes in the normal or wild-type genotype have all passed through many generations of natural selection; they have been screened for maximum adaptive value. And changes in such genes would be expected to be, mainly, changes for the worse, in the same way that random tinkering with a clock is more apt to "impair its functional efficiency" than to result in improvements.

Gene mutations are frequently said to be "random" changes in the genes. The term random requires qualification in this context. Mutational changes may not, in fact, be random at the molecular level. Certain alterations in the nucleotide order in a DNA chain may occur more frequently than others. The so-called randomness of the mutation process has reference not to the molecular arrangement, but to the adaptive properties of the mutant genes. Mutations are random in the sense that they are not oriented in the direction of any present or future state of adaptation of the organism.

Nevertheless, a small fraction of the total array of gene mutations in genetically well-studied organisms prove to be superior to the standard type in one way or another. Between 0.1 and 0.2% of a sample of mutants in barley showed an increased yield in the standard or parental-type environment (Gustafsson, 1951).

A mutant type that is inferior in the standard environment may be adaptively superior in a different environment. The mutant *eversae* in *Drosophila funebris* has a reduced viability, 98% of that of wild-type flies, at 15°C, but superior viability (104%) at 24°C (Dobzhansky 1951:84). Six mutant types in the snapdragon *(Antirrhinum majus)* were all inferior to the parental strain in a normal greenhouse environment, but the mutants showed superior growth as compared with the parental strain in various abnormal greenhouse environments (Brücher 1943; Gustafsson 1951).

It is important to know that *some* new mutations are adaptively superior to the wild type, in either the standard or a new environment, for this indicates that the mutation process can be the starting point of evolutionary advances.

Minor Mutations vs. Macromutations in Evolution

Most evolutionary geneticists have stressed the importance of minor mutations in evolution. A minority viewpoint, expressed by Goldschmidt (1940, 1952, 1953, 1955) and others, holds that macromutations are of primary importance in evolution. Controversies between these opposing viewpoints have developed in the past, but are unnecessary, since the two views are not mutually exclusive but on the contrary are complementary. Both minor mutations and macromutations play a role in evolution.

Minor mutations have certain distinct advantages as raw materials for evolutionary change. Each minor mutant produces only a slight phenotypic effect, for better or for worse. A slightly superior minor mutant allele can therefore be fitted into the pre-existing genotype without bringing about any drastic disharmonies. By employing a series of minor mutations at different loci, it is possible to built up to some adaptively quantitative effect without disrupting the functional efficiency of the organism during the interim stages.

Occasionally, a single mutation affecting a key structure or function can open up new possibilities for its possessor. Resistance to specific toxins is known to be caused by single-gene mutations in organisms as diverse as bacteria and mammals. The mutant toxin-resistant strain might be able to invade a toxic environment barred to the parental susceptible type. A mutant wingless fly might be favored in a windy island habitat where the normal winged form could not survive.

The single macromutation does not necessarily have to act alone. Its action can be regulated by a series of modifier genes with individually small effects. In such a case we have a combination of macromutations and minor mutations in evolution.

Direct evidence bearing on the question of the types of gene changes that are important in evolution is provided by genetic studies of interracial hybrids. The genetic evidence indicates that diverse gene systems play their respective parts in the differentiation of related races. Interracial character differences in plants and animals are commonly controlled by multiple gene systems, confirming the hypothesis of the evolutionary importance of minor mutations. But single-gene character differences are also found in sets of divergent natural populations. This situation is uncommon, however. A common type of gene system differentiating races in higher organisms is the combination of a major gene and several modifier genes. This type of gene system suggests an evolutionary change based on a conspicuous mutation associated with various minor mutations.

The species of columbine (Aquilegia) differ by floral characters which adapt them for pollination by different types of animals such as bumblebees, hawkmoths, or hummingbirds. Since the species are interfertile, a factorial analysis can be made of the interspecific floral differences. Prazmo (1965) found that the interspecific differences in some key floral characters, such as presence or absence of nectar-bearing spurs and erect versus nodding orientation of flowers, is determined by a single allele pair. Here is a clear case of macromutations in particular genes playing an important role in the evo-

CHAPTER 7
Dispersal and Gene Flow

A population may acquire a new allele either by mutation occurring in some individual of the same population, or by immigration of a carrier of this allele from another population. The latter process brings about gene flow.

The carrier of the new allele must have obtained it from some prior event of mutation, and gene flow can thus be regarded as a delayed effect of the mutation process. Nevertheless, the direct and immediate source of some variations in a population at any given moment may be the immigration of carriers of different alleles from other populations. Particular mutant alleles might never appear in the population, but might flow in from neighboring populations. The ultimate source of the variation in the recipient population is then gene mutation, occurring elsewhere at some previous time; but the effective cause of the variation, here and now, is gene flow.

Now gene flow depends on another process, dispersal, which provides its physical basis.

How effective a process is gene flow over space and time? How far can a particular gene spread through a population system, in a reasonable period of time, say 1,000 years, with a given dispersal potential and a normal migration rate, with or without the aid of selection? This is a fundamental question which will be discussed but not answered definitively in the present chapter.

Definitions

The relations between the processes mentioned above will be clarified by the following definitions.

Dispersal is the movement of individuals or their spores, gametes, or particular dispersal organs through space. It can take place in either occupied or unoccupied territory of the species.

Immigration is a special case of dispersal in which the movement is from one part of a population system to another part of that system. In short, immigration is dispersal within occupied territory of the species.

Gene flow is the movement of genes within and between populations. It includes immigration and interbreeding of the immigrants with the natives to introduce the foreign genes into the recipient gene pool.

Gene flow may lead to changes in gene frequency in the recipient population, but does not necessarily do so. It is desirable to distinguish the two cases terminologically (Grant 1980). Failure to do so has caused some confusion in the literature.

Isogenic gene flow is gene flow between populations with the same gene frequencies. Genetically effective gene flow is gene flow between populations differing in gene frequency (Grant 1980).

Dispersal

The dispersal units are individual organisms in motile higher animals, free-swimming larvae in sessile marine animals, pollen and seeds in higher plants, and spores in lower plants and fungi. Dispersal is a normal part of the life history. All organisms, including sessile types, appear to have some free-living and motile stage in the life cycle when dispersal takes place. Conversely, the vast majority of the dispersal units, including the individual organisms in motile animals, tend to remain close to their place of origin or home territory. Some degree of sedentariness is also normal. The two aspects—dispersal and sedentariness—work together to determine the spatial distribution of individuals, and their genes, from generation to generation.

The units of measurement of dispersal vary from case to case. The most widely useful data set is a frequency distribution of dispersal distances from birth, hatching, or seed setting to the adult reproductive stage for a good sample of individuals. Such data have been obtained for the rusty lizard (*Sceloporus olivaceus*) by marking the young animals externally and recapturing them as adults (Blair 1960). The use of genetic markers in *Drosophila pseudoobscura*, together with release and recapture after many months, revealed dispersal distances over several generations (Dobzhansky and Wright 1947). In many studies practical considerations require recapture of the marked individuals a short time after release, so as to reveal dispersal distances over a fraction of the active lifetime of the organism. Thus in a study of the *Drosophila pseudoobscura* group the marked flies were recaptured 1–2 days after release (Powell et al. 1976).

Two practical units of measurement of dispersal are advocated by Yablokov and his coworkers, and used by them in their studies of lizards and other animals. The first of these is the so-called radius of individual activity (Timofeeff-Ressovsky et al. 1977:35, 62–65). This is the distance traveled by an individual per day or per season. The second is the mean radius of individual activity (Yablokov et al. 1980). It is the arithmetic mean distance between point of release and point of recapture for a stated time interval.

In seed plants dispersal takes place in two stages, pollen transport and seed dispersal, which are independent of one another. The total dispersal distance per generation must therefore be the net difference between the two distances as often as it is their sum.

Dispersal Patterns

Data on dispersal distances of adult flies of *Drosophila pseudoobscura* and closely related species are summarized in table 7.1 (Powell et al. 1976). The flies were marked with ultraviolet-fluorescent dust, released, and recaptured 1–2 days later at intervals along three transects. The data for the three transects were pooled for our present purpose. The high frequency of short dispersal distances is evident. There is also an abundance of medium-range dispersal events and some moderately long-range dispersals.

Comparable data for the sand lizard (*Lacerta agilis*) in the Altai Mountains, USSR, are given in table 7.2 (Yablokov et al. 1980). Here the dispersal distances are for one year. The bulk of the population remained close to

Table 7.1. Number of marked flies of *Drosophila pseudoobscura* group recaptured at various distances from point of release on three transects (Compiled from table 3 in Powell et al. 1976)

Distance, M	No. Marked Flies	Distance, M	No. Marked Flies
Release point	123	440	19
40	115	480	11
80	116	520	9
120	91	560	2
160	67	600	4
200	65	640	3
240	67	680	11
280	55	720	6
320	45	760	6
360	29	800	2
400	35		

the release point, but a small proportion dispersed far and wide. The mean radius of individual activity for the population is 30.6 meters per individual per year.

In another study of lizard dispersal, a local population of the rusty lizard (*Sceloporus olivaceus*) in Texas was monitored over a six-year period (Blair 1960). The dispersal distances in this study refer to the individual life-span from hatching to maturity. The dispersal pattern is shown graphically in figure 7.1. It will be seen that most individuals remain within 200–300 feet of their origin, but some individuals range more widely, up to 1,650 feet from their hatching place.

Table 7.2. Frequency distribution of dispersal distances in the sand lizard, *Lacerta agilis* (Yablokov et al. 1980)

Distance, M	Frequency (%) All lizards	Males
0–5	60.1	61.2
6–10	25.9	19.7
11–30	10.0	13.7
31–60	2.5	3.3
61–100	1.0	1.1
101–200	0.4	0.7
201–400	0.07	0.2
401–700	0.03	0.1

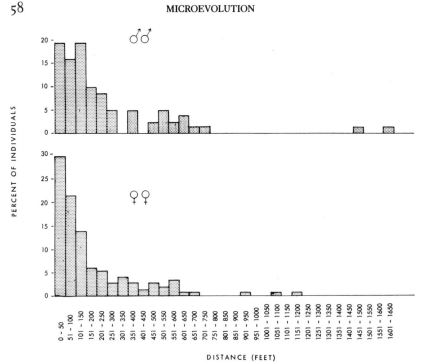

PERCENT OF INDIVIDUALS

DISTANCE (FEET)

Figure 7.1. Percentages of male and female rusty lizards dispersing various distances from point of hatching to place of sexual maturity. (Redrawn from Blair 1960)

The mean radius of individual activity for the various sex and age classes in the Altai population of the sand lizard is (Yabolkov et al. 1980):

Young	40.6 M
Adult males	34.8
Adult females	25.0

These results are typical. A greater dispersability in males than in females is found also in the rusty lizard (figure 7.1), the *Drosophila pseudoobscura* group (Powell et al. 1976), and various mammals. A greater dispersability of young than of adults has also been observed in flies and mammals.

There is some evidence in voles *(Microtus)* to indicate that the behavioral difference between wide-ranging and sedentary individuals of the same sex and age class has a genetic basis (Krebs et al. 1973).

An experimental set-up was used to reveal the dispersal pattern of pine pollen (Colwell 1951). Pollen of Coulter pine *(Pinus coulteri)* was tagged with radioactive phosphorus and, after wind dispersal, was traced with a Geiger-Müller counter and by radioautographs. One-liter lots of the pollen were released from inverted jars suspended at a height of 12 feet. Pollen release took place slowly on a fair day with gentle breezes. Pollen traps were placed at regular intervals along various radii from the point of release. The amount of pollen in the various traps could be estimated from the intensity of the radioactivity.

The main bulk of the pollen was found to be dispersed downwind to a distance of 10 to 30 feet. Beyond this zone of maximum concentration, the amount of dispersed pollen fell off rapidly. Only small amounts of pollen reached distances of 150 feet or more from the source point. The implication of these results is that a female cone in a pine forest will be swamped with pollen from neighboring trees, and will receive some pollen, but not very much, from trees standing a few hundred feet away (Colwell 1951).

The dispersal pattern described above has been found to be widespread in nature. The examples given are typical; other examples occur in rodents, birds, honeybees, butterflies, insect-pollinated plants, bird-pollinated plants, and wind-borne seeds. In general, the frequency distribution of dispersal distances shows a peak near the home base and a falling-off away from the home base. Much sedentariness is combined with some moderately long-range dispersal.

The expected result of this pattern is that interbreeding in an outcrossing organism will occur mostly between close neighbors. In the course of generations this will lead to a considerable amount of inbreeding despite an outcrossing breeding system. The close inbreeding is supplemented, however, by some wide outcrosses involving immigrant individuals or gametes. The wide outcrosses provide for a small but probably significant amount of gene flow over larger distances in the population.

Dispersal Ranges

Although the frequency distribution of dispersal distances is similar in different groups of organisms, the average dispersal distance is not. There are great differences between major groups in the typical dispersal range.

These differences are shown by table 7.3. The average dispersal distances are for one generation in most of the organisms listed and for a fraction of a generation in Drosophila. The results agree with expectations. Highly motile flies and sessile plants represent opposite extremes, separated by two or-

Table 7.3. Average dispersal distances in several organisms

Organism	Average Dispersal Distance, M	Source
Flies		
Drosophila pseudoobscura group, California	219, 361 in different populations	Dobzhansky and Powell 1974; Powell et al. 1976
D. pseudoobscura group, Colorado	176, 202 in different populations	Crumpacker and Williams 1973
Lizards		
Sceloporus olivaceus	49, 71 in males and females	Blair 1960
Lacerta agilis	31 (mean radius of activity)	Yablokov et al. 1980
Snail		
Cepaea nemoralis	8.1	Lamotte 1951
Herbaceous plants		
Senecio jacobaea	6.3 for achenes	Poole and Cairns 1940
Liatris aspera	2.5 for seeds	Levin and Kerster 1969
Phlox pilosa	1.1 for seeds	Levin and Kerster 1968

ders of magnitude. Terrestrial lizards are substantially less vagile than winged flies. Snails, on the other hand, do not range much more widely than plants.

Rate of Immigration

The immigration rate *(m)* varies quantitatively from 0 to 1, where $m = 0$ signifies no immigrant alleles, and $m = 1$ signifies swamping by immigrants in a single generation. In general, $m =$ number of immigrants per generation/number of natives plus immigrants in that generation; or $m =$ number of immigrant gametes per generation/total.

One of the best quantitative studies of immigration rates is that of Johnston and Heed (1976) on *Drosophila nigrospiracula* in Arizona. They present data on the percentage of marked flies in samples taken at varying dis-

Table 7.4. Immigration rates in *Drosophila nigrospiracula* (from table 3 in Johnston and Heed 1976, and Johnston, pers. comm.)

Dispersal Distance, M	Immigration Rate (m)
Release point	.517
19–30	.168
50–60	.084
90–100	.039
128–158	.041
234–274	.028
378–408	.029
443–512	.036
638–768	.052
871–932	.075

tances from the release point. Values of m calculated from these data are shown in table 7.4. These values represent weighted means for several different capture days at each distance, and were calculated by Johnston (pers. comm.)

The overall aveage immigration rate in these flies is $m = 0.037$. The value of m is not a constant, however, but falls off with distance in much the same way as dispersal distances in other studies that we have examined. The small excess of migrant flies at far-out distances is due to the irregular distribution of the flies' natural food sources (Johnston and Heed 1976).

Gene Flow

Gene flow is the genetic counterpart of migration. Migration is a factor affecting population variation if and only if the immigrants are genetically different from the natives. This is likely to be the condition where the migration takes place between populations or subpopulations that differ in allele frequency. Immigration in such a case will lead to what we have called genetically effective gene flow (in contrast to isogenic gene flow).

The relation between migration rate and allele frequency change is as follows (Falconer 1960, 1981). Let the frequency of an allele among the natives be symbolized as q_o, among the immigrants by q_m, and in the resulting mixed population by q. Then the change in allele frequency per generation (Δq) due to the gene flow is:

$$\triangle q = m(q_{\mathrm{m}} - q_{\mathrm{o}})$$

And the new allele frequency in the native population after one generation of immigration will be:

$$q = m(q_{\mathrm{m}} - q_{\mathrm{o}}) + q_{\mathrm{o}}$$

For example, if the frequency of an allele among the immigrants is $q_{\mathrm{m}} = 0.5$, and among the natives is $q_{\mathrm{o}} = 0.3$, and the immigration rate is $m = 0.01$, then the change in allele frequency resulting from one generation of immigration is $\triangle q = 0.002$, and the new allele frequency is $q = 0.302$.

The formula given above considers only m, the fraction of the population that migrates in each generation, and ignores population size, N. In studies of models of gene flow Slatkin (1980, 1981a, pers. comm.) finds that the effective amount of gene flow is measured better by Nm than by m alone.

In general, the rate of change in gene frequency in a population due to gene flow depends on two factors: (a) the immigration rate however expressed, and (b) the gene frequency difference between the donor and the recipient population or subpopulation (Falconer 1960, 1981).

Dispersal Though Time

Most studies of dispersal and gene flow, both empirical and theoretical ones, deal with single-generation changes. If our objective is to understand the dynamics of natural populations, however, it is clearly necessary to consider these processes over a long series of generations. We will attempt to put the time dimension into the picture in this and the next section (see also Grant 1980).

In an organism which disperses in all compass directions, some of the migratory moves in generation 2 will be an extension in the same direction of moves made in generation 1. A fraction of the migratory moves in subsequent generations will extend the range of the organism in this same direction. Average dispersal distance per generation can be multiplied by number of generations to give a reasonable numerical estimate of the cumulative linear distance which this organism could cover by stepwise migration through biological time. Such estimates for different organisms can then be put on a common comparative basis of chronological time by factoring in the generation time of each organism.

Table 7.5. Dispersal distances reached by stepwise dispersal through time in several types of organism (Further explanation in text)

Organism	Avg. Dispersal Distance per Generation, M	No. Generations Per Year	Extrapolated Dispersal Distance in 1,000 Years, Km
Drosophila pseudoobscura et aff.	176, 361	7	1,232, 2,527
Sceloporus olivaceus	49, 71	1	49, 71
Cepaea nemoralis	8.1	1	8.1
Senecio jacobaea	6.3	0.5	3.2
Liatris aspera	2.5, 3.5	0.2	0.5, 0.7
Phlox pilosa	1.1, 2.4	0.3	0.3, 0.7

The organisms considered in Table 7.3 are shown again in table 7.5, and annotated as to generation time. Values for average dispersal distance are used in four of the species. These are taken directly from table 7.3 in the case of *Drosophila* and *Sceloporus*; in the case of *Liatris* and *Phlox* the lower values represent seed dispersal, and the upper ones represent the sums of seed and pollen dispersal. The last column gives estimates of the cumulative dispersal or migration distance which each organism could reach in 1,000 years by a series of average dispersal moves in a rectilinear course.

We note that the highly vagile *Drosophila* can be expected to disperse or migrate 1,200 to 2,500 km. in 1000 years. The larger distance is about equal to the length or breadth of this organism's species area in western North America. The sessile plants, by contrast, would be expected to disperse or migrate only a few kilometers in 1,000 to 10,000 years. This distance is an insignificant fraction of the species area in these plants. Intermediate degrees of dispersability are exhibited by the lizard and the snail.

Gene Flow Through Time

We are concerned here with genetically effective gene flow occurring in a series of generations and involving three or more populations or subpopulations. We will apply the formula given earlier for a simple case to a slightly more complex one involving three generations and four populations.

Assume that four semi-isolated populations (A,B,C, and D) are distributed along a west-east transect. Population A has a new allele (G_2) at a frequency of 1.0 in generation 0; the other populations possess the old allele (G_1) at an initial frequency of 100%. Allele G_2 has no selective advantage or disadvantage relative to G_1. Migration occurs between neighboring populations in both directions at the rate of $m = 0.1$. After three generations the frequency of the new allele G_2 in the four populations will be:

Population A, $q = 0.755$
Population B, $q = 0.219$
Population C, $q = 0.025$
Population D, $q = 0.001$

The new allele clearly undergoes drastic reductions in frequency at each step along its pathway of migration. This is true even though the initial differential in allele frequency between population A and the other populations in our example is great, in fact maximal, and the migration rate is relatively high. Over a limited series of generations longer than postulated here, the new migrating G_2 allele will become so rare at some migrational step that it has only a negligible chance of being included in the next sample of emigrants. The process of genetically effective gene flow then peters out temporarily.

Over a longer series of generations, with continuing gene flow, the allele frequencies will tend to approach an equilibrium among all four populations, but this will require much time.

In the preceding section we concluded that dispersal distances have a substantial additive component in multi-generation time. Single-generation dispersal distances can be extrapolated to give reasonable estimates of the dispersal range attainable by the organism through time, as in table 7.5. Now we see that the migration of a new allele through space and time does not proceed in the same fashion. The new allele is normally carried by only a fraction of the emigrants in each generation, and moreover, the size of this fraction decreases in each successive generation. Genetically effective gene flow, in so far as it is determined by migration rates alone, is therefore much more restricted in space than stepwise dispersal. There is much slippage between dispersal and genetically effective gene flow (see also Grant 1980).

Let us relate these conclusions to the problem of gene migration throughout an extensive population system. We will assume that the population system is 1,000 km. across and that the time frame is 1,000 years long. Can a particular gene with no selective advantage migrate across this population system in the time allotted?

If the gene is carried by a sessile plant, or animal of low vagility such as a snail, the answer would appear to be no; the dispersal rate is too low, as shown in table 7.5. A highly motile, fast-breeding animal like *Drosophila* can readily extend its range across the 1,000 km. area, by stepwise dispersal, in the time available (table 7.5). Its dispersal potential is adequate for the task at hand. But it is not safe to assume that genetically effective gene flow, which is but a fraction of the dispersal potential, is adequate for the same task in the same organism.

It has been assumed up to now that the migrating allele G_2 is selectively neutral. Let us change this assumption and give it a selective advantage over the common and widespread allele(s) in the population system. This sets up a combination of forces, gene flow and selection, which will promote the spread of the allele.

But the migration of G_2 will still be slow because selection takes time. As G_2 enters a new population at a low initial frequency, it will require many generations of selection to raise its frequency to a level which ensures its passage to the next population, and once there the process must repeat itself again. In the case of stepwise gene flow under the control of the combined forces of migration and selection, we have to allow for many generations of selection at each migrational step.

The treatment of the problem presented here is qualitative and provisional. A quantitative treatment is highly desirable. This, however, is difficult since stepwise gene flow is a stochastic process, or to make matters more complicated, a series of separate stochastic processes, each of which can have a wide range of outcomes (Slatkin, pers. comm.).

Are We Underestimating Dispersal?

The quantitative data on dispersal, a sample of which was presented earlier, provides the basis for our extrapolations concerning gene flow through space and time. A definite possibility exists that the existing dispersal data are biased in favor of short and medium range dispersal.

Quantitative dispersal studies are, of necessity, carried out in a circumscribed study area. Individuals, spores, or other dispersal units that range beyond the limits of the study area are not caught or tabulated. Again, quantitative dispersal studies are conducted under "normal" weather conditions, but much long-distance dispersal can be accomplished during sporadic periods of high winds or floods.

Long-range dispersal is inferred from the colonization of oceanic islands and from some types of disjunct distribution patterns. It is documented by wind-borne pollen and insects over the oceans and stray individuals of bird species thousands of miles out of their normal range. It can be observed in action during periods of high winds in certain types of country. Savile (1972, pers. comm.) reports that a single gale in the Arctic can carry seeds, fruits, and some vegetative propagules hundreds of kilometers and as much as 2,000 kilometers.

Coyne et al. (1982) carried out release and recapture studies of *Drosophila pseudoobscura* in a large study area in desert mountains. They found that individual flies traveled up to 15 km.

There is, then, reason to believe that the available quantitative estimates of dispersal ranges are too low. Furthermore, evidence from a different source, namely, an inverse correlation between vagility and geographical race differentiation in certain groups, indicates that gene flow is indeed effective over areas of geographical extent. This line of evidence is presented in chapter 23.

Conclusions

Does gene flow play a significant role on more than a local scale in natural populations? Can genetically effective gene flow spread variations throughout an extensive population system? At present there are two contrasting viewpoints on this question.

Mayr (1954, 1963, 1970) is the chief spokesman for the view that gene flow does take place on a scale and at a rate such as to homogenize extensive population systems. The opposite viewpoint, expressed especially by Ehrlich and Raven (1969), and also others (Levin 1979, 1981; Ehrlich and White 1980) is that gene flow is too restricted to have any significant effects beyond the local population.

The first view is based on the morphological uniformity of large population systems in widespread species; the second on quantitative dispersal studies. In neither case is the evidence decisive. Morphological uniformity can be due to descent from a common ancestor as well as to gene flow; and the quantitative dispersal studies mostly do not go beyond single-generation changes.

The analysis presented in this chapter and elsewhere (Grant 1980; Slatkin 1981b) would suggest that highly motile animals like *Drosophila* conform to the Mayr viewpoint, and sessile plants to the viewpoint of Ehrlich and Raven, with other types of organisms falling in between. However, unknown quantities in the evidence render the analysis inconclusive. If genetically effective gene flow through time, with or without selection, turns out to be too small a fraction of dispersal, then populations of motile animals would conform to the Ehrlich and Raven model. On the other hand, if dispersal distances are significantly greater than current estimates, the populations of sessile plants would approach the Mayr model.

CHAPTER 8
Recombination

The Process
Amount of Genotypic Variability
Recombination and Mutation
Adaptive Value
Recombination and Complexity
Recombination and Evolution

Mutation and gene flow can produce variability in a population with respect to single genes. When allelic variation arises in two or more genes, as a result of these primary processes, the stage is set for the action of a second-order process, recombination. Recombination can combine the novel alleles, which at first are likely to be carried by different individuals, in a single genotype. Furthermore, recombination can multiply the number of different genotypes in the population. It converts a small initial stock of multiple-gene variation into a much greater amount of genotypic variation.

The Process

Assume that new mutations arise in two independent genes, A and B, in a diploid sexual population. Assume further that the mutant alleles *(a* and *b)* are originally carried by different individuals with the genotypes *AaBB* and *AABb*, respectively. The process of recombination can now proceed through the following series of steps: (a) crossing of the carriers of the different mutant alleles: *AaBB* × *AABb*; (b) production of the double heterozygote *AaBb* in F_1 (along with other types); (c) independent assortment of A and B at gamete formation to produce four classes of gametes: AB, Ab, aB, and *ab*; (d) production of nine classes of genotypes in F_2: AABB, . . . , *aabb*.

Most of the nine genotypes are new. The population at the beginning of the process contained three genotypes (*AABB*, *AaBB*, and *AABb*); two generations later it contains nine genotypes, including such new recombination types as *aaBb* and *aabb*.

Independence of the genes *A* and *B* is not a necessary assumption for recombination. The genes *A* and *B* can recombine whether they are borne on different chromosomes or at separate loci on the same chromosome. Linkage lowers the frequency of the recombination types but does not prevent their formation unless the genes involved are very tightly linked.

Amount of Genotypic Variability

Let the number of separate genes present in two allelic forms increase on an arithmetic scale $(2, 3, \ldots, n)$. The number of diploid genotypes then increases exponentially $(3^2, 3^3, \ldots, 3^n)$. In general, the number of possible diploid genotypes (g) is $g = 3^n$.

We saw above that two separate genes (*A* and *B*) present in two allelic forms each can produce nine genotypes, which is $3^2 = g$. It is well known in Mendelian genetics that a tri-hybrid cross, involving three genes (*A*, *B*, and *C*), can yield 27 genotypes $(g = 3^3)$.

Linkage disturbs the frequencies but not the total possible numbers of recombination types. If the separate genes are unlinked, their double or multiple heterozygote will produce the various recombination types in characteristic frequencies. If the genes are linked but separable by crossing-over, the recombinants will still be produced, but in lower frequencies proportionate to the strength of the linkage.

Polymorphic genes are commonly represented by multiple alleles in natural populations. In such cases, therefore, we do not raise 3, but rather some larger number, to the *n*th power in order to determine the number of genotypes. The general formula for the possible number of diploid genotypes (g), involving n (number of separate genes) and r (number of alleles of each gene), is

$$g = \left[\frac{r(r+1)}{2}\right]^n$$

Consider the application of this formula to the case of just two separate genes with various numbers of alleles. The results are shown graphically in

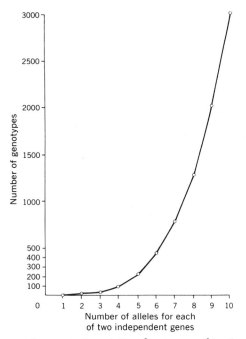

Figure 8.1. Increase in genotypic variation due to recombination with increase in number of alleles at each of two separate loci. (From V. Grant, *The Origin of Adaptations*, © 1963 Columbia University Press, New York; reproduced by permission)

figure 8.1. Individual variability due to recombination is seen to rise rapidly with an arithmetic increase in the number of alleles at the two loci.

Consider next the genotypic variability possible for multiple alleles at more than two loci. Some examples are listed in table 8.1. We see that a half million genotypes can be produced by recombination between 5 genes containing 10 alleles each. Going beyond the table, if a series of 10 or more alleles exists at each of 6 separable loci, the number of diploid recombination types is in the billions.

Most natural populations of higher animals and plants that have been studied genetically do contain polymorphisms for different genes. The foregoing numerical examples are not based on unrealistic assumptions; they are if anything too conservative.

Recombination is clearly a mechanism for generating tremendous amounts of individual genotypic variation. Given a moderate amount of polymorphism in just a few separable genes, this genic variation can be built up by recombination into astronomical numbers of genotypes. With only a mod-

Table 8.1. The number of diploid genotypes that can be produced by recombination between various numbers of separate genes each of which possesses various numbers of alleles (Grant 1963)

No. Alleles of Each Gene	Number of Genes				
	2	3	4	5	n
2	9	27	81	243	3^n
3	36	216	1,296	7,776	
4	100	1,000	10,000	100,000	
5	225	3,375	50,625	759,375	
6	441	9,261	194,481	4,084,101	
7	784	21,952	614,656	17,210,368	
8	1,296	46,656	1,679,616	60,466,176	
9	2,025	91,125	4,100,625	184,528,125	
10	3,025	166,375	9,150,625	503,284,375	
r	$\left[\dfrac{r(r+1)}{2}\right]^2$	$\left[\dfrac{r(r+1)}{2}\right]^3$	$\left[\dfrac{r(r+1)}{2}\right]^4$	$\left[\dfrac{r(r+1)}{2}\right]^5$	$\left[\dfrac{r(r+1)}{2}\right]^n$

erate amount of genic variation, the number of possible recombination types can easily exceed the total number of individuals in the species. Recombination is the reason why no two individuals developing from different zygotes are exactly alike genotypically in sexually reproducing organisms.

Recombination and Mutation

Linked genes are recombined following crossing-over. If the linked genes occupy closely neighboring loci, crossing-over will occur only rarely. A recombination type will then appear in the progeny as a rare event. The recombinant behaves like a mutant.

The similarity between rare recombination types and mutations can be illustrated by the following model. Two closely linked genes, A and B, control similar processes and can substitute for one another. In other words, a normal phenotype is produced by the dominant alleles $\dfrac{Ab}{aB}$ or by single dominant alleles (A or B). The double heterozygote $\dfrac{Ab}{aB}$ has a normal phenotype and usually breeds true. But crossing-over occurs rarely between A and B and yields some ab gametes. These produce aabb zygotes, which exhibit a "mutant" phenotype and breed true for the deviant character. If the crossing-over between A and B occurs at a rate comparable to mutation rates, it is not possible in ordinary practice to distinguish the rare recombinant from a mutant.

A number of cases conforming to the above general model have been revealed by refined genetic analysis in *Drosophila melanogaster, Zea mays,* and fungi. Crossover values as low as 0.00026, and possibly even lower, have been found in *Drosophila.* The compound gene A in corn *(Zea mays),* governing the color of the kernels and other plant parts, consists of two adjacent subgenes, which recombine rarely to yield types that superficially resemble mutants.[1]

The implication of these findings is that any sample of unanalyzed mutant types is likely to include a fraction of rare recombination types in addition to true gene mutations.

Adaptive Value

Certain recombination types among the large number produced may have a superior adaptive value. This is because the adaptive value of any given allele can be affected by other genes in the complement. Gene interaction is a factor determining how any particular gene performs.

Timofeeff-Ressovsky (1934a, 1934b) measured the viability of different mutant types in *Drosophila funebris* at 25°C and expressed the viabilities as a percentage of that of the wildtype. All single mutant types showed reduced viabilities. In some cases the recombination product, containing two mutant types, had a lower viability than either single mutant. In other cases the reverse was true. The recombination product of certain mutant types was adaptively superior to either single mutant and was nearly on a par with wildtype flies, as shown by the following comparison.

Miniature	69%	viability
Bobbed	85%	"
Miniature-bobbed	97%	"

Recombination and Complexity

Complex phenotypic characters are determined not by single genes, but by gene combinations. Such gene combinations are necessarily composed of

1. See Grant (1975: ch. 4) for review with references.

alleles that work harmoniously together. Recombination is the mechanism that assembles the gene combinations.

Recombination is important in nearly all organisms. Some means of exchanging genetic material exists in all organic kingdoms. Sexual reproduction, which is the chief means of bringing about recombination in eukaryotic organisms, has parasexual counterparts accomplishing the same net result in bacteria.

Although recombination is found in all major groups, its relative importance varies widely between groups. In higher animals there is the highest premium on recombination, which is promoted by obligatory sexual reproduction, high chromosome numbers, and other features of the genetic system. Bacteria, at the other extreme, manage to live successfully with a bare minimum of recombination. Higher plants are intermediate between these extremes in respect to the amount of recombinational variation normally generated.

These broad differences are correlated with the complexity of the organisms. Bacteria are relatively simple organisms with a relatively simple genotype. Important life functions, such as the ability to synthesize some vital metabolite, may be determined by single genes; the origin of new simple functions of this sort can often be brought about by gene mutation in combination with natural selection.

Single genes have relatively less individual importance in the enormously complex genotype of a higher animal. Here the adaptively valuable phenotypic characters are mostly determined by gene combinations, and usually by very complex gene combinations. Mechanisms for producing gene recombinations are therefore essential.

The Role of Recombination in Evolution

Evolutionary changes in single-gene characters can be brought about by the combination of mutation and selection. This combination of forces prevails in the bacteria. It operates on simple characters in multicellular organisms also, where, however, it plays only a secondary role in the overall process of change.

The evolution of new complex characters in multicellular organisms begins with the origin of multiple-gene variation and ends with the fixation of a new adaptive gene combination in a population. Recombination is an important midpoint in this process.

The process begins with mutations in two or more genes. The mutant alleles, if recessive, may be stored unexpressed for many generations in the diploid condition. The diploid state is a storehouse for mutational and multiple-gene variation. Sexual reproduction is the key to this storehouse. It produces the various possible recombination types out of the raw materials available in the gene pool.[2]

The biological function of sex is the production of an array of recombination types. The chances that any particular gene combination can be assembled in a single lineage by mutation alone and without sex are practically nil. Consider the formation of a derived genotype *abc* from an ancestral genotype ABC in a haploid organism. This transformation would be very slow if it had to depend on a series of mutations in asexual lines. If the population is sexual, however, the advanced genotype *abc* can be produced in just two generations by intercrossing between three lines carrying the alleles *a*, *b*, and *c* respectively (Wright 1931:145; Muller 1932).

The recombination process is blind as regards the adaptive value of its products. It mechanically churns out ill-adapted as well as adaptively useful recombination types. It is obviously desirable to minimize the proportion of the former. In any complex organism, foreign genes from a distantly related population are unlikely to work harmoniously in combination with the coadapted genes of the native population. One way of reducing the proportion of ill-adapted recombination types, therefore, is to erect barriers against wide hybridization.

The organization of populations into non-interbreeding biological species, each of which maintains its own separate pool of coadapted genes, is thus a corollary of sex. Biological species are a practical consequence of sexual reproduction. Recombination calls for the sexual mechanism, and sex in turn calls for a species organization (Dobzhansky 1937b).[3]

The recombination process, operating within species limits, produces a wide assortment of recombination types. Certain ones of these may be adaptively superior. The problem now shifts from the production of numerous recombination types to the preservation of certain superior ones. The sexual mechanism which puts together a valuable gene combination in one generation will inexorably break it up again in the next. The fixation of new superior recombination types now becomes the crucial step.

2. Recent treatments of the evolution of sex, with literature references, are given by Ghiselin (1974), Williams (1975), Maynard Smith (1978), and Shields (1982).
3. For a recent statement of this point see Grant (1981a:51–54).

Selection could in principle gradually replace an ancestral gene combination by a new one. But selection is a very inefficient and slow method of establishing a new gene combination if the population is large and outcrossing.

A more favorable condition for the fixation of a new adaptive gene combination is provided by inbreeding with selection of the inbred products (Grant 1963; Shields 1982).[4]

The inbreeding can be brought about in various ways. Small population size enforces inbreeding in an outcrossing organism. Localized dispersal patterns in a large population will also promote inbreeding (Bateman 1950; Shields 1982). Or a mating system that promotes consanguineous unions or self-fertilization will bring about inbreeding irrespective of population size. Two important special cases of fixation of gene combinations by inbreeding and selection will be discussed in later chapters: drift (chapter 16), and quantum speciation (chapters 24 and 25).

The formation and the establishment of recombination types require different, and indeed contradictory, conditions; outcrossing in the one case and inbreeding in the other. The contradiction can be resolved by alternating cycles of wide outcrossing and inbreeding. Thus a normally large population may pass through bottlenecks of small size. A predominantly self-fertilizing plant or animal group may have occasional episodes of outcrossing.

Inbreeding is a restricted form of sexual reproduction. Organisms can carry the restriction a step further by abandoning sexuality altogether during a short or long sequence of generations. An alternation of a sexual generation with a series of asexual generations is a favorable compromise solution permitting both the formation of new gene combinations and their subsequent replication, as suggested by Wright (1931:145) and a number of later authors. Life histories involving a balance between sexual and asexual generations are in fact more or less common in all eukaryotic kingdoms.[5]

---

4. The relation of inbreeding to homozygosity has of course been much discussed in works on population genetics and plant and animal breeding. The focus in these works is on single genes, quantitative characters, inbreeding depression, and other aspects. When I was writing *The Origin of Adaptations*, I could find nothing in the then existing literature on inbreeding in relation to the fixation of new adaptive recombination types, and so I developed this aspect myself (Grant 1963:278, 448–451, 569–570). Recently Shields (1982) has given an excellent book-length treatment of this subject.

5. The balance system of sexual-asexual reproduction in higher plants is discussed in Grant (1981a, esp. chs. 1, 5, 19, 31–33). Parthenogenesis in animals is discussed by White (1973).

In summary, we have one set of evolutionary forces, the mutation-selection combination, which provides an adequate explanation for primitive evolution and for simple character changes in higher organisms. Molecular evolutionists sometimes tend to emphasize the general importance of the mutation-selection system (e.g., Beadle 1963; Jukes 1966). They may also attempt to explain organic evolution in general in these terms (Beadle 1963).

The mutation-selection system does not, however, provide an adequate explanation for the evolution of multicellular organisms. The complex structures and functions of such organisms require correspondingly complex gene combinations. This requirement places a premium on recombination. A set of genetic systems then develops to promote and control recombination: sex, species, inbreeding, and secondary asexuality.

CHAPTER 9
Neighborhoods

One of the factors affecting population structure is migration. Population structure should be viewed in the light of dispersal and immigration patterns. The effect of these patterns is that the actual mating groups in a local population of cross-fertilizing organisms are smaller, and often much smaller, than the local population itself.

Two subpopulation units are recognized: the panmictic unit (Wright 1943), and the neighborhood (Wright 1946).

The Panmictic Unit

The panmictic unit is the group within which random mating occurs (Wright 1943, 1946). It is an actual unit in organisms which have obligate outcrossing and no preferential mating. Its size is usually symbolized by N_e. An important factor determining the size of N_e is migration rate (m). Another factor affecting the size of N_e is said to be population density; this relationship requires further study.

It should be noted that a certain proportion of inbreeding occurs in a panmictic unit. This proportion is (Wright 1943): $1/N_e$.

The Neighborhood Concept

A panmictic unit exists only if random mating takes place locally. But in many actual cases the breeding system brings about strong departures from local panmixia. A more general subpopulation unit, the neighborhood, was proposed to cover the wider range of breeding systems (Wright 1946).

Most subsequent authors have made no distinction between neighborhoods and panmictic units. For most practical purposes this is satisfactory. The size of the subpopulation is represented by the same symbol N_e in either case. However, there is a subtle difference depending on whether random mating occurs locally or not. If it does, the neighborhood equals the panmictic unit in all respects. If it does not, the neighborhood corresponds to a derivative of a purely theoretical panmictic unit. In that case one uses a theoretical panmictic unit to set up a standard from which to derive the size of the neighborhood.

The concept of the neighborhood starts with two assumptions. First, the chance of gamete union in a population falls off with distance between the parental individuals. This can be expressed as a mating probability distribution. Second, a proportion of inbreeding of $1/N_e$ occurs in a panmictic unit of size N_e. A large continuous population thus has a certain amount of local inbreeding per generation.

In dealing with a non-panmictic population, one could determine the actual proportion of inbreeding, then calculate the size of a panmictic unit with the same amount of inbreeding from the formula $1/N_e$, and the calculated value of N_e gives the neighborhood size. The factor $1/N_e$ can be used to derive neighborhood size in cases where local panmixia does not exist.

Neighborhood Area and Size

Neighborhood area in a continuous population is the size of the space occupied, on the average, by N_e individuals. An important factor affecting this quantity is the standard deviation of the dispersal distances in the population (σ). Data on the frequency distribution of dispersal distances, discussed in chapter 7, are important as a source for this statistic.

If a population has a two-dimensional spatial distribution, the neighborhood has a circular area with a radius of 2σ. Such a circle includes 86.5% of the parents of individuals at the center of the area (Wright 1946). In other words, if one takes an individual at the center of a circular area, and the radius of the circle is such that the individual's parents have a probability of 86.5% of coming from within this area, then the circumference of this circle outlines the neighborhood area.

If a population has a linear configuration, the neighborhood is a strip 3.5σ long. A strip of this size includes 92.4% of the parents of individuals at the center of the strip.

The values 2σ, 3.5σ, 86.5% and 92.4% relate to the properties of a normal frequency distribution. The statistics for a linear neighborhood are derived from a normal curve in which an area $\pm1.75\sigma$ ($=3.5\sigma$) long contains 92.4% of the variates. The circular neighborhood is related to a bivariate distribution where a circle 2σ in radius contains 86.5% of the variates (N. Fowler, pers. comm.).

Neighborhood size is the number of breeding individuals in the neighborhood area. A practical approach to the estimation of neighborhood size, therefore, is to obtain the standard deviation of dispersal distances (σ) from dispersal data. Then, for a two-dimensional population, one can stake out a circle of radius 2σ in the actual population, and count the number of individuals in this circular area.

Evolutionary Properties of Large Populations with Different Neighborhood Sizes

A large continuous population can be visualized as a series of overlapping neighborhoods. The size of the neighborhoods may be large or small in relation to the size of the overall population.

Migration rate is a factor affecting the size of the neighborhoods. As m increases, so does N_e. With high values of m, the large continuous population approaches, though does not necessarily attain, wide-scale panmixia. With low values of m, on the other hand, the large continuous population tends to break up into inbreeding groups, that is, into small neighborhoods.

On theoretical grounds it works out that a great deal of local differentiation can develop in a large continuous population if the size of the neighborhood is small. Assume a continuous, outcrossing, polymorphic population, with no selection. If $N_e = 1,000$ or more, little regional racial differentiation is expected to develop (apart from the effects of selection). But if $N_e = 100$, regional races can develop; and if $N_e = 10$, local races may develop without selection (Wright 1943).

The predicted result in the case of small neighborhoods in a large continuous outcrossing population has been confirmed in a computer simulation (Turner et al. 1982). A hypothetical population of 10,000 annual plants is assumed to be polymorphic for two alleles of a single gene, with no selective

GENERATION 1

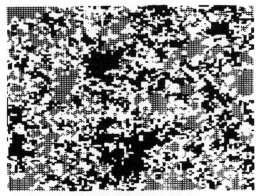

GENERATION 100

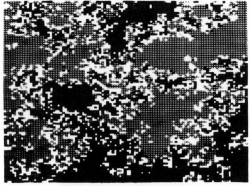

GENERATION 600

Figure 9.1. Spatial distribution of three genotypes at different times in a simulated plant population. Black = AA; gray = *aa*; white = A*a*. Further explanation in text. (Turner, Stephens, and Anderson 1982)

differential between the three genotypes. In the initial generation the three genotypes are randomly distributed throughout a large continuous population. The plants are outcrossing, but with nearest-neighbor pollination, so that neighborhood size is small ($N_e = 4.4 - 5.2$).

Over the course of one hundred to several hundred generations, with this neighborhood size and breeding structure, the originally homogenous population sorts itself out into patches of like genotypes or local races, as shown in figure 9.1 (Turner et al. 1982).

Table 9.1. Estimates of neighborhood size in several organisms

Organism	N_e	Source
Flies		
Drosophila pseudoobscura, California	4,922, 6,436	Crumpacker and Williams 1973
D. pseudoobscura, Colorado	3,239, 6,479	Crumpacker and Williams 1973
D. nigrospiracula	21,256 55,483 145,780	Johnston and Heed 1976
Birds		
Parus major	770, 1,806	Barrowclough 1980
Melospiza melodia	892	Barrowclough 1980
Larus argentatus	1,453	Barrowclough 1980
Puffinus puffinus	52,157	Barrowclough 1980
Rodents		
Peromyscus polionotus	240–360	Shields 1982
Mus musculus	12	Shields 1982
Lizard and frog		
Sceloporus olivaceus	225–270	Blair 1960; Kerster 1964
Rana pipiens	112–446	Shields 1982
Snail		
Cepaea nemoralis	190–2,850, sometimes up to 12,000	Greenwood 1974, 1976
Herbaceous plants		
Phlox pilosa	75–282	Levin and Kerster 1968, 1974
Liatris aspera	30–191	Levin and Kerster 1969, 1974
Lithospermum caroliniense	ca. 4	Kerster and Levin 1968
Lupinus texensis	42–94	Schaal 1980

Estimates of Neighborhood Size

Attempts to measure the evolutionarily important quanitity of neighborhood size in actual populations have been made in various organisms. Some representative estimates are given in table 9.1.

We see that neighborhoods are effectively large, with $N_e > 1,000$, in *Drosophila*, some birds, and some snail populations.

Moderately small neighborhoods representing a considerable departure from wide-scale panmixia are also found in various species. Thus in the rusty lizard *(Sceloporus olivaceus)* $N_e = 225-270$, but probably closer to 225, and neighborhood area is a circle with a radius of 213 meters (Kerster 1964). In the phlox *(Phlox pilosa)* $N_e = 75-282$ and the neighborhood area is $4.4 - 5.2$ meters in diameter (Levin and Kerster 1968, 1974).

In some wild populations of the house mouse *(Mus musculus)*, the breeding groups are very small, semi-isolated subpopulations consisting of several individuals (Selander 1970; DeFries and McClearn 1972; Shields 1982). A wild mouse population in a barn in Texas was polymorphic for an enzyme gene, esterase-3. Mouse traps were set in a grid pattern in one side of this barn and the trapped mice were scored for this genetic marker. The location of individuals carrying the three genotypes of esterase-3 at the time of capture is shown in figure 9.2. Individuals of like genotype are often clustered in a small area (Selander 1970).

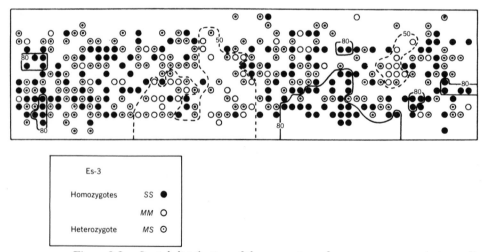

Figure 9.2. Spatial distribution of three genotypes for an enzyme gene (esterase-3) in mice *(Mus musculus)* in one side of a barn (Selander 1970)

This is not always the case in the house mouse, however, for Baker (1981), using another genetic marker, has found evidence for substantial amounts of migration and gene flow between local areas in mouse populations on farms in New York.

An extreme departure from wide-scale panmixia is found in the boraginaceous plant, *Lithospermum caroliniense*. Here the neighborhood area is about 4 meters in diameter and the effective size is about 4 plants (Kerster and Levin 1968).

With very small breeding groups, like those found in some natural populations as cited above, the situation is favorable for local racial differentiation without any action by selection.

PART III
Natural Selection

CHAPTER 10
Basic Theory of Selection

The Single-Gene Model
The Selection Coefficient
Rate and Extent of Change
Fitness
The Darwinian Concept of Selection
Differences Between the Darwinian and Population-Genetical Concepts of Selection
Components of Fitness

Natural selection comprises a diverse array of processes. The basic process is selective discrimination between different individual organisms. This process, though recognized by both nineteenth-century Darwinism and the modern synthetic theory, is conceived in different terms by the two schools of thought, as we shall see in this chapter. There are, in addition, various modes of individual selection and various selective processes, some of them controversial, that operate at levels of organization other than the individual. These will be discussed in later chapters.

The simplest form of selection is that envisioned in the single-gene model of the synthetic theory, with which we begin our discussion.

The Single-Gene Model

A large population is assumed to be variable for a gene A. The population contains an ancestral or wild-type allele A in high frequency, and a new mutant allele *a* in low frequency. If the carriers of *a* contribute more progeny to the next generation than do the carriers of A, and if this differential reproduction of the two alleles continues systematically generation after generation, then *a* will gradually rise in frequency in the population and A will decline in frequency.

This is the simplest form of natural selection. It suggests a definition applicable to the single-gene model. Natural selection in this sense is the differential and non-random reproduction of different alternative alleles in a population.

Natural selection takes place when the carriers of one allele (a) are more successful in reproduction than the carriers of an alternative allele (A), consistently and systematically during successive generations. The differential reproduction of the alternative alleles is non-random.

The selective advantage of a over A does not have to be great in order to lead to a change in allele frequency. The favored allele a may have only a slight advantage over other competing alleles, and a will still increase in frequency.

The Selection Coefficient

The selective advantage of one allele over the alternative allele (or alleles) can be expressed as a percentage or as a coefficient, the selection coefficient (s); s can have a range of values from 0 to 1.

The quantitative value of the selection coefficient is derived from the relative rates of reproduction of the alternative alleles. Assume that a is the favored allele and A the unfavored allele in a large population. For every 100 a alleles passed on to the next generation in this population, some number of A alleles (from 100 to 0) will also be passed on. The selection coefficient is a function of this ratio. The formula for s is: $s = 1 -$ reproductive rate of unfavored allele/reproductive rate of favored allele.

Consider the following numerical examples:

1. The relative rate of reproduction of a and A per generation is 100 a : 99 A. Hence $s = 1 - 99/100 = 0.01$. Or a can be said to have a 1% selective advantage.
2. The contributions of the alternative alleles to the next generation are in the ratio 1000 a : 999 A. Then $s = 0.001$, and the selective advantage of a is 0.1%.
3. The ratio is 100 a : 50 A per generation; $s = 0.5$.
4. The extreme case is 100 a : 0 A. Here $s = 1$. This means complete gene substitution in a single generation, which means in turn that A is a lethal gene.
5. The opposite extreme is 100 a : 100 A. In this case $s = 0$. No selection is taking place.

It is important to note that the selective advantage of *a* is not an all-or-none advantage, except in case 4. In every other possible case the selective advantage of the favored type is a statistical difference in reproductive rates. In any given generation in the breeding population, some individual carriers of *a* may fail to reproduce, and some individual carriers of A may be more successful reproductively than the average *a* types. But what counts, as regards changes in the gene pool, is the net reproductive contribution of all *a* carriers and all A carriers combined in the whole large population.

There is a random component in reproduction in the population. This random component may affect the relative rate of production of A and *a* either locally or in particular generations. But these random differences in reproduction of A and *a* are not selection. Selection occurs only if there is also a non-random component in the differential reproduction of the alternative alleles.

This is why the definition of selection given earlier specifies both "differential and non-random," following Lerner (1958) but not some other authors. It is also why we specified a large size in the population.

Rate and Extent of Change

The changes in allele frequency due to selection can run full course to complete replacement. The new allele *a* may be present initially at a low frequency as a result of mutation or gene flow. But so long as *a* has a selective advantage over the pre-existing allele A, even a slight advantage, *a* will continue to increase in frequency, and A will decline. The ultimate result, if enough time is available, may be the complete substitution of *a* for A in the population.

It is evident that these changes will take place faster with a high selection coefficient and slower with a low coefficient. The rate of change in allele frequency is directly proportional to the value of *s*.

However, for any given value of *s*, the rate of change in allele frequency is not constant at all allele frequencies. If the population is diploid, and the favored allele is dominant, change in allele frequency is rapid at low and intermediate frequencies, but falls off at high frequencies. For instance, a dominant allele with a 1% selective advantage ($s = 0.01$) can increase in frequency from 0.01% to 98% in about 6,000 generations; but it requires over 5000 generations to increase from 98% to 99%, and about 1,000,000 generations to increase from 99% to 100%. If the favored allele is recessive, on the other hand, change in frequency is very slow at low frequencies, but picks up at intermediate frequencies.

It has been assumed in the preceding discussion that the favored allele will progress all the way to complete fixation in the population. This is a simplifying assumption, but not necessarily a realistic one, for several reasons.

The selection coefficient may not remain constant throughout the long course of selection, but could undergo fluctuations and reversals. A given gene might then have a selective advantage at one stage in the history of the population and a selective disadvantage at another stage. Furthermore, during the course of time, other alleles are likely to be introduced into the population by mutation and gene flow. In this connection, the assumption of just two competing alleles in the population is unrealistic in itself.

Finally, if the population is diploid, the two or more alleles are incorporated into the respective homozygotes and heterozygotes, e.g., A and *a* in AA, *aa*, and A*a*; the real discriminating effect of selection will then be not among the alleles, but among their genotypes. The genotype *aa* might well be superior to AA, but the heterozygote A*a* might be superior to *aa*, and then the population will remain permanently polymorphic for A and *a*. In this case *a* will never reach complete fixation as a result of selection. We will discuss this situation in more detail in a later chapter. Meanwhile let us note that population geneticists are more and more coming to the conclusion that many alleles controlled by selection do tend to level off at varying frequencies lower than 100%.

Fitness

What is the basis for these changes in allele frequency? Why does *a* undergo systematic increases in frequency relative to A? In the first place, the basis of selective advantage must lie in some phenotypic character difference controlled by the A gene. In the second place, the phenotypic character involved must have some importance in survival and reproduction. The carriers of *a* must be superior to the carriers of A in some respect of life affecting reproductive success.

The question of the selective advantage or disadvantage of any given allele thus opens up large questions of gene action, gene-controlled development, gene-gene interaction, and genotype-environment interaction. We will explore some of these questions in later chapters. For the present we will simply assume that the *a* types have some superiority over the A carriers in respect to viability or fecundity.

This brings us to the concept of fitness in population genetics, also known as adaptive value or selective value. The fitness of a genotype is the average number of proleny left by that genotype relative to the average number of progency of other, competing genotypes. Fitness is thus a purely quantitative and operational concept. It is a quantitative measure of reproductive success.

Fitness (symbolized by *w*) is a function of the selection of coefficient $(w = 1 - s)$. If the ratio of the reproductive rates of A and *a* is 99 A : 100 *a*, then $s = 0.01$, as already noted, and we can now add that $w_A = 0.99$ ($w_A = 99$ A/100 *a* = 0.99).

High fitness, as the term is used in population genetics, does not necessarily carry the connotation of "survival of the fittest"; the synonymous term, adaptive value, does not necessarily mean that the genotype with a high value is especially well adapted to its environment. This may be the case. But, alternatively, the high fitness or high adaptive value might result solely from high fecundity. High fitness refers to genetically determined success in reproduction; it does not refer to the specific reasons for that success, which can vary widely from case to case.

The Darwinian Concept of Selection

Population genetics and modern evolutionary theory equate natural selection with differential reproduction of alternative forms of genes, genotypes, or other reproducible units. The original Darwinian concept was somewhat different. Darwin and his followers emphasized the life-and-death value of some variable characteristics of a species in "the struggle for existence."

Darwin's thesis in *The Origin of Species* (1859, 1872) can be summed up as follows. The individuals of a species compete for the means of life. These individuals differ with respect to minor variations in their characteristics and the variations are often hereditary. Some variant forms are better adapted for survival in the struggle for existence than others. Consequently the former will reproduce preferentially and pass their favorable characters on to future generations.

Darwin used the giraffe to illustrate his thesis. He postulated, plausibly, that the individuals in an ancestral population of giraffes differed slightly in the length of the neck and forelegs. The taller and the shorter animals would have entered into competition for leaves in times of food scarcity in their native savanna. In such times the taller individuals would have been able to reach leaves high in the trees that were not accessible to the shorter individuals. The latter therefore perished and their short-necked and short-legged characters disappeared with them. Conversely, the long neck and forelegs of the modern giraffe are a result of the preferential survival and reproduction, generation after generation, of the taller individuals (Darwin 1872 ch. 7).

In Darwin's own words (1859 ch. 4):

> Can it . . . be thought improbable [that] variations useful in some way to each being in the great and complex battle of life should sometimes occur in the course of thousands of generations? If such do occur, can we doubt (remembering that many more individuals are born than can possibly survive) that individuals having any advantage, however slight, over others, would have the best chance of surviving and of procreating their kind? On the other hand, we may feel sure that any variation in the least degree injurious would be rigidly destroyed. This preservation of favourable variations and the rejection of injurious variations I call Natural Selection.

Differences Between the Darwinian and Population-Genetical Concepts of Selection

There are two concepts of natural selection at the individual level. They differ in several respects. The population-genetical concept of selection has a well-developed mathematical basis. This enables it to show that, in theory, even slight differences in selective value can be effective in producing large changes in gene frequency over time. The nineteenth-century concept of course lacked this formulation.

Furthermore, the two concepts differ regarding the basis of the selective advantage of one type over another. According to the original Darwinian view, individuals of the superior type are better adapted to critical environmental conditions than individuals of the inferior type. The population-genetical theory of selection, by comparison, states that the type with the higher fitness or selective value is the one which leaves more progeny than its competitors, whether it is better adapted to its environment or not.

Imagine, for example, a plant or animal population containing two alternative genotypes, A and B. Type A is well-adapted to the physical environment, but is semi-sterile. Type B, on the other hand, is fully fertile but, while viable, is less well-adapted to the environment than type A. Type B will increase in frequency relative to A. Therefore it has the higher fitness or selective value according to the population-genetical concept of selection.

This example points up another difference between the two concepts. Natural selection in the Darwinian sense is a mechanism for improving the state of adaptedness in a population. This is not necessarily the outcome of population-genetical selection, where a given type is favored solely on its relative reproductive success. The reproductive success can indeed stem from secular adaptations, which are incidentally good for the population, but it can also stem from traits which, like social parasitism, benefit the individuals possessing them but are detrimental to the population.

The differences between the two schools of thought reflect differences in professional background. Darwin, Wallace, and their followers were field naturalists. The authors of the population-genetical theory of selection were statisticians and laboratory workers. I would suggest that the twentieth-century concept is not superior in all respects to the nineteenth-century one. The field naturalists had a perspective on the problem, which in the last analysis is a field problem, the problem of evolutionary change in natural populations, that is still needed today. In the next section we will suggest a way of integrating the two concepts of selection.

As noted in chapter 2, a number of current authors are referring to the synthetic theory as neo-Darwinism. This terminology obscures the real differences between nineteenth-century Darwinism and the modern theory as regards the nature of selection.

Components of Fitness

Differential reproduction is the end result of various factors. Fitness can be broken down into components, as follows:

> Differential mortality in early and juvenile stages
> Differential mortality in adult stage
> Differential viability
> Differential mating drive and mating success
> Differential fertility
> Differential fecundity

The original Darwinian theory of natural selection emphasized the first three components. Darwin later (1871) described the fourth component as a special category, sexual selection, which he regarded as distinct from natural selection. The population-genetical theory of selection, on the other hand, explicitly defines natural selection in terms of the end result, differential reproduction; and in practice it often places the emphasis on the last three components, the reproductive processes involved themselves.

There is an important difference between the first three components of fitness listed above, and the last three, as we have noted earlier. The first three set up a selective process leading to a better adaptedness to critical environmental conditions, whereas the last three components are not necessarily connected with improvements in secular adaptations.

Since the term fitness does not in itself specify the cause of differential reproductive success, it is useful to recognize two broad classes of fitness, corresponding to the two sets of components in the above list. The two classes of fitness are: (a) adaptedness, or the degree of adaptation of the individual or population to its environmental conditions, and the ability to leave more progeny for this reason; and (b) reproductive success per se.

CHAPTER 11
Gene Expression in Relation to Selection

The population-genetical theory of selection emphasizes genes, and the single-gene model focuses on the relative selective values of alternative alleles of one gene. But does natural selection really act on genes? In fact, selection acts directly on phenotypes and only indirectly on the underlying genes. Between the direct action of selection and the resulting change in gene frequency, therefore, lies the whole complex train of events by which gene action is translated into phenotypic characteristics.

Consequently, in order to understand the workings of natural selection, we must take the phenotypic expression of genes into consideration. This chapter will consider a series of different aspects of gene expression. One set of questions centers on the kind of phenotypic expression in relation to the environmental background; another on the degree of phenotypic modifiability; and still another on the multifarious modes of action and interaction of genes. An additional aspect of gene expression, the role of rate genes in phyletic change, will be examined in chapter 30.

The Relativity of the Selective Value

Let us consider the implications of a very obvious relationship. A given genotype determines one and the same phenotypic character expression in a variety of environments. The phenotype in question has a certain selective advantage in some environments; in other environments it might have a different selective advantage, or no advantage, or even a selective disadvantage. Under such conditions selection will favor the genotype determining that phenotype in some environments, but not to the same degree, or not at all, in other environments. In short, the selective value of a particular allele or genotype is not one of its intrinsic properties. It is instead a function of the phenotype-environment interrelationship.

A simple example is provided by shell color in the European land snail, *Cepaea nemoralis*. The snail lives in a variety of habitats, from closed dark beech woods to open sunny meadows. It is polymorphic for shell color, with brown, pink, and yellow forms. A single polymorphic gene determines these shell-color differences. The brown allele produces brown shells, and the yellow allele produces yellow shells, in the full range of environments inhabited by the snail.

The snails are subject to predation by thrushes and other birds that rely on their visual senses. The brown shells provide concealing coloration against the bird predators in beech woods, and the yellow snails are concealingly colored in meadows. As a result of the visual selective component in bird predation, the brown shell color is prevalent in snail populations in woodlands, and the yellow color is prevalent in meadow populations. The selective values of the brown and yellow alleles are thus correlated with the type of habitat, and become reversed with change of habitat (Cain and Sheppard 1952, 1954; Sheppard 1959; Lamotte 1959; Jones 1973).

Drosophila pseudoobscura in western North America is polymorphic for inversions on chromosome III. The types of inversions are assigned names and code letters: Standard *(ST)*, Chiricahua *(CH)*, Arrowhead *(AR)*, etc., as noted in chapter 5. It has been found that certain inversion heterozygotes have a selective advantage over the corresponding homozygotes. But this adaptive superiority of the heterozygous genotypes is manifested only under particular environmental conditions. Thus *ST/CH* has a higher selective value than either *ST/ST* or *CH/CH* at warm temperatures (21–25°C), but not at cool temperatures (16°C). The high selective value of *ST/CH* is manifested only when the flies are reared in crowded population cages. At a given warm temperature (21°C), *ST/CH* flies are superior when feeding on one kind of yeast *(Kloeckera)*, but not on another kind of yeast *(Zygosaccharomyces)* (see Dobzhansky 1970 ch. 5, for review).

Phenotypic Plasticity

The phenotype is the product of genotype-environment interaction during the course of development. A particular phenotypic characteristic is a product of environmental influences during development, as well as of gene action. Furthermore, the role of the environmental component in the formation of a phenotypic character varies among different characters and among different organisms. In man, for example, the environmental component is slight for the blood types, considerable for body weight, and very large in the sphere of mental and behavioral traits.

Different families of brome grass *(Bromus mollis)* were compared for range of phenotypic variation in panicle length and flowering time in a series of controlled environments (Jain 1978). Large differences occur between families in the amount of phenotypic plasticity. This result points to genetic control of phenotypic modifiability.

The proportion of the variation in a phenotypic character that is due to genotypic differences, as opposed to the proportion caused by environmental influences, is known as heritability. The effectiveness of selection is directly correlated with the degree of heritability.

If selection is working on a phenotypic character that is determined predominantly by the genotypic component in development, with only a slight environmental influence, i.e., a character with high heritability, the effect of selection on the composition of the gene pool will be immediate and relatively direct. But selection will be much less effective—its effects will be more delayed—in the case of phenotypic characters that are molded by the environment to a large extent. The capacity of a genotype to respond by appropriate phenotypic modifications to a wide range of environmental conditions hampers the effectiveness of environmental selection.

Pleiotropy

A gene usually has different and often unrelated phenotypic effects, that is, the gene affects more than one phenotypic character, a condition known as pleiotropy. Pleiotropy complicates the action of selection.

Suppose that a particular allele has two pleiotropic effects, one of which is favorable and the other unfavorable. The selective value of the allele for all practical purposes will then be the net value when its selective advantages are weighed against its disadvantages. At the point when the selective disadvantages outweigh the advantages, selection ceases to favor the allele and begins to discriminate against it.

Plant and animal breeders confront this problem in artificial selection constantly. Some economically useful characters, such as very high yield, simply cannot be fixed in a population of domesticated plants or animals if they are developmentally correlated with strongly disadvantageous characters, such as reduced fertility. Pleiotropic effects undoubtedly hold back the effects of natural selection in many natural populations in a similar way.

Expressivity Modifiers

The phenotypic expression of a particular gene or gene combination is affected by other genes of the complement, known as modifiers. In this and the next section we will discuss the roles of two classes of modifier genes, expressivity modifiers and dominance modifiers.

Expressivity is the degree of phenotypic expression of a gene or gene combination. A series of individuals with the same constitution for a given gene, when grown or reared in a standard environment, may exhibit different degrees of phenotypic expression of that gene. Thus different individual *Drosophila* flies, and lines of flies, may have different numbers of body bristles, although they have the same constitution for the main bristle-controlling gene and have been reared in the same environment. The gene is said to have variable expressivity.

The variable expressivity may be caused by expressivity modifiers. Plus modifiers enhance the phenotypic expression of the major gene and minus modifiers suppress it. The individual organisms, though genotypically uniform with respect to the major gene, are genetically different in their sets of expressivity modifiers. Furthermore, the action of the modifier genes may be affected by environmental factors, thus introducing another complicating factor. The variable expressivity will then be a result of complex interactions between the plus and minus modifiers and the environmental factors. This in turn complicates the action of selection.

Let us assume that a population is monomorphic for a major gene or gene combination determining a phenotypic character, but contains latent variability in the system of modifier genes. As a result of minus expressivity modifiers, the phenotypic character is not expressed in one environment (E_1), in which it would be adaptively valuable, and is only weakly expressed in another environment (E_2), in which it has a positive selective advantage. Of course selection will be ineffectual in environment E_1. But selection can be expected to build up systems of modifiers that enhance the phenotypic expression in environment E_2. The new set of modifiers may then have effects that carry over to environment E_1. The character comes to expression and can be stabilized by selection in E_1 as well as in E_2.

Experiments carried out with certain abnormal phenotypes in *Drosophila melanogaster* (Waddington 1953, 1956, 1957; Bateman 1959) seem to conform to the above model. The experiments made use of the known fact that the abnormal phenotypes in question (crossveinless wings, the bithorax condition of the thorax) can be induced in some adult flies by treating populations of eggs or pupae with environmental shocks (high temperatures, ether). The induced phenotypic alterations are not ordinarily hereditary, but they eventually became so in the experiments.

The environmental shock—the heat treatment or ether treatment—was applied to starting populations of flies in the egg or pupal stages; the appropriate phenotypic responses were obtained in some of the adult flies, and selection was practiced for the abnormal phenotypes. This process was repeated each generation for 24 to 29 generations. It was, in effect, artificial selection for an abnormal phenotypic trait in an abnormal environment in which that trait is expressed.

The selection was effective. The descendent generations of flies at the end of the experiment contained a significantly higher frequency of individuals that exhibited the phenotypic response to the environmental shock than did the ancestral generations at the beginning of the experiment. Furthermore, some of the derived flies exhibited the abnormal phenotype not only in the abnormal environment in which the selection was carried out, but also in the normal environment, for which they were not selected. The advanced generations contain some flies that develop abnormal phenotypes without the stimulus of the abnormal environment.

The experimental results can be explained on the following hypothesis. First, selection built up sets of modifiers that enhanced the expressivity of the altered wing or thorax character in the abnormal environment. And next, these new, strong modifiers produced their effects on phenotypic expression, not only in the abnormal environment, but also and as a by-product, in the normal environment (Stern 1958, 1959; Bateman 1959; Milkman 1960a, 1960b, 1961; Grant 1963).

The model involving expressivity modifiers has implications for evolutionary theory. The model indicates that selection for a set of modifier genes in one environment may produce a genotype that gives rise to unpredicted phenotypic expressions in other environments. New potentialities for phenotypic expression can sometimes be created inadvertently by selection for expressivity modifiers. In this way a population being selected for one new environment could acquire genotypically controlled phenotypes that, as a side effect, preadapt it for still other new environments.

Dominance Modifiers

The relationship of dominance and recessiveness in an allele pair in a diploid may stem from the relative strengths of action of the two alleles themselves, but this is not the whole story, for dominance and recessiveness are also brought about by the action of other genes, known as dominance modifiers. A particular heterozygous allele pair may produce a dominant phenotype when in one genetic background and an intermediate phenotype in another background. One genetic background has a set of strong dominance modifiers, modifier genes that enhance the phenotypic expression of the dominant allele, while the other genetic background has a weak or otherwise different set of modifiers.[1]

It will be recalled that the majority of new mutations are both deleterious and recessive. The usually deleterious effects of mutations are an inevitable consequence of the randomness of the mutation process in an organism that has previously gone through many generations of selection. But the recessiveness is not an intrinsic property of the mutation process. Why then should most new mutations be recessive?

Fisher (1930, 1958) suggested that natural selection in diploid organisms has built up systems of dominance modifiers that strengthen and stabilize the action of the normal, wild-type alleles. The dominance modifiers would be advantageous in protecting the diploid organism from the immediate ill effects of most new mutations. Dominance modifiers enable the diploid population to store mutations, deleterious or otherwise, in the recessive state, and to expose them to selection slowly in a proportionately small number of homozygous recessive segregates.

Fisher's theory of the origin of dominance is consistent with much evidence concerning both dominance and mutation, and provides a unifying explanation of diverse observed facts, but remains controversial. It has been

1. For a review of the genetics of dominance modifiers in plants see Grant 1975: 135–140.

defended by some later authors (Mayo 1966; Sheppard and Ford 1966), and criticized by others (Crosby 1963; Ewens 1965; Wright 1977; Wagner 1981).[2]

A difficulty of the theory is that selection for dominance modifiers would be operative only on wild-type/mutant heterozygotes, and these would be very rare in the population after the mutation occurs. Moreover, the selection, to act favorably, would require heterozygous mutant genotypes containing also plus dominance modifiers, and these would be even more rare. Consequently, the process of selection for dominance modifiers would be extremely slow.

Wright's (1977) criticism of Fisher's theory is that selection would be too slow to be effective in developing dominance modifiers, and that dominance could develop more easily in other ways. Wright's account is partly an explanation of other causes of dominance which do not involve dominance modifiers. Dominance modifiers are in effect argued away. But dominance modifiers do exist and require an explanation.

Wagner (1981) has suggested a way whereby the selection process would be speeded up. An allele that is advantageous will spread more rapidly if it is also dominant. A dominance modifier will enhance the selection of the major gene. This in turn enhances the spread of the modifier gene. Positive feedback thus occurs between selection for the major gene and selection for the modifier. The feedback loop will enhance the selection process to the maximum extent if the major gene and modifier are linked.

The Genotype as a Unit of Selection

The applicability of the single-gene model of selection to the real world lies mainly in the bacteria. In multicellular organisms, the phenotypic characters exposed to the action of selection are, with rare exceptions, not single-gene characters, but products of gene combinations. Selection is not given the opportunity of discriminating between the alleles of a single gene, but must operate on the alternative forms of a gene system composed of many components.

The primary form of selection takes place not among genes themselves, or even gene systems, but among individuals in a population. And the in-

2. For reviews with more references see Wright (1977:508–526) and Wallace (1981 ch. 21).

dividuals in the case of a sexually reproducing species normally differ with respect to many genes.

A view put forward by Dawkins (1976) that individuals in higher organisms are vehicles for "selfish genes," out to multiply themselves, is unrealistically reductionist and simplistic.

The individual organism, particularly in the more advanced forms of life, is a complex machine composed of many organs with different functional roles. The diverse organs and functions must be coordinated and harmonized. A change in one character may well be advantageous in relation to its own particular function, but have disadvantageous side effects on other functions of the organism. Selection will then favor or reject the new character on the basis of its net advantage or disadvantage to the individual organism as a whole.

The end result in cases where selection has opposing tendencies is often a compromise. Life is full of such compromises. The colorful plumage of birds of paradise serves a useful role in courtship, but also advertises the presence of the birds to predators. The conflict is resolved by a compromise. The brilliant plumage is confined to the male birds, which are expendable, while the females remain plain-colored.

CHAPTER 12
Examples of Selection

The ability of natural selection to bring about changes as predicted by selection theory has been tested and confirmed in countless studies over the years. The studies have been carried out with diverse species of animals, plants, and microorganisms. The studies can be grouped, according to the approach, into three classes: selection experiments in the laboratory or breeding plot; observations in uncontrolled natural populations; and the tracing of historical changes due to artificial selection in domesticated plants and animals. A few examples representing each of these three approaches will be presented in this chapter.

The Illinois Corn Selection Experiment

A long-term selection experiment with corn *(Zea mays)* has been carried on at the Illinois Agricultural Experiment Station since 1896. Artificial selection was carried out for several characters in different lines derived from a variable foundation population. The characters are protein content of kernels, oil content of kernels, and height of ear. The results have been reported by Winter (1929), Woodworth, Leng, and Jugenheimer (1952), Bonnett (1954), and Leng (1960).

Protein content of the grains was subject to selection in two parallel lines: a high line (for high protein content) and a low line, derived from the same foundation stock in 1896 with 10.9% protein. In each subsequent generation, or year since corn is an annual plant, a large number of ears were harvested and some of their kernels assayed for protein content. The fraction of the ears with the highest protein was then assigned to the high line and used as seeds for the next generation. Similarly, the fraction of ears with the lowest protein was used as seeds for each new generation in the low line. This fraction was 24 out of 120 ears in each line in the early years of the experiment, and 12 out of 60 ears in the later years. The selection experiment reached generation 60 in 1959.

Response to the selection was immediate and long-continued. In the high line, protein content rose from 10.9% in generation 0 to 13.8% (in generation 5) and 19.4% (in generation 50). In the low line, protein content went down from 10.9% (generation 0) to 9.6% (generation 5) and 4.9% (generation 50). Curves showing the progress of selection in the two lines up to generation 60 are presented in figure 12.1. Response to selection for oil content of the kernels was even more marked than that for protein (Woodworth, Leng, and Jugenheimer 1952; Leng 1960).

The selection for protein or oil content led to unexpected correlated changes in various morphological features. Thus the derivative high-protein line had small ears with translucent flinty kernels, and the low-protein line had large ears with long starchy kernels (Woodworth, Leng, and Jugenheimer 1952).

Selection for height of ears above ground started with a foundation pop-

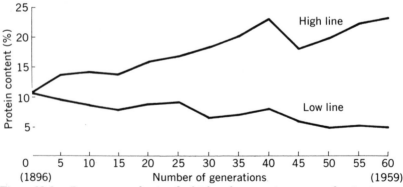

Figure 12.1. Response to selection for high or low protein content of grains in corn (*Zea mays*) during a 60-generation period. (Drawn from data of Woodworth, Leng, and Jugenheimer 1952, and Leng 1960)

ulation grown in 1903 in which the average height varied from 43 to 56 inches in different individuals. The plants with the ears closest to the ground in each generation were chosen as parents of the next generation in the low line. A parallel high line was continued by a similar procedure. The two lines were continued for 25 generations until 1928.

Here again the response to selection was marked. In generation 24 (in 1927) the average height of ears was 8.1 inches in the low line and 120.5 inches in the high line. As before, unpredicted but correlated changes took place in the highly selected lines. The low-ear strain came into flowering early in the season, whereas the high-ear strain was 10 to 14 days later in flowering (Bonnett 1954).

Zea mays is an outcrossing and highly heterozygous plant whose populations contain large reserves of genetic variability. The gradual and long-continued responses to selection in the Illinois experiment are consistent with the idea that the characters selected are determined by many genes that were polymorphic in the foundation populations. It is probable that new recombinational variation was released during the long course of the experiment, permitting the selection to continue to produce upward or downward changes, and it is quite possible that new mutational variation could have arisen too. The selection brought about correlated changes in other characters, thus demonstrating the complex and integrated nature of the genotype.

Viability in *Drosophila*

Dobzhansky and Spassky (1947) designed an experiment with *Drosophila* so that the changes in the populations would be guided by natural selection rather than by the conscious will of the experimenter. It is thus complementary to the Illinois corn selection experiment, in which artificial selection was the guiding factor.

The foundation populations were strains of *Drosophila pseudoobscura* with reduced viability. Their viability was expressed quantitatively as a percent of normal viability. Seven related strains were used as foundation populations for seven replicate cultures. Each replicate was subdivided into four subcultures which were treated differently. The treatments are designated briefly as follows: (a) irradiated and crowded; (b) crowded but not irradiated; (c) irradiated but not crowded; (d) not crowded and not irradiated.

The lines given treatments (a) and (b) were made homozygous in certain chromosomes to facilitate the expression of new mutations; the lines under treatments (c) and (d) were maintained in heterozygous condition for the same chromosomes.

Treatments (c) and (d) serve as controls for treatments (a) and (b), respectively. All lines were maintained for 50 generations in the laboratory. Viability and other characteristics were tested every few generations and at the end of the experiment.

Let us first consider the changes in one line of the seven given treatment (b). The viability of this line at the start of the experiment was 29% (of normal viability). Males and females were placed in a bottle containing food material and allowed to lay unlimited numbers of eggs, which soon resulted in the container's becoming overpopulated. A random sample of the offspring was then transferred to a new culture bottle and allowed to breed again. The line was continued in this fashion for 50 generations.

It was assumed that the crowded conditions in the bottles and the strong competition for food would lead to an increase in the frequency of any mutant types possessing greater vigor than their siblings. This expectation was realized. The viability of the strain rose from 29% to 90% in the course of the experiment.

Most of the other replicate lines of treatment (b) also underwent improvements in viability. One line showed viability increases from 60% to nearly 100%, another from 30% to over 80%. Viability rose in five of the seven lines under treatment (b).

The seven lines given treatment (a) were handled generally in the same way as the foregoing series, except that the male parents of each new generation were treated with X-rays before they were introduced into the fresh culture bottle. The purpose of the X-ray treatment was to increase the number of new mutations. A gain in viability was exhibited by six of the seven irradiated lines. Some of the viability gains were quite large, as from an initial 29% to a final 103% in one line, and from 65% to 115% in another line.

Altogether 11 of the 14 lines maintained under the conditions of intense competition for food in crowded culture bottles underwent marked improvements in viability during 50 generations. Similar improvements did not occur in the control lines (treatments (c) and (d)). The microevolutionary changes in the first two series of lines (treatments (a) and (b)) are associated with a partially homozygous constitution facilitating the expression of new genetic variations, and with differential morality in relation to food getting. In short, natural selection was at work and produced observable effects in the crowded lines but not in the controls.

Melanism in the Peppered Moth

The peppered moth *(Biston betularia)* occurs widely in England. It flies by night but spends the day at rest on the trunk or branches of trees, where it is vulnerable to the attacks of robins, thrushes, and other insectivorous birds. These birds hunt by sight. Two polymorphic forms of the moth concern us here: the so-called typical form, which is speckled gray, and a melanic form with black wings and body, known as carbonaria. The gray moths blend well into a background of lichen-covered bark, and bird predation is reduced by this concealing coloration. The melanic form, on the other hand, is conspicuous against a background of gray bark but is concealingly colored on soot-covered bark (Kettlewell 1956; Ford 1964).[1]

The typical and carbonaria forms differ with respect to a single gene, C. The carbonaria form is the result of a rare dominant mutation in this gene. The gray form is cc and the melanic type, Cc or CC.

The background environment of the peppered moth was changed drastically by the Industrial Revolution in England. One side effect of industrialism was an outpouring of smoke and soot over the formerly unpolluted rural English countryside. The lichens disappeared and the tree trunks and branches turned black with soot. These changes affected profoundly the composition of *Biston* populations in the industrialized areas.

The peppered moth populations consisted almost exclusively of the typical gray form in pre-industrial England, and this situation still exists in unpolluted parts of England. The melanic carbonaria form was first noticed as a rare mutant in 1848 at Manchester. From 1848 to 1898 the carbonaria form rapidly increased in frequency in industrial districts; it became the common type, while the typical gray form became rare. The frequency of the carbonaria allele C is estimated to have increased from 1% to 99% during the 50 generations from 1848 to 1898. A very high selective advantage of the carbonaria allele is required to account for the observed change in gene frequency (Kettlewell 1956, 1973; Ford 1964).

The selective factor is visually oriented predation by insectivorous birds. The typical form is protected on lichen-covered bark and the carbonaria form on soot-covered bark. Release and recapture studies show that the survival rate of the typical form is twice as high as that of the carbonaria form in an unpolluted area; conversely, the survival rate of carbonaria exceeds that of the typical form by a factor of about two near industrial Birmingham (Ket-

1. For review see Kettlewell 1973.

tlewell 1956). The historical change in substrate brought about by industrialization thus reversed the relative selective values of the c and C alleles.

The C gene has pleiotropic effects on physiological vigor that may lead to
secondary selective values in relation to other factors. The carbonaria type
is more viable than the typical form in stressful environments in the laboratory and probably also in nature. This favorable pleiotropic effect of the
carbonaria allele is overruled by strong counterselection in the form of bird
predation in unpolluted areas. In sooty areas, however, the moth populations can exploit the physiological advantages of the carbonaria allele.

Parallel changes from a gray or brown to a melanic form took place in
several other species of moths in England during the same period; a parallel
evolution of melanism in moths occurred again on a somewhat later time
schedule in industrial central Europe and the northeastern United States
(Owen 1961; Ford 1964; Kettlewell 1973).

Shell Color in Land Snails

The European land snail *(Cepaea nemoralis)* is common in a variety of
habitats—woods, meadows, hedgerows—in England and western Europe. The
populations are usually polymorphic for the color of the shell, which may
be brown, pink, or yellow (figure 12.2). Shell color is determined by a series
of multiple alleles, in which brown is top dominant and yellow the bottom

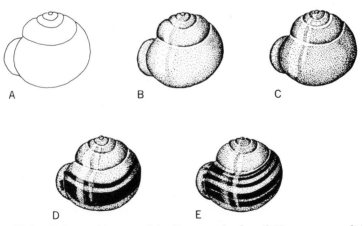

Figure 12.2. Polymorphic types of the European land snail *(Cepaea nemoralis)*. (A)
Yellow unbanded. (B) Pink. (C) Brown. (D) Two-banded. (E) Five-banded.

recessive (Cain and Sheppard 1954; Sheppard 1959; Lamotte 1959; Jones 1973).

Shell color plays the central role in the concealing coloration of the snails. Thrushes and other birds hunt the snails by sight, and take a higher proportion of contrastingly colored snails than of cryptically colored ones. What constitutes contrasting or cryptic coloration depends, of course, entirely on the background.

One of the habitats of *Cepaea nemoralis* is dense beech woods. These woods have a carpet of red-brown leaf litter, which remains about the same throughout the year. Here the brown and pink snails are protectively colored, and here these color forms are common. Yellow snails are rare in beech woods but common in green meadows (Cain and Sheppard 1954; Sheppard 1959).

Still another habitat is mixed deciduous woods. The character of the background changes with the seasons in these woods. In early spring the floor is brown with leaf litter, but in summer the floor turns green. And here the selective values of the different shell types also change with the seasons. In early spring the brown and pink snails have an advantage in relation to bird predation, whereas in summer the yellow snails have the advantage. Populations in deciduous woods undergo cyclical changes in the relative frequencies of the different polymorphic forms throughout the year (Cain and Sheppard 1954; Sheppard 1959).

Concealment is not the only function of shell color. There is some evidence indicating differences in physiological hardiness between the various color forms. The yellow form seems to possess better tolerance for temperature extremes, for both heat and cold, than the brown and pink forms. The greater temperature tolerance of the yellow type would of course be advantageous in open meadow habitats and in warm-summer areas. The yellow form does attain a high frequency in meadow habitats, as already mentioned, and also increases in frequency on a geographical transect from northern to southern Europe (Lamotte 1959; Jones 1973).

Another feature of the shells is the banding pattern. The shells may be banded or unbanded, and, if banded, they can have from 1 to 5 bands (figure 12.2). The presence or absence of bands is determined by a gene (*B*) closely linked to the shell-color gene. At this *B* locus, unbanded is dominant over banded. A separate locus controls the number of bands. Recombination produces the various possible combinations of shell color and banding pattern (Cain and Sheppard 1954; Sheppard 1959).

The role of natural selection is less clear in the case of banding pattern than in that of shell color. In other animals dark bands around the body are known to contribute to concealing coloration in semi-shady habitats with broken patterns of light and shadow. A similar function probably exists in *Cepaea nemoralis*. In some habitats, at least, the banded snails are at an advantage in relation to bird predation. It is also probable that some of the variation in banding pattern is controlled by genetic drift; we will return to this aspect of the problem in a later chapter.

Resistance to Toxins

Penicillin, streptomycin, and other antibiotics, when first introduced into medical use in the 1940s and 1950s, were effective in small doses in controlling disease bacteria. Shortly after the medical use of these antibiotics became widespread, however, their effectiveness in controlling bacterial infections began to decline, and higher doses had to be employed to achieve the desired results. There are antibiotic-resistant and antibiotic-susceptible strains of bacteria. The resistant types arise by spontaneous mutations at a given low rate. The application of antibiotics in light or moderate doses then sets in motion a selection process in favor of the resistant strains.

Such microevolutionary changes have been reported in laboratory experiments. An example is a selection experiment carried out with a strain (no. 209P) of *Micrococcus pyrogenes aureus (Staphylococcus aureus)*, a pathogenic bacterium that causes suppuration in wounds and food poisoning (McVeigh and Hobdy 1952). The original population of this strain was susceptible to various antibiotics and its growth was blocked by minute concentrations of these agents. Isolates from the original population were subcultured and grown in a succession of growth media containing increasing concentrations of penicillin and other specific antibiotics. Resistance to these antibiotics developed in the various lines. The greatest increase in resistance occurred during the first 10 to 20 subcultures; increase thereafter was slower. The increase in resistance to the various antibiotics was as follows:

Chloromycetin	193-fold
Aureomycin	210-fold
Sodium penicillin	187,000-fold
Streptomycin	250,000-fold

The derived resistant types exhibit other characteristics. They generally grow more slowly than the original, susceptible strain. The penicillin-resistant type had lost much of its pathogenicity and its ability to grow anaerobically. The susceptible form is thus superior to the resistant types in the ordinary antibiotic-free environment. Removal of antibiotics from the environment of the bacteria leads to selection in the reverse direction in favor of the susceptible form.

Parallel microevolutionary changes have been observed in several species of insect pests in relation to the widespread application of insecticides. The development of DDT-resistant strains of the common housefly *(Musca domestica)* is one well-known example. The evolution of cyanide-resistant strains of scale insects (*Aonidiella aurantii* and other species) in California citrus orchards is another.[2]

Various toxins occur in the natural world, and the process of evolutionary adjustment to them takes place in nature as well as in agriculture and medicine.[3] For example, the senita cactus *(Lophocereus schottii)* of Mexico contains alkaloids that are toxic to most species of *Drosophila*. But *Drosophila pachea* is resistant to this alkaloid and breeds in the rotting stems of the senita cactus. In its unique breeding sites, *D. pachea* is free from competition from other species of *Drosophila* (Kircher and Heed 1970).

Domestication

Scores of species of plants and animals have been transformed greatly by a combination of artificial and natural selection in the course of domestication. The histories of domesticated plants and animals, where known, provide numerous good examples of the efficacy of selection. Darwin, in *The Origin of Species*, cited the differences between the wild rock pigeon and the domestic pigeon, including such specialized breeds as the tumbler and pouter, as an example of the power of artificial selection to bring about profound evolutionary changes. Other cases come readily to mind: the dog, cattle, tomato, squash, etc.[4] In the following section we will consider one such case, maize.

2. For review see Dobzhansky 1970: ch. 7.
3. For more on this subject see Rosenthal and Janzen (1979) and Futuyma and Slatkin (1983).
4. For reviews see Darwin (1875), Zeuner (1963), Protsch and Berger (1973), Heiser (1973), Harlan (1975), and Simmonds (1976).

Evolution of Corn

Indian corn or maize (*Zea mays*), a member of the tribe Maydeae of the grass family (Gramineae), is strikingly different from all other members of its tribe and family in a number of characters, particularly in the distribution of the sex organs and in the structure of the corn ear or cob. The corncob is a very complex and highly specialized plant part; nothing like it is found anywhere else in the grass family. Nevertheless, modern corn with its unique cob and other features has evolved from some wild grass ancestor in tropical and subtropical America.

The evidence has been gained from comparative morphological, taxonomic, genetic, cytogenetic, ethnobotanical, archeological, and palynological studies. Numerous workers have participated in the gathering of this evidence. These workers agree on the general conclusion stated above, but fall into several different schools of thought with regard to the specific phylogeny, a matter to which we will return later.[5]

In *Zea mays* the staminate and the pistillate spikelets are segregated into different inflorescences, the tassels and ears, respectively; and corn is thus monoecious, an unusual arrangement in the Gramineae. The pollen-bearing tassel develops at the top of the corn stalk, while the grain-bearing ears develop in the axils of leaves in the mid-region of the stalk (figure 12.3). The corn ear is a spike consisting of a stout central axis bearing many rows of pistillate spikelets and, later, as many rows of mature kernels. The whole ear is enveloped by modified leaf sheaths (the husk). The grains are firmly attached to the cob and are naked, that is, not covered by bracts (glumes or chaff) as in other grasses (figure 12.4A, B).

In other grasses the grains are protected individually by the glumes or chaff, and separate individually from their inflorescence, so that they can function as units of seed dispersal. The opposite character combination in the corn ear—the firmly attached naked grains and the husk surrounding the whole ear—represents a loss of a mechanism for seed dispersal. But these same characters are useful in agriculture in that they make it possible to harvest

5. The evidence and various interpretations are presented in the following publications: Mangelsdorf (1952, 1958, 1959, 1974, 1983), Mangelsdorf and Smith (1949), Mangelsdorf and Reeves (1959), Mangelsdorf, MacNeish, and Galinat (1964), Mangelsdorf, Dick, and Camara-Hernandez (1967), Mangelsdorf, Barghoorn, and Banerjee (1978), Anderson and Brown (1952), Galinat (1970, 1971a, 1971b, 1977), De Wet, Harlan, and Grant (1971), De Wet and Harlan (1972), Wilkes (1967, 1972, 1982), Beadle (1972, 1980), Iltis, Doebley, Guzman, and Pazy (1979), Doebley and Iltis (1980), and Iltis (1981, 1983a, 1983b).

Figure 12.3. Form of the shoot in two varieties of corn *(Zea mays)*. (A) Dent corn. (B) Popcorn.

the corn efficiently and with minimum loss of kernels. A feature of the more highly developed varieties of modern corn is the large size of the individual grains and of the whole ear (figure 12.4A). In some primitive varieties of *Zea mays*, as in its wild relatives, the grains are tiny and the ears are small (figure 12.4B). The grains are hard and flinty as well as small in primitive corn, but are floury or sugary and thus easier to grind or chew in advanced corn.

The increased yield of modern corn requires a large leafy shoot to produce the stored food materials. This requirement is met; modern corn is a towering plant for an annual herb, as large as some bamboos, and is much taller than the primitive varieties of corn (figure 12.3). These and other features of the advanced varieties of corn are useful to man and have been bred into corn during many generations of artificial selection.

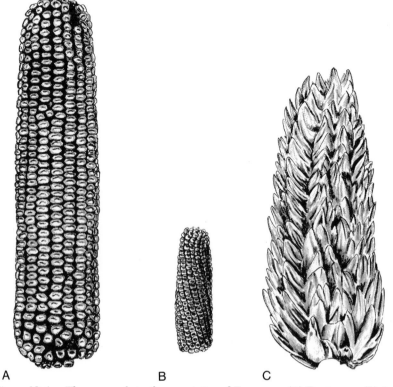

A   B   C

Figure 12.4.   The corn cob in three varieties of *Zea mays*. (A) Dent corn. (B) Argentine popcorn. (C) Pod corn. (Redrawn from Mangelsdorf 1958)

Corn has been cultivated by various agricultural groups of American Indians for at least 7,000 years (Mangelsdorf 1974, 1983; Wilkes 1972). Centers of corn culture developed in eastern North America, the Southwest, Mexico, Central America, and the Andean region. Some primitive characters are found in varieties of corn still being grown in these centers; others are found in long-extinct varieties preserved in archeological deposits.

Thus popcorn is relatively small in stature, with several stalks from the base of the plant, and has small ears bearing tiny hard kernels (figures 12.3B and 12.4B). Pod corn exhibits another primitive character, namely, the envelopment of the kernels by chaff, as in other grasses (figure 12.4C).

An important fossil corncob is that from Bat Cave, New Mexico, a cave inhabited by an agricultural people who cultivated a primitive type of corn. The Bat Cave corn ear was small, smaller than an ear of popcorn. The ker-

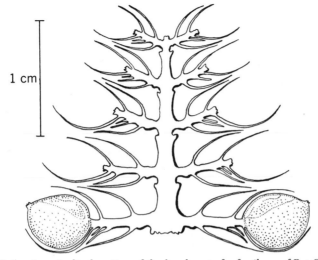

Figure 12.5. Longitudinal section of the basal part of a fossil ear of Bat Cave corn. The tip is missing. (Mangelsdorf 1958)

nels were tiny, as in popcorn, were borne on long pedicels from a relatively slender central axis, and were enclosed by chaffy bracts, as in pod corn (figure 12.5). The tip of the spike may have borne staminate spikelets. The oldest specimens of Bat Cave corn are dated by radiocarbon and other methods as 3,000–4,000 years old (Mangelsdorf and Smith 1949; Mangelsdorf 1958, 1974). Older fossil corn ears, from Tehuacan Valley, Mexico, about 7,000 years old, also possessed the combined features of popcorn and pod corn, and were bisexual with a staminate tip (Mangelsdorf et al. 1964; Mangelsdorf 1974, 1983; Wilkes 1972).

A still older fossil deposit from Mexico City contains pollen grains like those of modern corn. Although identifications on pollen grains alone, without other plant parts, are usually uncertain, the Mexico City fossil pollen has been shown by careful measurements to be corn and not teosinte or some other grass. It probably represents an ancestral wild form of corn. Its age is estimated to be 25,000–80,000 years old (Mangelsdorf 1974; Mangelsdorf et al. 1978).

Mangelsdorf (1958) crossed modern popcorn with pod corn to produce a synthetic pod-popcorn containing the primitive characteristics of each parental type. The synthetic pod-popcorn has a slender, bisexual, grasslike inflorescence with female flowers in the lower part and male flowers above. The grains are enveloped by chaffy bracts and separate readily from the in-

5 cm

Figure 12.6. Grain-bearing inflorescences of three relatives of modern corn with primitive characteristics. (A) *Tripsacum dactyloides*. Upper spikelets staminate and lower spikelets pistillate. (B) Synthetic pod-popcorn. Upper spikelets staminate and lower spikelets pistillate. (C) *Zea mexicana*. Branched series of pistillate spikes. (Redrawn from Mangelsdorf 1958, and Wilkes 1967)

florescence when ripe. Synthetic pod-popcorn thus has a grasslike inflorescence (figure 12.6B). It exemplifies a possible early stage in corn evolution.

The closest relatives of *Zea mays* in the American tropics and subtropics fall into three groups: (a) the small wild genus *Tripsacum* (figure 12.6A); (b) the weedy *Zea mexicana* or annual teosinte (figure 12.6C); and (c) the recently discovered wild perennial teosintes, *Zea perennis* and *Z. diploperennis*. These have all been considered as the ancestors of *Zea mays*.

The several species of *Tripsacum*, including *T. dactyloides*, are widely distributed in the warm parts of the New World. *Tripsacum* differs from *Zea mays* and *Z. mexicana* in being perennial and having bisexual inflorescences (figure 12.6A). The inflorescence of *Tripsacum* is not too different from that of synthetic pod-popcorn. *Tripsacum* can be crossed successfully with *Zea mays*, but not with *Z. mexicana*, and the F_1 hybrids, when obtained, are more or less sterile.

Teosinte *(Z. mexicana)* is a widespread annual plant in Mexico and Central America that frequently occurs as a weed in or near cornfields. *Zea mays* and *Z. mexicana* can be crossed artificially, and hybridize naturally in Mexico, and the F_1 hybrids are fertile. *Zea mexicana* has separate pistillate and staminate inflorescences, like *Z. mays*, but in the former the pistillate inflorescence is a branched system of slender spikes surrounded by a husk. In *Z. mexicana* the pistillate spikelets on a spike are protected individually by special envelopes (cupules); the cupules are present but reduced and serve a strengthening rather than protective function in *Z. mays*. In *Z. mexicana* the axis of the spike disarticulates, so that the grains separate freely from one another. Teosinte does not have a cob (figure 12.6C).

Great interest attaches to two recently discovered wild perennial forms of teosinte in Jalisco, Mexico, since the perennial herbaceous habit is generally primitive as compared with the annual habit in plant groups containing both life forms. The first perennial form, *Z. perennis*, a tetraploid, had been known previously but considered extinct, and so was actually rediscovered. The second, *Z. diploperennis*, a diploid, was new to science when discovered in 1978 (Iltis et al. 1979; Doebley and Iltis 1980).

There is general agreement that the ancestral characters of corn can be found among its relatives, *Tripsacum*, annual teosinte, and perennial *Zea*. When one tries to draw a specific phylogeny, however, the task becomes more difficult and controversial. One plausible phylogenetic hypothesis is that *Tripsacum* is the ancestor of the genus *Zea* (Doebley and Iltis 1980). A second is that *Zea diploperennis* in hybridization with wild corn produced teosinte (Mangelsdorf et al. 1978; Wilkes 1982; Mangelsdorf 1983).

Two additional and much debated hypotheses concern the ancestor of modern corn. One hypothesis sees teosinte as the ancestor of modern corn (Galinat, Beadle, Harlan, De Wet, Iltis). The other considers pod-popcorn to be the ancestral type (Mangelsdorf).

A bold recent version of the teosinte hypothesis holds that the maize ear is derived, not from the teosinte ear, but by a transformation and sex change of the central spike of a male tassel (Iltis 1983a, 1983b).

Opposed to the teosinte hypothesis, according to Mangelsdorf, is the ethnobotanical and archeological evidence. Teosinte is not an important food plant for humans, and as far as known, was never cultivated by Indians (Mangelsdorf et al. 1978; Mangelsdorf 1983). Furthermore, the first appearance of teosinte comes after that of corn in all archeological sites where both types can be reliably identified (Mangelsdorf 1983).[6]

Archeological evidence does favor the pod-popcorn hypothesis. The Bat Cave and Tehuacan Valley fossils were types of pod-popcorn.

Two primitive races of corn, still living in Mexico, are phylogenetically significant. The Chapalote race, a pod corn, is adapted to intermediate altitudes, and the Palomero Toluqueño race, a popcorn, is adapted to higher altitudes. Hybridization between these races may have given rise to the direct ancestor of modern corn (Mangelsdorf 1983).

The characters of domesticated corn were later affected by hybridization with *Zea diploperennis* and *Z. mexicana* and introgression of particular genes and chromosome segments from these species into *Z. mays*. South American corn was also affected by introgression from *Tripsacum* (Mangelsdorf 1983).

The morphological differences between *Zea mays* and its wild ancestors are very great. Nevertheless, these differences can be accounted for as a result of interaction between genetic variability and natural and artificial selection. Modern races of corn contain large stores of genetic variation. Some of this variation is mutational in origin; some stems from spontaneous hybridization between the races of corn and between corn and its relatives, teosinte, and *Tripsacum* (Mangelsdorf 1974).

Some large changes in particular morphological features in *Zea mays* are known to be produced by single genes and simple gene systems (Mangelsdorf 1958, 1974; Galinat 1971a; Beadle 1972, 1980). Pod corn, with its chaffy bracts, for example, is determined by a mutant allele (*Tu*, tunicate) of the gene *Tu* on chromosome 4. The difference between sessile and stalked pistillate spikelets is another example of a genetically simple but phylogenetically significant character difference in corn. The difference between flinty and relatively soft kernels is a single-gene difference. Perennial vs. annual habit may also be a single-gene difference. Some of the morphological changes in the evolution of corn could then be due to the artificial selection of a relatively small number of major genes and their modifier genes. Other morphological and physiological changes in corn evolution are due to the selection of more complex systems of multiple factors.

6. For more details on teosinte see inter alia Wilkes (1967), De Wet et al. (1971), Mangelsdorf (1974), and Yamakake (1975).

Conclusions

We have examined cases where selection of mutations in single genes has brought about relatively simple evolutionary changes (black peppered moths, toxin-resistant bacteria). Other cases involve selection for multifactorial characters (protein content of grain in corn). But in the evolution of corn itself from its wild ancestors we are confronted with changes of greater magnitude, changes in a complex of characters determined by a combination of genes. A black peppered moth is still a peppered moth, but modern corn represents an essentially new type of plant in the grass family.

Critics of selection theory have long argued that selection is a negative force; it can eliminate the unfit, but, in combination with blind mutations, it cannot be expected to produce anything new (e.g., Koestler and Smythies 1969).

This criticism misses the point that character complexes are determined not by simple genes, but by gene combinations. Sexual reproduction is a mechanism for assembling the combinations of alleles that determine new character complexes, as we have seen earlier. And selection is a mechanism that can then establish the new allele combinations in populations if they are favorable. Seen in this light, selection is a mechanism for bringing about highly improbable events, as pointed out by Fisher (1930, 1958) and Huxley (1943). A corn ear is an improbable phenomenon in the grass family. What are the chances of a corn ear developing *without* the agency of selection?

CHAPTER 13
Effects of Density and Frequency

Competition
Mixtures of Genotypes Under Competition
Density-Dependent Factors
Frequency-Dependent Selection
Hard and Soft Selection

Competition

Interactions of various sorts between individuals affect the process of selection. The most general of these interactions is competition for necessary raw materials, energy sources, or living space. The competition can take place between individuals of the same species or between species. Here we are concerned with intraspecific competition.

Competition is not necessary for the effective action of selection. A selective differential between genotypes, with respect to physical environmental features, can show up in a sparsely populated habitat in which all genotypes are potentially able to multiply. However, the selective pressures are more intense and the effectiveness of selection is greater under conditions of competition.

In selection experiments with *Drosophila pseudoobscura*, for example, the selective advantages of the various homozygotes and heterozygotes (*ST/ST, ST/CH, ST/AR*, etc.) appear only when the population cages are crowded, and not in thinly populated cages (see chapter 11).

Sukatchev (1928) compared the viabilities of three asexually reproducing (apomictic) strains of the dandelion *(Taraxacum officinale)* in experimental plots in Leningrad. The strains are designated as biotypes A, B, and C. They were grown in pure open stands (with plants spaced 18 cm apart), in pure dense stands (individuals 3 cm apart), and in mixed stands (A, B, and C).

Natural mortality in the different plantings was measured at the end of two years. We are concerned here with the two types of pure stands.

The observed viabilities of the three biotypes, as represented by the percentage of individuals surviving through two years, was as follows:

Biotype	Survival, open stand	Survival, dense stand
A	77%	27%
B	69%	49%
C	90%	24%

Survivability was lowered by competition in all biotypes, but the biotypes differed in their response to the competition. In open stands biotype C was superior. In dense stands B was superior to both C and A.[1] Selection would favor one type in a sparsely populated habitat and a different type in a dense stand.

The above results suggest that the genotypically determined ability to weather physical environmental factors is different from the ability to succeed in a strong competition (Sukatchev 1928). Subsequent experiments along similar lines in both plants and insects have confirmed this conclusion and led to the concept of competitive ability as a special gene-controlled property of the organism. In barley *(Hordeum sativum)* and flour beetles *(Tribolium castaneum* and *T. confusum)*, different genotypes differ in their degree of competitive ability (Sakai and Gotoh 1955; Lerner and Ho 1961). The barley studies show further that competitive ability is a character separate from general vigor.

In the higher vertebrates, competitive ability often assumes the form of aggressive behavior. Aggressive behavior is usually associated with competition, is manifested most strongly and sometimes exclusively under conditions of crowding, and is a technique for coping with competition (Wilson 1971).

Recent experiments with two species of filaree *(Erodium)* indicate that intraspecific competitive ability is different from interspecific competitive ability (Martin and Harding 1982).

---

1. See also the more recent competition experiment in dandelion by Solbrig and Simpson (1977), which is different but yields comparable results.

Mixtures of Genotypes Under Competition

In wheat *(Triticum aestivum)* and oats *(Avena sativa)*, two varieties grown together in competition give a larger number of plants at harvest and a higher total yield than either variety sown alone (Gustafsson 1951). Chromosomally polymorphic experimental populations of *Drosophila pseudoobscura* likewise exceed monomorphic populations in total biomass of flies produced from a given amount of food (Beardmore, Dobzhansky, and Pavlovsky 1960).

Furthermore, a given genotype often reacts very differently, as regards competitive ability, in a pure dense culture and in a mixed dense culture. The genotype that succeeds well in competition with other individuals of the same genotype is not necessarily a superior competitor in a polymorphic mixture of genotypes. And conversely, a relatively poor competitor in a dense pure stand may become the dominant member in a mixed stand. Gustafsson (1951) calls this the Montgomery effect.

The Montgomery effect is exhibited in another part of Sukatchev's (1928) experiment with *Taraxacum*. He grew biotypes A, B, and C in pure dense stands and in a mixed dense stand in the experimental garden, and counted the number of flowering heads per individual, a good measure of plant vigor and fruitfulness. The results were as follows:

Biotype	Heads/individual, pure dense stand	Heads/individual, mixed dense stand
A	20–35	1–8
B	34–43	12–20
C	8–11	16–23

The relative competitive ability of genotype C shows a reversal in going from a pure culture to a polymorphic mixture.

Similar results have been observed repeatedly in single-variety vs. mixed plantings of barley, wheat, timothy *(Phleum pratense)*, and other grasses (Gustafsson 1951). And again, parallel results are obtained in experimental monomorphic and polymorphic populations of *Drosophila melanogaster* and *Tribolium castaneum* (Lewontin 1955; Sokal and Karten 1964).

Density-Dependent Factors

Some factors in the environment that operate as selective agents become more severe as population density increases. Those environmental factors whose severity of action on the population increases with population density are called density-dependent (or density-regulated) factors. They stand in contrast to density-independent factors, the strength of action of which is not correlated with population density.[2]

In general, physical factors in the environment, such as temperature, moisture, floods, lava flows, etc., tend to be density-independent, while biotic factors tend to be density-dependent.

Infectious diseases provide a good example of a density-dependent factor. As the population becomes more dense, infections and mortality increase; and at some point in the rising population density, epidemics may develop. As population size decreases, on the other hand, the disease factor becomes less active, and, at the lower extreme of population dispersion, may become quite inactive.

Competition is a density-dependent factor. The action of herbivorous animals on a plant population, and of predators on a population of prey animals, are other such density-dependent factors.

Density-dependent factors tend to have a stabilizing effect on population number. When the population becomes very dense, the factor reduces its size; but when the population is thinned down, the factor plays a permissive role, allowing the numbers to build up again.

Density-dependent factors generally exercise their selective role at and above certain threshold population densities. Disease organisms set in motion a selection for disease resistance when the population reaches some level of density, and this selective process increases in intensity at higher densities. The selective value of a disease-resistant gene is not constant, in other words, but varies with population density. The variation in selective value of the gene may span a wide range, being virtually nil in a dispersed state of the population, acquiring a positive selective value at a threshold density, and having a high selective advantage at high population densities.

Frequency-Dependent Selection

A number of cases have been found in both insects and plants where the selective value of a gene or genotype varies with the frequency of that gene or genotype. This situation is known as frequency-dependent selection. The selective value usually varies with frequency in a negatively correlated way. That is to say, the gene or genotype usually has a higher selective advantage

2. Review of density-dependent selection in Roughgarden (1979: ch. 17).

when at low frequency than it has at higher frequencies. The opposite situation of low-frequency selective disadvantage has also been reported.

Good examples of frequency-dependent selective values in nature are provided by Batesian mimics in various species of butterflies. The success of the mimicry as a means of protection against bird predators depends on the mimics' being relatively uncommon in comparison with the number of the models. The insectivorous bird learns to avoid the model species of butterfly, which it recognizes by sight, through experience with its noxious or poisonous qualities. The mimic species benefits indirectly from its resemblance to the model. If the Batesian mimic becomes abundant, however, it confuses the learned reaction of the bird predator to its own disadvantage. Therefore the selective advantage of the mimetic form is high when it is rare and drops when it becomes common.

In several species of *Drosophila*, including *D. melanogaster*, *D. pseudoobscura*, *D. paulistorum*, and *D. willistoni*, mutant males and wild-type males have been compared with respect to mating success in experimental populations. It has been found in these cases that the mating success of a type of male varies with the frequency of that type in the experimental population. The type of male that is rare, whether it is the mutant or normal type, has the advantage in mating (Ehrman and Spiess 1969).

Other examples of frequency-dependent selective values in insects are an enzyme locus and an inversion type in *Drosophila melanogaster*, and the black mutant in *Tribolium castaneum* (Kojima and Yarbrough 1967; Nassar, Muhs and Cook 1973; Sokal and Karten 1964). Parallel examples in plants include a gene for seed-coat pattern in *Phaseolus lunatus* and variant corolla types in *Phlox drummondii* (Harding, Allard, and Smeltzer 1966; Levin 1972).

Hard and Soft Selection

Wallace (1968, 1981) recognizes two contrasting modes, hard vs. soft selection. With hard selection, there is strong selection for individuals possessing a given character state, and elimination of those lacking it. Hard selection makes absolute discriminations. Soft selection, by contrast is moderate and relative. In a series of populations exposed to soft selection, the selection process favors the relatively superior individuals in each population.

These modes correlate with the modes described in the two preceding sections. Soft selection is density-dependent and frequency-dependent. Hard selection is independent of frequency and density (Wallace 1981).

CHAPTER 14

Modes of Individual Selection

Natural selection at the individual level takes different forms depending on the types of variants eliminated and preserved from generation to generation. Four such modes can be recognized: directional, stabilizing, disruptive (diversifying), and balancing selection.

Directional Selection

This is the mode discussed in chapters 10 and 12. Directional selection brings about a progressive or unidirectional change in the genetic composition of a population (figure 14.1). It occurs when a population is becoming adapted to a new environment or, conversely, when the environment is changing progressively and the population is keeping pace with it. A good example of the first situation is the rise in frequency of the melanic form of *Biston betularia* and other moths when confronted with a sudden change in environment, namely, industrial soot (chapter 12). Environmental changes that occur as gradual trends, such as many climatic changes, illustrate the second situation that evokes directional selection.

MODE OF SELECTION

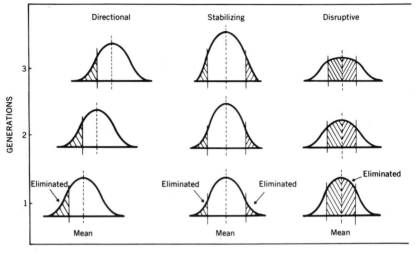

Figure 14.1. The effect of three modes of selection on the genetic variation of a population.

Stabilizing Selection

Stabilizing selection can be contrasted with directional selection. If a population is well adapted to a given environment that remains stable, the main action of selection is to eliminate such ill-adapted and peripheral variants as arise by mutation, gene flow, segregation, and recombination. A range of genotypes of proven adaptedness is thus preserved while the ill-adapted types are weeded out. This is stabilizing selection.

The effect of stabilizing selection is shown graphically in figure 14.1. The genetic variation in a population is assumed to have a normal distribution, with numerous individuals near the mean and a few individuals at the extremes for any given variable and measurable character. Under stabilizing selection the peripheral variants on both extremes of the normal curve are eliminated, generation after generation. The preferential reproduction of the individuals possessing characteristics around the mean for the population results in the preservation of a constant modal condition through time. This is contrasted with directional selection, in which the elimination of the genetic variation is one-sided and the mean consequently shifts during a succession of generations (figure 14.1, left).

Prout (1962) carried out an experiment on stabilizing selection in *Drosophila melanogaster*. The variable character used was time of development of the flies. In the line subjected to stabilizing selection, the individuals close to the mean for development time were selected in each generation. After a number of generations of such stabilizing selection, the variance of development time decreased in this line relative to that in the starting generation and in unselected control lines.

Stabilizing selection is ubiquitous, though unspectacular, and is the most common mode of selection in nature. We have previously referred to the striking example of directional selection in *Biston betularia*, but note the complementary role of stabilizing selection in the moth populations. Prior to industrialization, the rare melanic mutants were weeded out of the populations by stabilizing selection, a selective process that probably went on for many centuries; and in the altered industrial environment occupied by the new, predominantly melanic populations, stabilizing selection is again at work, weeding out the occasional gray types.

Disruptive Selection

In disruptive selection the extreme types in a polymorphic population are selected for, and the intermediate types are selected against. The result is to preserve and accentuate the polymorphism (figure 14.1, right).

Consider again Prout's (1962) selection experiment on development time in *Drospophila melanogaster*. Disruptive selection was practiced in one line by mating the earliest flies with the latest flies to emerge in each generation. The result, after a number of generations, was an increase in the variance of development time as compared with the unselected control lines.

A series of experiments on disruptive selection in *Drosophila melanogaster* has been carried out by Thoday and his coworkers. The experiment to be described here (Thoday and Boam 1959) involves selection for number of body bristles (chaeta number).[1]

Selection was carried out concurrently for high bristle number in a set of high lines and for low bristle number in a set of low lines. This disruptive selection tended to subdivide the original population into different high and low subpopulations. But the high and low lines were constantly intercrossed

1. Other experiments on disruptive selection in *Drosophila* are described by Millicent and Thoday (1961), Thoday and Gibson (1970), and Soans, Pimentel, and Soans (1974). For a review see Thoday (1972).

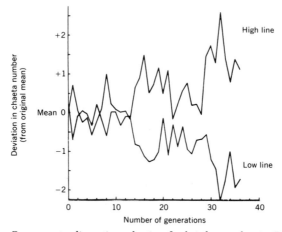

Figure 14.2. Response to disruptive selection for bristle number in *Drosophila melanogaster* under conditions of maximum gene flow. Further explanation in text. (Redrawn from Thoday and Boam 1959)

so that gene flow between them was always present and was working in opposition to the disruptive selection. The experiment was continued in this manner to generation 36.

The high and low lines did not diverge significantly during the early generations. They began to diverge in generation 14 and the divergence became large after generation 30. The divergent selection curves for one high and one low line are shown in figure 14.2. It should be noted that the high and low lines here represent a different situation from that of the high and low lines in the corn-selection experiment described earlier (figure 12.1). The high and low lines were kept isolated in the corn experiment but were cross-mated in the present experiment. This experiment shows, therefore, that disruptive selection can prevail over gene flow.

Experiments of Streams and Pimental (1961) with *Drosophila melanogaster* are interesting in quantifying the interaction between disruptive selection and gene flow. Four levels of gene flow are used (6%, 20%, 50%, and 0% in the control lines) and two intensities of disruptive selection (moderate and strong). Strong disruptive selection produced its effects in spite of 20% gene flow but was ineffective with 50% gene flow. Moderate disruptive selection, however, was overcome by even slight (6%) gene flow.

Disruptive Selection in Nature

The efficacy of disruptive selection is now well established under experimental conditions. The next question is to assess the role of this force in natural populations. Here we are on less certain ground and must speak in terms of probabilities rather than certainties.

One situation in nature in which disruptive selection probably comes into play is that where well-differentiated polymorphic types have a definite selective advantage over poorly differentiated polymorphic types. This situation is realized in the case of sexual dimorphism. Females and males with well-differentiated secondary sexual characters have greater success in mating and reproduction than various types of intermediates (intersexes, homosexuals, etc.).

A second possible situation arises where a polymorphic population occupies a heterogeneous habitat (Levene 1953). The polymorphic types are assumed to be specialized for different niches or subniches in the habitat and to live mainly in their respective special niches. Thus polymorphic type A might be adaptively superior in subniche A' but inferior in subniche B', whereas morph B would flourish in subniche B' but not in A'. The correspondence between polymorphic types and the array of subniches in a heterogeneous habitat could be brought about by disruptive selection.

An example that apparently conforms to the above model occurs in the sulphur butterfly, *Colias eurytheme*. The females of this North American species are polymorphic for wing color; there are orange-winged females and white females. This polymorphism is controlled by a single gene.

In midsummer at several localities in California the white and the orange forms reach their peaks of activity at different times of day. The white form is more active in the early morning and late afternoon, and the orange form in midday, suggesting that the two polymorphic types have different temperature and humidity preferences. There is also a seasonal change in the frequency of the two color forms in California populations, the white form rising in frequency as temperatures decline in the fall (Hovanitz 1953; Remington 1954). Physiological studies show that wing pigmentation plays an important role in temperature regulation in *Colias eurytheme* (Watt 1969). The color polymorphism in this species thus appears to broaden the temperature range and perhaps the length of the season in which the populations can remain active.

A third situation in which disruptive selection operates in nature is found in plants under certain special conditions. The plant population is sedentary but outcrossing and stands astride two ecologically different zones—for instance, two soil types or two topographic zones. It is not uncommon to find a different set of adaptive characteristics developed in the two halves of the plant population. This differentiation is maintained in spite of interbreeding. The chief controlling factor involved is almost certainly disruptive selection.

A high-montane species of pine, *Pinus albicaulis*, occurs at and just above tree line in the Sierra Nevada of California. The populations on mountain slopes up to timberline consist of erect trees, the common growth form of the species. Above timberline the species is represented by a low, horizontal, elfinwood form. The arboreal and the elfinwood subpopulations are contiguous and are cross-pollinated by wind, as shown by the presence of some intermediate individuals, but maintain different growth forms in their respective subalpine and alpine ecological zones under the influence of divergent selective pressures. Parallel racial variation across the narrow timberline belt occurs in other associated high-montane species of conifers and willows (Clausen 1965).

Another set of examples illustrating the same set of forces involves pasture grasses in lead-mining districts in Wales. In such districts the soil changes abruptly from ordinary pasture soils without significant quantities of lead to mine soils containing much lead. *Agrostis tenuis* and other species of grasses in the region form continuous populations cutting across the different adjacent soil types. The populations are held together genetically by wind cross-pollination. Nevertheless, minor racial divergences have developed between lead-intolerant subpopulations on the pasture soils and lead-tolerant subpopulations on the mine soils (Jain and Bradshaw 1966; Antonovics 1971).

Balancing Selection

Heterozygotes are often superior to the corresponding homozygous types in general vigor or in some specific component of viability, such as competitive ability or disease resistance. This superiority can occur in genotypes that are heterozygous for either a single gene or a block of genes.

When a heterozygote *Aa* has a selective advantage over one or both homozygous types, an effect of selection will be to preserve both alleles (A and *a*) in the population. An equilibrium frequency of A and *a* will develop in the gene pool, the exact level of which will depend upon the relative selective values of the alternative polymorphic types. But neither the A nor the *a* allele will go to extinction in the population insofar as the allele frequencies are controlled by selection. A state of balanced polymorphism will be maintained.

Selection in favor of heterozygotes is known as balancing selection (also heterozygote superiority or heterozygote advantage).

Under balancing selection, the alternative alleles or gene blocks do not go to complete fixation or extinction, as noted above, but instead the favored unit of selection is a heterozygous allele pair or gene combination. This has a bearing on the expected effects of directional selection in a balanced polymorphic system. Directional selection in combination with balancing selection is expected to bring about, not a replacement of one gene or gene block by another in the gene pool, but a series of shifts from one heterozygous combination to another, e.g., $A_1A_2 \rightarrow A_2A_3 \rightarrow A_3A_4$ (Lerner 1954:113–114).

Balanced polymorphism based on heterozygote superiority is a fairly widespread phenomenon. Among animals it is found, for example, in grasshoppers, chickens, mice, man, and various species of *Drosophila*; among plants it is found in corn *(Zea mays)*, barley *(Hordeum sativum)*, *Arabidopsis*, and *Oenothera*. The most thoroughly investigated case is that in *Drosophila pseudoobscura*, which will be described next.

Heterozygote Superiority in *Drosophila pseudoobscura*

The wild North American species *Drosophila pseudoobscura* is remarkable for the variation in its chromosome III. Some 16 types of chromosome III, differing from one another with respect to inversions, are known in the species (as was mentioned in chapter 5). The inversion types, which can be identified cytologically in the salivary gland chromosomes of the larvae, are designated by names (Standard, Arrowhead, Chiricahua, Timberline, Pikes Peak, etc.) or by letter symbols (*ST*, *AR*, *CH*, *TL*, *PP*, etc.). Most populations are polymorphic for some of the inversion types and produce the various possible homozygotes and heterozygotes.

The inverted chromosome segments differ in their genic contents as well as in their cytological features. A population polymorphic for *ST*, *CH*, and *AR* contains not only three types of chromosomes but also three different sets of genes on the inverted segments; and the *ST/CH*, *ST/AR*, and *CH/AR* genotypes are genic heterozygotes as well as inversion heterozygotes. Now inversions prevent effective gene recombination in inversion heterozygotes.

Consequently the inversion polymorphism in *Drosophila pseudoobscura* is at the same time a polymorphism for blocks of genes that remain intact from generation to generation.

Dobzhansky and his school have explored the inversion polymorphisms in natural populations of *D. pseudoobscura* throughout the range of the species. The most extensive sampling has been carried out in certain populations in the Sierra Nevada and San Jacinto Mountains in California that have been studied over a period of many years. Much experimental work, using population cages in the laboratory, has also been carried out with strains of flies collected from the same natural populations in California.

The common inversion types in these California mountain populations are *ST*, *CH*, and *AR*; also present in low frequencies are *TL*, *PP*, and *SC*. The populations often exhibit an excess of inversion heterozygotes over expectation on the basis of a Hardy-Weinberg equilibrium. One determines the actual frequency of the various inversion types in the gamete pool, by appropriate methods of population sampling; and from this information one calculates, by the Hardy-Weinberg formula, the expected frequency of heterozygous genotypes. The latter is then compared with the known actual frequency of inversion heterozygotes. A statistically significant excess of heterozygotes is regularly found in the natural populations at certain seasons of the year.

Thus a population in the San Jacinto Mountains, polymorphic for five inversion types, was sampled in May 1952, and found to have the following observed and expected frequencies of inversion heterozygotes and homozygotes (data of Epling, Mitchell, and Mattoni 1953):

	Observed	Expected
ST/CH	0.232	0.168
ST/AR	0.232	0.168
All heterozygotes	0.830	0.724
ST/ST	0.134	0.184
All homozygotes	0.170	0.276

There is a consistent excess of heterozygotes and a corresponding deficiency of homozygotes as compared with the expected Hardy-Weinberg proportions.

The observed deviations from the Hardy-Weinberg equilibrium could be explained on the basis of selection for heterozygotes. One way to test this hypothesis is to compare the frequencies of inversion heterozygotes and homozygotes in an egg sample taken from the natural population with the frequencies in a sample of adult flies. The egg samples were found to contain the heterozygous and homozygous genotypes in Hardy-Weinberg proportions. The deficiency of the homozygous classes in the adult samples must therefore be due to a differential mortality during development between the egg and adult stages, operating in favor of the heterozygotes (Dobzhansky and Levene 1948).

Parallel evidence for heterozygote superiority was obtained in artificial populations reared in population cages. The founders of the artificial populations were strains of flies from the San Jacinto Mountain natural population carrying ST, CH, and AR chromosomes. The population cages were maintained at warm temperatures and allowed to become overcrowded. In the egg stage the inversion homozygotes and heterozygotes were found to be present in Hardy-Weinberg proportions. But the same artificial populations exhibited a significant excess of inversion heterozygotes in the adult stage (Dobzhansky 1947a).

Developmental and behavioral studies reveal that the superior fitness of the inversion heterozygotes can be resolved into a number of different components: greater pre-adult viability, development rate, longevity, fecundity, and mating speed (Moos 1955; Dobzhansky 1970:137–138).

It should be stressed again that the heterozygote superiority is manifested only under certain conditions of temperature, food medium, and population density; when these conditions are not fulfilled, in either the natural or artificial populations, the inversion heterozygotes cease to have a selective advantage over the homozygotes.

The important point is that the environmental conditions bringing heterozygote superiority to expression do exist periodically in natural populations. Balancing selection acts intermittently if not continuously in these populations. Its intermittent action is sufficient to maintain a permanent balanced polymorphism. The AR and CH chromosomes rise in frequency in the natural populations in the cool early summer (May and June), and the ST chromosomes increase in the hot summer and fall (July to September or October), but the changes never go all the way to complete fixation or extinction (Dobzhansky 1943, 1947b, 1948a). The balancing selection thus plays a role in increasing the seasonal range of a population over what it would presumably be if it were monomorphic.

Factors Favoring Polymorphism

Three modes of selection discussed in this and the preceding chapter have a direct bearing on the problem of the extensive polymorphisms in natural populations.

The hypothesis was advanced in the 1950s that polymorphism should be favored by environmental heterogeneity. A habitat containing numerous niches and subniches can support a larger array of polymorphic types than a more homogeneous habitat (da Cunha et al. 1950; da Cunha and Dobzhansky 1954; Dobzhansky 1950b; Ludwig 1950; Levene 1953).[2] There is empirical evidence for this in several organisms, for example in the tropical *Drosophila willistoni* (da Cunha et al. 1950, 1959; da Cunha and Dobzhansky 1954), in fungus-feeding *Drosophilas* in the eastern United States (Lacy 1982), and in marine molluscs in the Mediterranean (Lavie and Nevo 1981).

The enhanced polymorphism in heterogeneous environments was at first attributed to balancing selection, and much of it is indeed due to this mode, but Levene (1953) showed that a polymorphism could also be maintained in a heterogeneous environment by disruptive selection.

A third selection mode that promotes polymorphism is frequency-dependent selection (Clarke 1979). Frequency-dependent selection favoring the rare type commonly occurs in predation and parasitism. Since these ecological interactions are general in nature, it is argued that frequency-dependent selection could be a primary source of polymorphic variation in natural populations (Clarke 1979).

The occurrence of selectively neutral alleles in populations provides a fourth source of polymorphic variation (Kimura and Ohta 1971; Kimura 1979). Undoubtedly such selectively neutral alleles exist. How common they are in natural populations, however, is at present a highly controversial issue. We will return to this question in a later chapter.

2. For reviews see Mayr (1963:237–251), Hedrick et al. (1976), and Lacy (1982).

CHAPTER 15
Levels of Selection

Selection at Subindividual Levels
Sexual Selection
Reproductive Behavior in Bighorn Sheep and Red Deer
Interdeme Selection
Social-group Selection
Altruism
Kin Selection
Selection at the Species Level

Living material exists in a hierarchy of units of organization from genes and chromosome segments through individual organisms to populations and species. The entities at the various levels are also units of reproduction. A population of reproducible units at any level can, if genetically variable, undergo systematic changes in the relative frequencies of the alternative component types. Selection, which up to now has been considered at the individual level, its primary form, can also occur at various subindividual and supra-individual levels.[1]

Selection at Subindividual Levels

Single-gene selection, hitherto considered as an oversimplified but useful abstraction, was presumably a real process, and indeed the main selective process, in primitive stages of organic evolution when the prevailing unit of organization was gene-like particles. An approximation to gene selection is found in the modern world in the differential reproduction of virus particles or bacterial cells differing with respect to a single viral or bacterial gene.

1. Previous treatments of selection as a multilevel set of processes are given by Wright (1956), Grant (1963: ch. 10) and Lewontin (1970).

Certain gene-controlled aberrations of meiosis in *Drosophila melanogaster* cause one chromosome of a homologous pair to become included in more functioning sperms than the other. The distorted chromosome segregation at meiosis leads to an increase in frequency of one chromosome type over that of its homolog in the gamete pool. This process is known as meiotic drive. Corresponding changes in frequency take place in the genes borne on the two homologous chromosomes and in the phenotypes that they determine. In one case the sex ratio is altered in the direction of an excess of females; in another case a recessive lethal is caused to rise in frequency (Sandler and Novitski 1957; Hiraizumi, Sandler, and Crow 1960). Meiotic drive is essentially a process of differential reproduction of homologous chromosomes. It therefore represents selection at the chromosome level.

In flowering plants the male gametes are carried by an independent unit, the male gametophyte, consisting of the pollen grain and pollen tube. A heterozygous plant produces genetically different classes of pollen, and, in some cases, the pollen segregates for genetic factors affecting the viability, germination ability, or growth rate of the pollen. Furthermore, pollen is normally produced and delivered to the stigma in excess of the numbers required for fertilization, so that a competition among pollen grains or tubes ensues. The result of this competition, where the pollen is segregating for growth factors, is that some classes of male gametes are more effective than others in fertilization and hence in the production of embryos or endosperm in the seeds. Selection at the gamete level takes place.

Let some marker gene determining a morphologically visible trait be linked to a pollen growth gene. Then, instead of the expected Mendelian ratio for the marker gene in the progeny, one observes an altered ratio. The morphological type determined by the marker allele borne by the superior class of pollen is present in excess, and the opposite morphological type is deficient, in the seed or seedling generation. Altered segregation ratios due to selective discrimination against certain classes of pollen are well known in various species of flowering plants.

A well-analyzed case in corn *(Zea mays)* concerns the linked genes *Su* (for type of endosperm) and *Ga* (controlling pollen tube growth). The *Su/su* heterozygote usually produces corn kernels segregating into starchy vs. sugary endosperm classes in Mendelian ratios. But let the sugary allele *(su)* be linked to the *ga* allele (for slow pollen tube growth) in a heterozygote *(Su Ga / su ga)*, and let the heterozygote be used as a male parent, so that the linked segment *su-ga* is transmitted through the pollen. Then a marked deficiency of sugary types appears in the kernels in the next generation (Mangelsdorf and Jones 1926).

Sexual Selection

Sexual selection is selective discrimination between the members of one sex in a dioecious organism, usually animals. It is individual selection, but a special form of that process involving individuals of one sex class within the population, usually the males, and it is therefore also referred to as inter-male selection.

The starting point for sexual selection is the division of labor between the gametes and their bearers which developed early in the history of sexual reproduction. The eggs and eventually the female individuals became specialized for nourishment and protection of the zygote and embryo, the sperms and eventually the males for finding and fertilizing the eggs. This division of labor led early on to a numerical preponderance of sperms over eggs, to a greater motility of sperms or males than of eggs or females, and to a stronger sex drive in males than in females. These conditions set the stage for inter-male competition and selection. Sexual selection as narrowly defined is inter-male selection. To be sure, some current authors broaden the concept of sexual selection to include the characteristics associated with the basic division of labor between the sexes, but this is lumping the results with the preconditions.

The phenomena which the theory of sexual selection addresses are secondary sexual characters as found in many groups of animals. Sexual dimorphism in animals often consists of two sets of character differences. The primary sex characters are involved directly in reproduction; they are not the issue here. The secondary sexual characters, developed in the males, which enable them to secure mates, are the issue. Such secondary sexual characters fall into two broad classes: characters of size, strength, and armature (e.g., the superior size of male sea lions, the antlers of male deer); and characters of ornamentation and display (e.g., colorful plumage of male ducks, gorgets of male hummingbirds, special songs, courtship displays).

In *The Descent of Man and Selection in Relation to Sex* (1871), Darwin reviewed the then known instances of secondary sexual characters in male animals, belonging to a wide range of animal groups, and advanced the theory of sexual selection to account for them.

Darwin conceived of sexual selection as a process supplementary to the more general and pervasive natural selection, which builds up the adaptive features of a species as a whole, including the secular adaptations common to the two sexes and the primary sex differences related directly to reproduction. Natural selection in the original Darwinian sense of the term did not, however, account for secondary sexual characters. Such characters as the antlers of male deer and the colorful plumage of male ducks are not adaptively beneficial to the species as a whole, nor are they necessary for reproduction, but they appear to enhance the mating success of the male individuals that possess them. The theory of sexual selection was introduced to explain the development of such special male characters.

Darwin describes the proposed process as follows (1871: ch. 8): "sexual selection . . . depends on the advantage which certain individuals have over others of the same sex and species solely in respect of reproduction." He goes on to explain (and I am paraphrasing his statement here while preserving his phraseology). When the two sexes follow exactly the same habits of life, yet the male has sensory or locomotive organs more highly developed than those of the female, the males have probably often acquired their characters, not from being better fitted to survive in the struggle for existence, but from having gained an advantage over other males in reproduction. "Sexual selection must here have come into action."

The theory of sexual selection has had a tortuous history since 1871. It was controversial in Darwin's time. The subject later fell into a state of neglect. When revived in the early modern period of evolution studies, by Fisher (1930) and Huxley (1938), the subject entered upon a new scene, one in which the same phenomena were being viewed from other alternative standpoints, and the sexual selection theory had to make a new debut. This second debut was successful and the subject is now an active research field with a voluminous literature.[2]

To evaluate the status of sexual selection today we have to begin by recognizing that secondary sexual characters constitute a highly varied assemblage of features. These characters serve diverse functions in the life of the animals. Their formation probably cannot be attributed to any single mode of selection. Size differences between sexes may be due to selection for ecological divergence (see Part 5) as well as to sexual selection. The weapons of males may serve to establish territories or to establish social dominance as

2. Reviews by Campbell (1972), Maynard Smith (1978: ch. 10), and Clutton-Brock et al. (1982: ch. 1). See also Darwin (1874) and Wallace (1889: ch. 10). A historical review is given by Kottler (1980).

well as to secure females. Male adornment, displays, songs, and scents may play roles as courtship stimuli and as species-specific recognition signals. In the latter case they could be due to selection for reproductive isolation (see Part 5) as well as sexual selection.

In the whole broad range of secondary sexual characters we have a mixture of selective processes at work, which makes it difficult to sort out the effects of sexual selection alone. The role of sexual selection can be seen more clearly if we consider a narrower range of phenomena in which the circumstances are especially favorable for the operation of this process.

Sexual selection requires the following conditions to be effective: (a) competition between males for females; (b) genotypic differences between males in competitive ability in this struggle; and (c) successful males have a reproductive advantage over other males.

The first condition is met in mammals and birds with a polygynous (polygamous) mating system. In polygynous species the strongest males collect and guard a harem of females, or a territory containing their females, and drive off their weaker brothers, who consequently possess few or no mates.

Polygyny is common in mammals and occasional in birds. Among mammals polygyny occurs in deer, cattle, sheep, most antelope, elephants, seals, sea lions, walruses, and baboons; among birds it occurs in chickens, pheasants, and peacocks. Secondary sexual characters are well developed in the males of these animals, in marked contrast to the condition in related, non-polygynous groups. Thus in polygynous chickens, pheasants, and peacocks, the cocks are notably larger, more pugnacious, and better decorated than the hens; while in monogamous partridge, grouse, and ptarmigan the differences between the sexes are relatively slight. In walruses and sea lions, the males are huge; in many ungulates they possess antlers or horns; and in baboons they are large and aggressive. By contrast, the sexes are nearly equal in size and strength in monogamous wolves and certain monogamous species of monkeys, in the members of the cat family that form maternal families, and in colonial but non-polygynous rodents (Darwin 1871: ch. 8).

The tendency of characters of male combat to develop preferentially in polygynous groups suggests strongly that these characters are products of sexual selection. The correlation between size dimorphism and mating system is not perfect, however, and there are numerous exceptions and complications (Ralls 1977).

A second correlation, one between mating system and parental investment, comes into the picture in mammals and birds (Trivers 1972; Zeveloff and Boyce 1980).

In birds the males and females are involved more or less equally in the various aspects of parental care, including nest-building and foraging for nestlings. The reproductive strategy in most birds calls for cooperation of both parents. This results in monogamy which is the common and widespread condition in this class. Under monogamy, with the males all paired off, an inter-male competition for mates does not develop. Consequently, characters of male combat and dominance also do not develop.

Polygynous birds, which do have male dominance, are the exception that proves the rule.

The reproductive strategy in mammals, by contrast, is based on predominantly maternal care of the young, through internal gestation and provision of milk. The parental investment by the males is often negligible. Where the males are not tied down by parental care they are free to compete for females and build up harems. This sets the stage for sexual selection for characters of male dominance, which are common in mammals. Monogamous mammals, with their equal sexes, are the exception that proves the rule in this class.

The role of sexual selection in the origin of characters of male adornment and display in non-polygynous groups is less clear. Sexual selection for this type of character requires female choice at the intraspecific level, which is an uncertain quantity. The initial stages of development of male display characters are perhaps guided by ordinary individual selection to promote courtship and mating. A possibility suggested by Mayr (1972b) is that sexual selection joins ordinary individual selection at this point, and the two modes, acting jointly, carry the character development to a greater expression on the basis of the mating advantage of the more highly decorated males.

Breeding Behavior in Bighorn Sheep and Red Deer

Sexual selection will be effective only if the victorious males leave more surviving progeny than the unsuccessful males. Data on the relative reproductive success of different male individuals are hard to obtain in nature and scarce in the literature. This aspect is becoming better documented now, however; two examples will be given here.

The reproductive behavior of bighorn sheep (Ovis canadensis) has been carefully studied in the wild and in captivity by Geist (1971). The adult rams have large heavy horns and a thick skull, as is well known. The rams engage in butting contests during the rutting season, when they are competing for access to the ewes. The fights are often severe and may result in the wounding or death of the vanquished animals. The observed facts of rutting behavior in sheep and other ruminants are at variance with a commonly held opinion in animal behavior studies that male animals engage mainly in non-injurious bluff and other displays. Displays of symbols of rank in the dominance hierarchy of rams are present, to be sure, but are backed up by combat if necessary.

Some individual rams emerge from combat as dominant members of their social group. These dominant males then chase other males off from the ewes, but copulate freely with the ewes themselves. The mating system in bighorn sheep does not preclude some copulation between subordinate males and ewes, but it does ensure preferential breeding by the dominant males. The latter therefore apparently sire more lambs than the subordinate males (Geist 1971).

A long-term study of the red deer (*Cervus elaphus*) in Scotland by Clutton-Brock and coworkers (1982) provides data on the reproductive success of individual deer. The species is strongly dimorphic. The body weight of males is nearly twice (ave. $1.7 \times$) that of females. During the fall rut the males fight among themselves, using their size, strength, and antlers. Depending on their fighting ability they gain and hold harems of varying size, from 1–22 females, or remain solitary.

The successful males fertilize several or many females each breeding season, while the unsuccessful males may not breed at all. The lifetime reproductive success in a sample of 13 stags ranged from 0–25 surviving calves per stag. Reproductive output of stags correlates with harem size which, as we have seen, correlates with fighting ability. A large differential clearly exists between high-ranking and low-ranking males in reproductive success (Clutton-Brock et al. 1982).

Interdeme Selection

We consider next the controversial question of selection at the population level. Interdeme selection (also often referred to as group selection) is differential reproduction of different local populations. We will begin with interdeme selection in non-social organisms.

Wright (1931, 1960) compared two types of population systems, a large continuous population and a series of small semi-isolated colonies, with respect to the theoretical effectiveness of selection. The total size (N) is assumed to be the same in both population systems, and the organism is assumed to be outcrossing.

In the large continuous population, selection is relatively ineffective in raising the frequency of favorable but rare recessive mutations. Furthermore, any tendency of a favorable allele to rise in frequency in one part of the large population is opposed by interbreeding with neighboring subpopulations in which the allele is rare. In the same way, favorable new gene combinations that succeed in becoming assembled in one local segment of the population are broken apart and swamped out by interbreeding with neighboring segments.

These difficulties are largely obviated in the island-like population system. Here selection, or selection and drift jointly, can quickly and effectively raise the frequency of a rare favorable allele in one or more small colonies. Favorable new gene combinations can also be established readily in one or more small colonies. Isolation then protects the gene pools of these colonies from the swamping effects of migration from and interbreeding with other, less-favored colonies. The model up to this point invokes only individual selection or individual selection combined with drift, in certain colonies.

Now suppose that the environment of the population system has changed in such a way as to lower the adaptedness of the old genotypes. The new favorable genes or gene combinations established in certain colonies have a high potential adaptive value for the population system as a whole in the new environment. The stage is now set for interdeme selection to come into play. The less-fit colonies dwindle and become extinct, while the better-adapted colonies expand and replace them throughout the range of the population system.

The subdivided population system thus acquires a new set of adaptive characteristics as a result of individual selection within some colonies followed by the differential reproduction of different colonies. The combination of interdeme and individual selection can bring about evolutionary results which individual selection alone cannot accomplish.

It is recognized that interdeme selection is a second-order process which complements the primary process of individual selection. Since interdeme selection is a second-order effect it will be slower—probably much slower—than individual selection. The turnover of populations requires more time than that of individuals.

Interdeme selection has had wide acceptance in some circles but rejection in others. Williams (1966, 1975) and his followers consider this selection mode to be unnecessary. They argue that various possible models of individual selection can bring about all of the effects attributed to interdeme selection.[3] Wade (1978) argues in rebuttal that the models in question contain assumptions which are unnecessary and which bias them against interdeme selection.

Wade (1976, 1977, 1982) has performed a series of selection experiments with the flour beetle, *Tribolium castaneum*, to test the efficacy of interdeme selection, and he finds response to this selection mode. Furthermore, when individual and interdeme selection are both operating on a trait, and operating in the same direction, the rate of change in that trait is faster than it is under individual selection alone (Wade 1976, 1977). Even moderate amounts of immigration (6 and 12%) do not prevent interdeme selection from bringing about differentiation between populations (Wade 1982).

A feature in the organic world which is difficult to explain on the basis of individual selection, but readily explained as a product of interdeme selection, is sexual reproduction. Models have indeed been devised in which sexual reproduction is favored by individual selection (Williams 1975), but these seem unrealistic. Sexual reproduction is the process which generates recombinational variation in breeding populations. The beneficiary of sex is not the parental genotypes, which are broken up in the recombination process, but the population in future generations with its enhanced store of variations. This seems to implicate a selection process at the population level as one of the agents involved.[4]

Social-Group Selection

Wynne-Edwards (1962) introduced the term, group selection, for the special case of interdeme selection in which the demes are social groups. The term has since been used extensively for interdeme selection in general, so that its meaning has undergone a semantic shift. This leaves us without an unambiguous term for the original concept although such a term is still needed. The modified term, social-group selection, is adopted here for selective discrimination between social groups.

The adaptive characters of the neuter worker caste(s) in social insects af-

3. The opposing viewpoints are presented by Williams (1971).
4. For more on this subject see Williams (1975), Maynard Smith (1978), Shields (1982), and Nyberg (1982).

ford a clear example of social-group selection at work. The colony in the honeybee, *Apis mellifera*, for example, is a social population consisting of females (queens), males (drones), and neuters (workers). The queens and drones are sexually fertile and reproductively active but economically useless.

The important functions of food getting, maintenance of the colony, rearing of the broods, and defense are carried out by the workers, which possess bodily and psychic adaptations for performing these functions. The chief secular adaptations by which the colony lives reside in these worker bees. Yet the workers, being sexual neuters, do not reproduce as individuals, and consequently have no opportunity to pass the genes determining their adaptive characteristics on to the next generation. That task is performed by the queens and drones. If, therefore, the queens and drones do not carry genes making for adept and efficient worker bees, the hive will not thrive, and may be eliminated by competition from other hives containing better-adapted workers. Such a process of replacement of one strain of beehives by another has actually been observed in recent years; the aggressive African strain has replaced the older European strain in Brazil during the past twenty years (Michener 1975). The main unit of selection in honeybees definitely seems to be the colony.

Social-group selection in the social Hymenoptera is facilitated by the hymenopteran or haplodiploid method of sex determination in which males develop from unfertilized eggs, and genetic females—queens and workers—from fertilized eggs. The result is that queens and workers in the same colony during the same generation are sisters. Furthermore, sisters in a hymenopteran colony are especially closely related, more so than parents and offspring, in contrast to the symmetrical relationships within a family in a normal dioecious species. Because of the especially close relationship between workers and queens in social Hymenoptera, selective pressures brought to bear on the workers belonging to different colonies can be translated readily into selection between their respective queens (Hamilton 1964, 1972; Trivers and Hare 1976).

Social Hymenoptera represent one of the extremes of social integration in the animal kingdom, and they exhibit the effects of social-group selection in a pronounced fashion. This mode of selection probably operates also in less highly socialized animals such as birds and mammals. Some characteristics of the human species are probably products of social-group selection, as we will attempt to show in Part 8.

Altruism

Altruistic behavior in animals includes a wide range of specific behavior patterns. The common denominator is behavior that benefits other individuals. Beyond this, the degree of altruism ranges from behavior benefiting other individuals in a moderate way to self-sacrifice of suicidal proportions. And the beneficiaries of the altruistic behavior may be offspring, other kin, or social groups.

The existence of different kinds of altruistic behavior suggests that different modes of selection may be involved. Let us consider three different cases.

The first case is altruistic behavior of parental animals in relation to their offspring. One can think of the general phenomena of parental care in birds and mammals, or of specific behavior patterns such as injury-feigning and decoying in the vicinity of a nest by lapwings, killdeer, and other ground-nesting birds. Parental care is clearly a product of individual selection, since individual selection favors the genes of the parental individuals that leave the largest number of surviving offspring.

Consider next self-sacrificing defensive behavior of the workers in social bees such as *Apis mellifera*. The use of the stinger by the worker bee is suicidal to the individual, but valuable to the colony in defending it against intruders. The colonies with the best defenses tend to survive and multiply more successfully than those with less effective defenses. The self-sacrificing defensive behavior of worker bees, like other characteristics of the worker caste, can be adequately explained as a product of social-group selection, since the beneficiary is the bee colony.

Our third case is a primitive human group in the hunting-gathering stage, as exemplified by the Bushmen of southwestern Africa (Lee and Devore 1976). The social groups are bands composed of families, other blood relatives, in-laws, and occasional visitors from other bands. The habit of food-sharing is strongly entrenched. When a large animal is killed the meat is divided among the members of the band whether they are kinsmen or visitors. Other cooperative behavioral patterns are also developed.

Now let us assume for purposes of argument that food-sharing and other similar social behaviors have some genotypic basis, so that we can inquire into the modes of selection that are probably involved.

Individual selection for parental care would be powerful. It is difficult to visualize the food being shared with the offspring only, however, and not with other members and close relatives, because of normal plasticity in the behavioral phenotype and social pressures in the group. Rigid and narrowly oriented behavior patterns are not characteristic of social primates. The food-sharing behavior would naturally tend to overflow its primary objective, the offspring, and extend throughout the family and into the kinship group. Social-group selection would also be expected to promote food-sharing behavior. The band depends on cooperative group activity for food-getting and indeed for survival and would be benefited by broadly based food-sharing habits. The food-sharing behavior patterns promoted by social-group selection would tend to embrace all members of the band, blood relatives and in-laws alike, and would overlap with the behavior patterns produced by individual selection in the intermediate zone of kinships. In short, food-sharing in the band could be explained adequately as a product of the combined action of individual selection and social-group selection for a culturally modifiable behavior pattern.

Kin Selection

The term, kin selection, was introduced and defined by Maynard Smith (1964) as selection for characteristics which favor the survival of close relatives of a given individual. The relatives could be either offspring or siblings and other relatives. As examples of characteristics brought about by kin selection he cites parental care, injury-feigning, and sterile worker castes in social insects.

Maynard Smith and later authors treat kin selection as a process close to social-group selection, and perhaps intermediate between individual and group selection.

Kin selection has become a popular term and concept and is extensively used in the literature. Nevertheless, in my opinion, its explanatory power is slight and its ability to cause confusion of thinking is great (see Grant 1978). One part of kin selection is a later synonym of individual selection for parental care, another part is a later synonym of group selection.[5]

The phenomena attributed to kin selection are adequately explained by these other selection modes, that is, by individual selection for parental care, or social-group selection, or both processes acting in concert.

Some authors prefer the term kin selection over the prior term group se-

5. For further reading on kin selection, pro and con, see Williams (1971), Hamilton (1972), Alexander (1974), Wilson (1975: ch. 5), Maynard Smith (1976), Grant (1978), and Darlington (1980: ch. 7).

lection because the former emphasizes the role of kinship. However, kinship is strongly implicit in the group selection concept. All social groups in animals are kinship groups. A separate term, kin selection, is unnecessary.

Can we find any phenomena in nature that are determined by a mode of kin selection narrowly defined to exclude the previously recognized modes of individual selection for parental care and social-group selection? Certain phenomena commonly attributed to kin selection, such as altruism and worker castes in social insects, have already been discussed. Another set of characters, warning coloration in unpalatable insects, was for awhile thought to be a product of kin selection.

A predator kills a warningly colored insect, discovers that it is unpalatable, and learns the color signal. This favors siblings of the dead insect with similar color markings. The kin-selection explanation of warning coloration depends on the assumption that unpalatable insects are killed while the predator is tasting it. Wiklund and Järvi (1982) tested this assumption in tests with a wide variety of warningly colored insects exposed to four species of birds. They find that birds usually do not kill the insect. They taste it, reject it, and it often (in 84% of the trials) survives. Therefore unpalatability and warning coloration can be adequately explained on the basis of individual selection alone (Wiklund and Järvi 1982).

Selection at the Species Level

At least three modes of selection operate at the species level. They are: (a) selection for ecological divergence; (b) species replacement; and (c) selection for reproductive isolation. These modes are only mentioned here, to round out our survey, and they will be discussed in Part 5.

Stanley (1975, 1979) introduced the term "species selection" and Gould and Eldredge (1977) hailed it as a new concept, a "previously unrecognized mode of operation for natural selection." In fact species selection is a later synonym of species replacement.

A reductionist would point out that the selective process involved in species replacement actually involves individual organisms. So it does. But here the individuals involved in the selective process are representatives of different species. There is evidence in *Erodium*, as noted in chapter 13, that interspecific competitive ability is a different character than intraspecific competitive ability (Martin and Harding 1982).

There has been some discussion regarding the possibility that selective processes may operate among biotic communities. It has been suggested that more stable communities might outlast less stable ones through ecological time (MacArthur 1955; Hutchinson 1959; Dunbar 1960). Stability, however, seems to be an elusive or even inappropriate concept to apply to biotic communities (Symposium 1969; Connell and Sousa 1983). The question of selection at the community level remains open and needs further study (see Wilson 1980).

CHAPTER 16
Genetic Drift

General Considerations

According to the Hardy-Weinberg law a selectively neutral allele will tend to remain at a constant frequency from generation to generation. But it will be recalled (from chapter 4) that this law is applicable only to a very large population. The predicted constant allele frequency represents a statistical average for many trials; gene reproduction in a large population fulfills this condition of many trials. In any set of few trials, such as occurs in gene reproduction in a small population, one expects deviations from the average allele frequency due to chance alone.

Allele and genotype frequencies are subject to chance variation from generation to generation in a small population. Genetic drift refers to this random component in the rate of gene reproduction. In a small polymorphic population, drift leads, first, to fluctuations in allele frequency from generation to generation, and eventually, to complete fixation or extinction of the allele.

We have assumed that the allele is selectively neutral. This assumption is simplifying but unnecessary. Let us assume alternatively that the allele has a slight selective advantage or disadvantage. The predicted change in allele frequency from generation to generation is again a matter of statistical averages and is again subject to chance deviations.

If, for example, a population is polymorphic for A and a, and a has a 0.1% selective advantage($s = 0.001$), the predicted contributions of the two alleles to the gene pool in the next generation are in the ratio 1000 a:999 A. This average proportion will be realized in a large population. But significant deviations from the ratio due to random factors are expected to occur in a small population.

Thus the action of selection does not in itself rule out the possibility of the action of drift. In fact, there is reason to believe that the most important evolutionary role of genetic drift lies in its joint action with selection.

Effect of Population Size

Whether allele frequencies in a population will or will not be affected significantly by drift depends on four factors: (a) the size of the population (N); (b) the selective value of the allele (s); (c) the mutation pressure (u); and (d) the amount of gene flow (m). These four factors interact. Their interrelationships have been investigated and quantified by Wright (1931).

Let us first consider population size. As noted previously, random fluctuations in allele frequency will be negligible and insignificant in a large but not in a small population. In the small population, by chance alone, an allele may change from a low to a high frequency, or reach fixation, in one or a few successive generations.

The theoretical distribution of allele frequencies in very small isolated populations has been shown by Wright (1931) to be represented by a U-shaped curve (figure 16.1). Assume that a given allele (A_1) is selectively neutral ($s = 0$), and is a polymorph in a series of small isolated populations (of size N). This allele is expected to be approaching either extinction or fixation in most of the polymorphic populations, and to be present in medium frequencies in only a relatively few populations, as shown by the curve for N in figure 16.1.

The distribution of allele frequencies in small populations (of size N) can be compared with that in a series of large populations (of size 4N). Here most populations have the allele at medium frequencies (curve for 4N in figure 16.1).

The graph in figure 16.1 can also be read to indicate the theoretical distribution of allele frequencies (abscissa) in a series of polymorphic genes (ordinate) in a single population. Most polymorphic genes are represented at

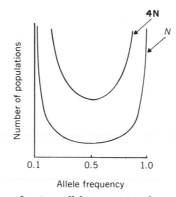

Figure 16.1. Frequency of a given allele in a series of populations of two size classes. The populations are either small *(N)* or large (4N). *s* = 0. (Redrawn from Wright 1931)

extreme frequencies in a small population (curve N), but at medium frequencies in a large population (curve 4N).

Effect of Selection

The polymorphic allele was assumed to be selectively neutral in the preceding discussion, and this assumption is reflected in the symmetry of the U-shaped curves. In a series of small populations the selectively neutral allele approaches either extinction or fixation in equal measure. If the allele has a small selective value in the small populations, however, the curve remains U-shaped but is skewed. The direction of the skewness depends on whether the allele is selectively advantageous or disadvantageous, and the degree of skewness on the magnitude of the selective value (figure 16.2A). An allele with a small selective advantage tends to be present at either high or low frequencies, but more often at high frequencies, in a series of small isolated populations.

Here again sets of small populations differ from sets of large populations with respect to the expected behavior of genes. The gene frequencies are controlled more effectively by selection in large populations than in small ones at equivalent values of *s* (figure 16.2).

Another way of verbalizing the curves in figure 16.2 is to say that selection, theoretically, has a relatively small effect on gene frequencies in populations below a certain critical size, whereas chance alone is likely to be

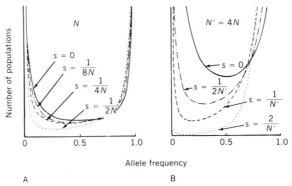

Figure 16.2. Expected distribution of allele frequencies for different values of s and two values of N. (A) In a small population. (B) In a population 4 times larger, with equivalent intensities of selection. (Wright 1931)

very effective in controlling gene frequencies under these same conditions (Wright 1931). Weak selective pressures are likely to be overruled by drift in a small population.

This brings us to the question of how small is "small" and how large is "large" for the action of drift. The critical value of N at which drift becomes effective is a function of s. The relations between N and s and drift are as shown in table 16.1.

These relations can be visualized when plotted graphically on a linear scale (Figure 16.3). We see that when N is low relative to s, drift prevails; that when N is relatively large, selection prevails; and that a range of overlap exists where drift and selection may act jointly.

These general relations can readily be translated into concrete figures. Suppose an allele has a selective value of $s = 0.01$. Its frequency is controlled by drift if $N \leqq 50$. But if the selective value of the allele is $s = 0.001$, its fre-

Table 16.1. Relationships between population size, strength of selection, and drift (From Wright 1931)

Allele frequencies controlled by:	When N is	When s is
Drift	$N \leqq \dfrac{1}{2\,s}$	$s \leqq \dfrac{1}{2\,N}$
Selection	$N \geqq \dfrac{1}{4\,s}$	$s \geqq \dfrac{1}{4\,N}$
Selection and drift	$N = \dfrac{1}{4\,s}$ to $\dfrac{1}{2\,s}$	$s = \dfrac{1}{4\,N}$ to $\dfrac{1}{2\,N}$

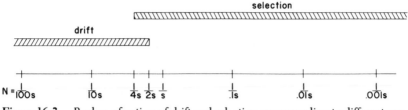

Figure 16.3. Realms of action of drift and selection corresponding to different proportions of N and s. (From V. Grant, *The Origin of Adaptations*, © 1963 Columbia University Press, New York; reproduced by permission)

quency is controlled by drift in a population of $N \leqq 500$. Thus, in general, if the selective value of the allele is *fairly* low, it can be fixed or lost by drift in a quite small population; but if the selective value is *very* low, drift can control its frequency in a medium-sized population.

The range of overlap in which both drift and selection are operative also varies with the magnitude of s. In the numerical examples given above, that range is $N = 25 - 50$ for $s = 0.01$, and $N = 250 - 500$ for $s = 0.001$.

The possibility of joint action of selection and drift can, theoretically, be very important in evolution. Wright (1931, 1949, 1960) points out that a favorable gene can be established much more rapidly by selection and drift in an islandlike population system than by selection alone in a single large population of the same overall size.

It was desirable to quantify Wright's conclusion. Assume that an initially rare gene with a slight selective advantage occurs in two types of populations, a single continuous and a subdivided population, each containing a total of 10^6 individuals. What is the comparative probability of fixation of this gene? The probability of fixation was found to be an order of magnitude greater in the subdivided population (Flake and Grant 1974).

Effect of Gene Flow

Wright's (1931) equations describing the interaction between gene flow and drift are as follows. Allele frequencies are controlled by drift when

$$N \leqq \frac{1}{2m} \quad \text{or} \quad m \leqq \frac{1}{2N}$$

Allele frequencies are controlled by gene flow when

$$N \geqq \frac{1}{4\,m} \quad \text{or} \quad m \geqq \frac{1}{4\,N}$$

and by both forces in the intermediate range,

$$N = \frac{1}{4\,m} \quad \text{to} \quad \frac{1}{2\,m}$$

For example, if $N = 100$, the critical value of m is $\frac{1}{2\,N} = 0.005$. Thus immigration of foreign alleles at rates greater than $m = 0.005$ can prevent drift from occurring in a population of 100 or less breeding individuals.

In general, small amounts of gene flow can prevent drift. A small population must be rather well isolated for drift to occur. But when this condition exists, when N and m are both low, drift can affect gene frequencies significantly.

The effect of mutation rate is described by equations parallel to those for gene flow. Allele frequencies are controlled by mutation pressure when $N \geqq \frac{1}{4\,u}$ and by drift when $N \leqq \frac{1}{2\,u}$. High mutation rates could prevent drift in a small population.

The Fixation of Gene Combinations

The joint action of selection and drift in small populations can promote the fixation not only of single genes, but also of gene combinations. The latter case can be very important in evolution.

Suppose that a large population contains two rare mutant alleles, a and b, of the independent genes A and B. Most individuals in this population have the diploid genotype AABB: in addition there are a few mutant-carrying individuals, AaBB and AABb. Suppose further that the gene combination aabb has a high adaptive value in relation to new environmental conditions.

In the large population, the sexual process will produce aabb only rarely, and will then immediately break it up again. It is difficult for selection to "get a hold on it," and therefore its rise in frequency occurs at a slow rate.

But a small isolate of this population might have an intermediate or high frequency of the otherwise rare alleles, *a* and *b*, as a result of chance. Proportionally more *aabb* zygotes would be produced and exposed to selection in each generation. Selection could then act effectively to raise the frequency of *aabb* further. Thus the new gene combination can be established more rapidly by selection and drift in certain small colonies than by selection alone in the large population.

Experimental Evidence

Kerr and Wright (1954) tested the theory of drift in a series of very small populations of *Drosophila melanogaster* polymorphic for a body-bristle gene, forked *(f)*. They started 96 replicate lines with four female and four male flies each. The initial frequency of the mutant allele forked was 0.5 in each line. The lines were continued by a random choice of the parental individuals (four females and four males) for the next generation. The 96 lines were maintained in this way to generation 16.

By generation 16, the wild-type allele had become fixed in 41 lines; forked was fixed in 29 lines; the remaining 26 lines were still polymorphic.

A parallel experiment made use of the eye mutant, Bar *(B)*, in *Drosophila melanogaster*. Here 108 lines were started and continued as before. The Bar type is selectively disadvantageous. As of generation 10, the wild-type was fixed in 95 lines; Bar was fixed in 3 lines; 10 lines were still polymorphic (Wright and Kerr 1954).

It is apparent that the polymorphic genes do drift into fixation in a high proportion of the small experimental populations. Sometimes they become fixed in spite of counterselection. More often they become fixed by the combination of drift and selection, as in the case of the 95 ultimately wild-type lines in the experiment with Bar.[1]

The joint action of selection and drift has been demonstrated in other experimental studies. One set of experiments involves inversion types in laboratory populations of *Drosophila pseudoobscura* (Dobzhansky and Pavlovsky 1957; Dobzhansky and Spassky 1962). Another experiment carried out over a 17-year period with an annual herbaceous plant, *Gilia*, involves vigor and fertility in a series of related inbred lines (Grant 1966a).

1. Other drift experiments in *Drosophila melanogaster* are those of Merrell (1953), Buri (1956), and Katz and Young (1975).

Drift in Natural Populations

Population size occasionally becomes favorable for the effective action of drift, with or without selection, under certain conditions in nature. Three such situations exist; they are not uncommon: (a) a population system consists of a series of permanently small, isolated colonies; (b) the population is usually large, but is decimated periodically, and then builds up again from a few survivors; (c) a large population gives rise to isolated daughter colonies, the latter being founded by one or a few founder individuals. The new colonies thus pass through a bottleneck of small size in their early generations, although they may build up in size later. The last-mentioned case is the founder effect or founder principle of Mayr (1942, 1963).

If drift is playing an effective role (again with or without selection, but probably with), we would expect the variation pattern of the colonies to show the following characteristic features. First, the small colonies—the sister colonies in mode a and the daughter colonies in the early generations in modes b and c—should be relatively uniform genetically. Second, there should be a marked colony-to-colony variation in genetically determined characters. This local racial differentiation is expected to be displayed most clearly in series of small sister colonies (mode a), but should also be evident in some series of larger derived populations descended from small colonies (modes b and c). And third, the intercolonial variation, when plotted on a distribution map, should be somewhat irregular and haphazard.

Variation patterns have been carefully studied in a number of plant groups with colonial population systems. In some of these groups the variation pattern conforms to the above expectations and thus suggests the effective action of drift.

Species of cypress trees (*Cupressus spp.*) in California exist as series of isolated groves; the groves are uniform, but each grove has its distinctive morphological characteristics (Wolf 1948; Grant 1958). The same type of variation pattern is found in herbaceous plants with colonial population systems, such as *Gilia achilleaefolia* in California, the *Erysimum candicum* group in the Aegean Islands, and the *Nigella arvensis* group in the Aegean Islands (Grant 1958; Snogerup 1967; Strid 1970). The case for drift in these examples is strengthened by the fact that related species of *Juniperus, Gilia,* and *Nigella* in other areas form large continuous populations with a different type of variation pattern, namely, gradual intergradation along geographical transects.

Haphazard local variation is found again in some population systems of the European land snail, *Cepaea nemoralis*, for the presence or absence of bands on the shells (figure 12.2). This shell character is determined by a single allele pair, with bandless dominant over banded, as noted in chapter 12. The frequency of the bandless phenotype and bandless allele fluctuates widely from one colony to the next in areas of France where *Cepaea nemoralis* has a colonial population structure (figure 16.4). But the bandless allele varies in frequency in a more gradual manner along geographical transects through large populations of the European land snail (Lamotte 1951, 1959).

The Torrey pine *(Pinus torreyana)* is a narrow endemic species in southern California consisting of two small isolated populations. A mainland population near San Diego contains just over 3,400 trees. A second population on Santa Rosa Island, 280 km. from the first, contains 2,000 trees. Each population is uniform in morphological characters. There are slight morphological differences between the populations (Ledig and Conkle 1983).

Biochemical evidence, derived from an electrophoretic assay of 25 enzyme systems representing about 59 loci, agrees with the morphological evidence. All individuals sampled in each population are homozygous for all enzyme loci. The individuals are also genetically identical in each popula-

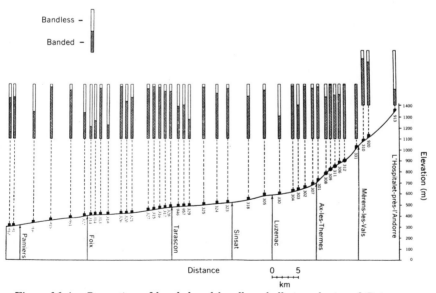

Figure 16.4. Proportion of banded and bandless shells in colonies of *Cepaea nemoralis* in the Ariège Valley of the Pyrenees, France. (Lamotte 1951)

tion. The populations differ in two of the 59 enzyme loci (Ledig and Conkle 1983).

This uniformity and slight interdeme differentiation are probably due to genetic drift. The San Diego population of Torrey pine was probably reduced to a smaller size during the xerothermic period 8,500–3,000 years ago at which time gene fixation could have taken place. The Santa Rosa Island population could have been founded by a small number of seeds, or only one seed, from the mainland population (Ledig and Conkle 1983).

Blood-Group Alleles in Human Populations

Some of the strongest evidence for drift in natural populations is provided by the human species. Population size has been favorable for drift in many parts of the world throughout human history. Small isolated or semi-isolated populations consisting of 200–500 adult individuals were common in the food-gathering and hunting stages of culture. Small isolated farming and fishing communities exist in various parts of the world. Some religious sects form small isolated breeding populations because of religious beliefs prohibiting intermarriage with outsiders.

The large body of data available on the ABO and other blood groups in human populations, both small and large, together with the simple genic basis of the blood types, makes the blood types a useful indicator of the genetic similarities or differences between populations. The polymorphism in the ABO blood groups was described briefly in chapter 3.

An interesting case is that of the Polar Eskimos near Thule in northern Greenland. This small band of 271 or fewer individuals existed in complete isolation for generations. When first contacted by another band of Eskimos from northern Baffin Island, who spent several years trying to reach them, they believed they were the only people in the world.

The Polar Eskimos were found to differ markedly from the main populations of Eskimos in the frequencies of the blood-group alleles. The I^A allele varies from 27–40% in the larger Eskimo populations of Greenland (table 16.2). Similar frequencies of I^A are found in the Eskimos of Baffin Island, Labrador, and Alaska. But the Polar Eskimos deviate from this norm with an allele frequency of I^A of 9% (table 16.2). On the other hand, the Polar Eskimos have a particularly high frequency of I^O as compared with those of other Eskimo populations in Greenland and elsewhere (Laughlin 1950).

Other small isolated human populations showing a marked local differ-

Table 16.2. Frequency of the ABO blood-group alleles in Eskimo populations in Greenland (Laughlin 1950)

	Allele Frequency (%)		
Region	I^A	I^B	I^O
Nanortalik, Julianehaab District, southern Greenland	27	3	70
South of Nanortalik	35	5	60
Cape Farewell	33	3	64
Jakobshavn	29	5	66
Angmagssalik, eastern Greenland	40	11	49
Thule, northern Greenland	9	3	84

entiation in respect to the ABO blood types are the aboriginal tribes of southern Australia, the Dunker religious sect in eastern North America, the Jewish community of Rome, and certain village communities of islands and mountains in Japan (Birdsell 1950; Glass et al. 1952; Dunn and Dunn 1957; Nei and Imaizumi 1966).

The Dunker religious sect was founded in Germany in the early 1700s and later immigrated to the eastern United States. The members of this sect marry largely among themselves and have thus remained reproductively isolated for generations from the surrounding German and American populations. Some Dunker communities are quite small; one in southern Pennsylvania consisted of about 90 adult individuals when studied in the early 1950s. It is significant that the Pennsylvania Dunkers deviate from the general German and American populations in blood types and other characteristics (Glass et al. 1952).

Table 16.3 gives the frequencies of the I alleles in the Pennsylvania Dunkers and in the racially similar large populations of West Germany and the eastern United States. It is apparent that the general German and American populations are similar in gene frequency. But the Dunkers deviate from their German ancestors and modern American neighbors in this respect, the I^A

Table 16.3. Frequency of the ABO blood-group alleles in three related Caucasian populations (Glass et al. 1952)

	Allele Frequency (%)		
Population	I^A	I^B	I^O
West Germans	29	7	64
Eastern Americans	26	4	70
Dunkers in Pennsylvania	38	2	60

allele having a significantly higher frequency and the I^B allele being close to extinction. Other characters in the Pennsylvania Dunkers, such as type of ear lobe or hair, showed similar deviations from those in the ancestral or neighboring populations. In five different genes the allele frequencies found in this small breeding group diverged significantly from those typical of the surrounding populations (Glass et al. 1952).

The I alleles are not selectively neutral. Some human populations show a positive correlation between the incidence of stomach ulcers and an $I^O I^O$ genotype, and between the incidence of stomach cancer and the genotypes $I^A I^A$ and $I^A I^O$. The indication that the I alleles have a selective value is sometimes cited as an argument for a role of drift in this gene. This argument is based on the misconception that drift and selection are mutually exclusive forces.

Conclusions

The conclusion that gene frequencies are controlled to a significant extent by drift in small populations is demanded by the laws of probability and confirmed by experimental studies. The next question is the probable role of drift in natural populations. Does drift play a significant role in evolution?

This question was the subject of a long controversy which has apparently died down in recent years. Opponents of genetic drift contended that, if selection can be shown to be operative in a given case, drift is thereby ruled out as unnecessary (Fisher and Ford 1947; Ford 1955, 1964, 1971; Mayr 1963; 207–210). The implication is that the issue is drift vs. selection with the latter always winning. As Wright and his followers, including myself, see it, however, the evolutionarily important issue is the selection-drift combination vs. selection alone.

Mayr (1942, 1954, 1963, 1970) has long advocated the founder principle, which has become generally accepted. A section in his 1963 book is highly critical of drift (Mayr 1963: 203–214; also 1970). In at least two discussions Mayr treats the founder effect as different from drift (Mayr 1954, 1963: 211–212). Other evolutionists (e.g., Dobzhansky and Spassky 1962; Grant 1963) see the founder effect as a special case of drift, and this is how we presented it in the preceding section.

An argument used against drift is that this factor has not been proven to be effective in natural populations. This argument misapplies the rules of evidence applicable to experimental populations to the vastly more complicated situation in nature. We cannot control the variables in natural populations, as we do in experiments, and therefore we cannot positively identify and quantify the various factors involved. But we can look for patterns in nature that are (or are not) consistent with theoretical and experimental findings. The best evidence for the effectiveness of drift in microevolution is the pattern of haphazard local differentiation in series of permanently or intermittently small isolated colonies. This pattern has been found repeatedly in various groups of plants and animals with a colonial population system. The pattern may not prove, but it does strongly suggest, that drift plays a significant role in such population systems.

CHAPTER 17

Cost of Selection

Genetic Load

Natural selection has both a positive and a negative aspect. It entails the preferential survival and reproduction of some genotypes and the preferential elimination and non-reproduction of others. In this chapter we are concerned with the second aspect.

The loss of a certain proportion of genotypes from a population as a result of the action of selection is obvious when the selective elimination takes the form of mortality in the Darwinian sense. Then there is a visible corpse. But selection also involves success or failure in the reproductive phase alone, as we have seen earlier. Individuals of a given genotype may be physically vigorous but still make a reduced contribution to the next generation because of low fecundity. In this case there is no visible corpse. But the genes for low fecundity decline in frequency and may eventually die out anyway.

The non-reproduction of an individual due to selection in its negative aspect, whether it comes as selective mortality or as failure of reproduction per se, is known as genetic death (or selective death). Genetic deaths lower the reproductive potential of the population.

The population would be stronger in numbers without its quota of genetic deaths. Therefore the latter represent a burden to the population, at least potentially. This burden is referred to as genetic load. Genetic load is defined by Wallace (1968) as 1 − (mean population fitness). It can also be thought of as the sum total of genetic deaths per generation.

Various types of genetic load can be recognized (Brues 1969; Crow 1970). The main types are the mutational load, segregational load, misplaced-individual load, and substitutional load. These types of load are correlated with different types of selection.

The mutational load is an inevitable by-product of the mutation process. This process generates deleterious mutations, which must be weeded out by stabilizing selection. A segregational load exists in populations that exploit the advantages of heterozygote superiority. The superior heterozygotes segregate the less fit homozygotes in each generation. The latter lower the average fitness of the population. They represent the cost of balancing selection.

A misplaced-individual load occurs in a polymorphic population in a heterogeneous environment, where some individuals will inevitably find themselves outside their special niche or subniche. Either disruptive selection or balancing selection or both could be involved in this type of load.

A substitutional load is involved in directional selection. The substitution of an old allele by a new, superior allele entails the genetic deaths of the carriers of the old allele. The total number of selective deaths involved in a complete gene substitution—the overall substitutional load—may be quite high, since the old allele, which is destined to be replaced, is normally present at a high frequency at the beginning of the substitution process.

Protein Polymorphism and the Segregational Load

Electrophoretic assays of enzyme loci in a number of populations of *Drosophila*, vertebrates, and plants have revealed that rather high proportions of these loci are polymorphic. Thus 42% of 19 enzyme loci in *D. melanogaster* are polymorphic, and the proportions are comparable in *D. pseudoobscura* and *D. willistoni* (table 3.3). It is often assumed that the percentage of polymorphic loci detected by electrophoretic methods can be taken as representative of the genotype as a whole, thus increasing the estimated total amount of gene polymorphism.

The occurrence of large numbers of polymorphic genes—some observed, others predicted by extrapolation from the observed ones—runs counter to expectations based on considerations of genetic load. Each balanced polymorphism in a population, since it entails the continual segregation of homozygotes of reduced fitness, produces a certain segregational load. With an increase in the number of independent balanced polymorphisms this segregational load goes up exponentially. At some point the population should be unable to bear the segregational load. Yet the highly polymorphic populations of *Drosophila* are flourishing.

One school of workers, led by Kimura, attempts to solve the dilemma by postulating that the protein polymorphisms are not in fact maintained by selection. The alleles of most enzyme genes are supposed to be selectively neutral and to be drifting at various frequencies in populations (Kimura 1968, 1983; Kimura and Ohta 1971b; King and Jukes 1969). Consequently there is no segregational load. The neutral theory of Kimura would indeed solve the problem if the enzyme genes are selectively neutral.

But are they neutral? There is evidence that some enzyme genes, at least, are not (see chapter 32). Where do we stand? The neutralism-selectionism controversy is unresolved at this writing. It seems probable that some polymorphic genes are selectively neutral or nearly so, while others are probably controlled by balancing selection.

The latter class of polymorphic genes must, therefore, be contributing to a segregational load. The problem as to how the population bears that load still has to be faced.

One possibility is that electrophoretically detectable enzyme loci are not representative of the genotype. This is contrary to a common but unstated assumption, but is in line with some evidence (see chapter 32). It may not be correct, therefore, to estimate the overall rate of gene polymorphism by simple extrapolation from the observed rate of protein polymorphism. The total amount of gene polymorphism could well be substantially less than it is estimated to be. In that case the segregational load would also be a less burdensome one for the population.

Haldane's Cost of Selection

The substitutional load is closely related to what Haldane (1957, 1960) called the cost of selection, the main subject of this chapter. Haldane's concept of cost of selection refers to the total number of selective deaths involved in one complete gene substitution. Hence cost of selection is synonymous with cumulative substitutional load over a series of generations; and what we can call incremental cost of selection is synonymous with substitutional load per generation.

Haldane's two pioneering papers (1957, 1960) contain a mathematical analysis of the total or cumulative substitutional load relative to the population size in any generation. He considered how this relation is affected by selection intensity and other factors. And he considered the bearing of the cost of selection on evolutionary rate. The problem has been restudied and reformulated by various later workers.[1] The presentation that follows is based on Haldane (1957, 1960), Crow and Kimura (1970), and Flake (pers. comm.).

The number of selective deaths required for the complete substitution of one allele by another (ΣD) will be many times the number of adult individuals in any single generation (N). There is a definite relation between ΣD and N. This relation is expressed by a cost factor (C). In general, if N is constant, $\Sigma D = C \cdot N$.

The formula for C for a single locus in a haploid system (which is much simpler algebraically than a diploid system) is as follows. Cost per generation is the ratio

$$\frac{w_1 s q}{\bar{w}} \quad \text{or} \quad \frac{w_1 - \bar{w}}{\bar{w}}$$

where w_1 is fitness of favored allele, $\bar{w}$ is mean population fitness, and q is frequency of inferior type. The total cost of substitution is then

$$C = \sum \frac{w_1 - \bar{w}}{\bar{w}}$$

The value of C thus depends on the initial frequency of the favored allele and the rate of replacement.

The cost factor is high when the favored allele is at a low initial frequency, and decreases as the initial frequency increases. Crow and Kimura (1970:250) give the following values of C for the following gene frequency changes in a diploid system with no dominance:

q	C
0.001 to 0.999	13.81
0.01 to 0.99	9.19
0.1 to 0.9	4.39
0.2 to 0.8	2.77

1. See inter alia Maynard Smith (1968), Crow (1970), Crow and Kimura (1970), Wallace (1970, 1981: ch. 12), Ewens (1972), and Flake and Grant (1974).

If the initial gene frequency is constant, the cost factor varies with the rate of replacement. The rate of replacement depends, in turn, on selection intensity, and therefore s or w enters into the formula for computing cost (C). The total number of selective deaths involved in the gene substitution turns out to be practically independent of selection intensity; but the number of generations over which these selective deaths are spread is measured by selection intensity (s).

Haldane (1957) obtained numerical estimates of the cost of selection in a typical situation. He assumed a large diploid population containing a favorable mutant allele at a low initial frequency. The new allele is supposed to have a moderate selective advantage over the old allele. Under these conditions, the total number of genetic deaths involved in the complete gene substitution is usually 10 to 20 times, or sometimes up to 100 times, the number of breeding individuals in one generation $(C = 10 - 100)$. Taking a single figure as representative, one could expect, on the average, about 30 times as many selective deaths during the course of gene substitution as there are adult individuals per generation $(\Sigma D = 30N; C = 30)$.

Selection Cost and Evolutionary Rate

If the rate of replacement (s) is high, the gene substitution could come fast theoretically. In actuality, however, the population may be unable to stand the cost of intense selection. Selective pressures of too great an intensity may exterminate the population.

The number of genetic deaths that can be tolerated by a population in any one generation is strongly limited. The process of gene substitution must be spread out over numerous generations, therefore, if the population is to maintain a sufficient strength in numbers continuously. This restriction places an upper limit on the rate of evolutionary change for a single gene. In the case where $\Sigma D = 30N$, the gene substitution can be accomplished in 300 generations, but not, obviously, in 30 generations (Haldane 1957).

The cost of selection for two or more independent genes is still higher. The cost rises with the number of independent genes undergoing selection simultaneously (Haldane 1957). In a continuously viable population, therefore, the theoretical maximum rate at which evolutionary change can take place in several or many genes concurrently must be much slower than that for any single-gene character alone. Thus, where a single gene is able to undergo substitution in as few as 100 generations, two independent genes being selected for concurrently might require a minimum of 200 generations for substitution.

There is a strong restriction on the rate of gene substitution for any given number of independent genes determining separate single-gene characters, as noted above. This restriction on the evolutionary rate becomes even more severe if the same number of independent genes form components of a new adaptive gene combination, for then there are two interrelated difficulties standing in the way of rapid multiple-gene replacement. To the high cost of selection for multiple-gene substitutions is added the continual breaking-up of the favored gene combination by the sexual process. Both of these factors are especially strong in their restrictive effects when the component alleles of the new gene combination are still rare in the population.

These considerations suggest that evolutionary changes in complex characters and character combinations will normally take place at moderate or slow rates. The empirical evidence concerning evolutionary rates in some low-rate and moderate-rate evolutionary lines is within the theoretical limits imposed by selection cost (Haldane 1957).

There are, however, some cases of rapid evolution that apparently exceed the theoretical limit. One example will be described.

Certain races of *Mimulus guttatus* (Scrophulariaceae) occupy habitats of post-Pleistocene age in Utah. Their habitats are dated as approximately 4,000 years old. Hence racial differentiation in *Mimulus guttatus* has taken place in 4,000 years in Utah. This herbaceous plant could pass through 4,000 generations in this time (Lindsay and Vickery, 1967). The character differences between the races of another herbaceous plant, *Potentilla glandulosa* (Rosaceae), are due to allelic differences in at least 100 genes (Clausen and Hiesey 1958). Assume that the genetic differences between races in *Mimulus* are of the same order as those in *Potentilla*. Then we have 100 genes undergoing substitution in 4,000 generations, or an average rate of one gene substitution per 40 generations. This estimate errs on the conservative side as regards both the time element and the gene number.

Some organisms have thus evolved at rates that apparently exceed the ceiling imposed by a tolerable cost of selection. The question naturally arises: how have they managed to do so?

Gene Linkage and Gene Interaction in Relation to the Cost Restriction

Haldane's original model postulated a large population size, independence of the two or more genes undergoing substitution simultaneously, and several other conditions. Undoubtedly these conditions are often realized in nature. It is equally true that deviations from the postulated conditions occur frequently in natural populations. Some of these deviations ease the cost restriction on evolutionary rate (cf. Grant and Flake 1974b).

Assume that the separate genes constituting a new adaptive gene combination are closely linked so as to form a supergene. The component genes in the supergene can then undergo substitution at the rate and cost of one gene. The cost of selection is no greater for a supergene than for a single Mendelian gene. If the component genes are closely linked, a population can acquire a new adaptive gene combination within the cost limitations applicable to a single-gene substitution (Grant and Flake 1974b).

Certain modes of gene interaction would also reduce the cost of multiple-gene substitution. This would be the case where the separate genes undergoing selection simultaneously have correlated selective advantages or disadvantages. Different favored alleles may sometimes occur together in the same genotype, and, conversely, different unfavored alleles may also occur together in the same alternative genotype. Then the genetic deaths for numerous separate genes will be pooled in a smaller number of individual deaths. This in turn would make possible a more rapid evolutionary rate for a multifactorial character (Mayr 1963; Maynard Smith 1968; Mettler and Gregg 1969). This result is also suggested by some other authors on the basis of different models and terminology (King 1966; Crow 1970; Felsenstein 1971).

Population Structure in Relation to the Cost Restriction

Deviations from the assumption of a single large continuous population can also reduce the cost limitation on the rate of multiple-gene substitution. Let us consider two alternative types of population structure: the series of founder populations and the subdivided or colonial population system (Grant and Flake 1974a).

It will be recalled that the cost is highest when the favored alleles are at low frequencies. If the favored alleles could be raised above the low level by random non-selective factors, they would escape the most burdensome part of the total substitutional load. A mechanism for such changes exists in founder populations.

In a species with colonizing habits, a large ancestral polymorphic population may give rise to series of daughter colonies founded by one or a few colonizing individuals. Such daughter colonies or founder populations begin with a non-random sample of the ancestral gene pool. Previously rare alleles can be shifted to middle or high frequencies by partly random factors during the establishment of some of the new daughter colonies. These colonies simply evade the heavy part of the substitutional load. And they can support the lighter cost of selection for all subsequent increases in gene frequency, up to complete replacement.

The selective deaths take their inexorable average toll in a population for any given selection intensity. The population has, however, a reproductive potential that allows for the normal proportion of accidental deaths as well as for some quota of selective deaths. In certain population structures the accidental mortality rate can vary upward or downward. Where it varies downward it provides leeway for a higher number of selective deaths.

Take a large continuous population as a standard. Its combined total of selective deaths and accidental deaths sets a limit on the rate of evolution. Now consider a subdivided population of the same total size as the standard population, and subject to the same environmental pressures. In the subdivided population there will be colony-to-colony variation in the accidental mortality rate. Those colonies with low accidental mortality rates can absorb a heavier than average burden of selective deaths and hence can afford a more rapid evolutionary rate than that of the standard large population (Grant and Flake 1974a).

Other Ways Around the Cost Restriction

Some other solutions to Haldane's cost restriction have been suggested. Haldane himself (1957) pointed out that the selective deaths would be less detrimental to the population if they came in embryonic or juvenile stages rather than in the form of adult mortality. Factors favoring the expression of genes and their exposure to selection in early developmental stages would therefore make the cost more tolerable.

Wallace (1968, 1970) suggested that his distinction of hard and soft selection (see chapter 13) could have a bearing on the cost restriction. In effect the cost restriction is a feature of hard selection. Under soft selection the carriers of an unfavorable allele might become selective deaths under stringent environmental conditions but not under permissive ones.

A condition that was implicit but undeveloped in Haldane's original paper (1957) was that the population has a fixed reproductive potential slightly but not greatly in excess of bare replacement. The reproductive excess is presumably kept down to a low level by strong interspecific competition. The selective deaths have to come out of this reproductive excess.

This condition can be altered. Compare a population living in a multi-species community and facing strong interspecific competition with a similar population living in an open habitat. The first population can stand only so much mortality and still maintain its place in the community. The pioneer population does not have this restriction. The absence of interspecific competition enables it to tolerate a greater number of selective deaths and thus to undergo a more rapid gene substitution. Haldane (1957) touched briefly on this possibility.

A variety of ways is thus available for rapid evolution by natural selection.

PART IV
Acquired Characters

CHAPTER 18
Phenotypic Modifications

What Is an Acquired Character?
Phenotypic Plasticity
Adaptive Properties of Phenotypic Modifications
The Role of Phenotypic Modifications in Evolution

What Is an Acquired Character?

The theory of inheritance of acquired characters was generally accepted through most of the history of biology from the Greeks to the nineteenth century (Zirkle 1946). It was accepted by both Lamarck and Darwin. The doctrine was shown to be false by Weismann in the late nineteenth century and by classical genetics in the early twentieth century. The traditional view did not accept defeat, however, but continued to hold sway in various quarters of the scientific world up until the mid-twentieth century. Controversy over the question prevailed from Weismann's key works (1889–1892, 1892) until recent times.

One factor helping to keep the controversy alive was a dualism embedded in the traditional concept of acquired characters. Some acquired characters are in fact hereditary, e.g., hereditary pathogenic diseases, while other acquired characters, now known as phenotypic modifications, are non-hereditary.

Weismann's (1889–1892, 1892) solution to the problem, which was a big step forward in its day, began with a division of the body into soma and germplasm. The latter is the exclusive vehicle of inheritance. Weismann recognized a category of so-called somatogenic variations. These are reactions of somatic tissues or of the body as a whole to external influences, including mutilations and the effects of use or disuse. Weismann argued that, since the soma does not enter the stream of heredity, neither can somatogenic variations. He cited observational and experimental evidence to support this argument.

It has often been pointed out that a clearcut morphological separation of the germplasm from the soma is far from universal. Such a separation does exist in various groups of invertebrate animals. It is less clear in other animal groups and nonexistent in other kingdoms (see Buss 1983). Thus in plants both vegetative and reproductive organs develop from a common growing point. Dissatisfaction with the germplasm theory arose over incongruities like this.

Weismann (1892) was aware of this difficulty and attempted to resolve it. He also (1889–1892) noted that the evidence for non-inheritance of somatogenic variations seemed to hold true in all groups including plants. In this he was not entirely correct.

In the early period of genetics, Weismann's distinction between germplasm and soma was superseded by the genotype-phenotype dichotomy of Johannsen (1911). The former is an anatomical distinction and is not general; the latter is a genetic distinction and is general. The genotype-phenotype dichotomy clarifies the issue of inheritance. Characters determined by newly acquired genetic material in somatic tissues can be hereditary if those tissues go on to produce reproductive cells, as they well may in plants. On the other hand, phenotypic characters are never inherited, and therefore phenotypic modifications are also non-hereditary.

Buss (1983) has recently called attention to an alleged flaw in the synthetic theory of evolution. His argument goes as follows. The synthetic theory focuses on inter-individual genetic variation, but disregards genetic variation in somatic tissues within individuals, on the assumption that these are never transmitted. This assumption, in turn, is based on the acceptance of Weismann's theory of a separate germline, which does not hold up in many major groups. The synthetic theory needs to be modified to accommodate the possibility of inheritance of certain kinds of somatic variation (Buss 1983).

The error lies in Buss' premise that the synthetic theory looks back to Weismannism for guidance on inheritance. The synthetic theory bases itself on the genotype-phenotype dichotomy.

In summary, we have to distinguish two types of acquired characters: (a) phenotypic characteristics determined by newly acquired genetic material; and (b) characteristics resulting from phenotypic reactions to environmental stimuli, in other words, phenotypic modifications. The first may be transmitted in inheritance; the second is non-hereditary. The two types of acquired characteristics play entirely different roles in evolution.

We will deal with them separately. Phenotypic modifications will be discussed in this chapter, and the effects of acquired genetic material in the next.

Phenotypic Plasticity

The genotype is the sum total of genes of an organism, and the phenotype is its particular set of character expressions. Between genotype and phenotype lie the long and complex processes of gene action and development. These processes take place in an environment and are influenced by that environment. The phenotypic character expression is then a product of two sets of factors: the genotypic determinants and the environmental conditions in which development takes place.

Each genotype has the capacity to give rise to a certain range of phenotypes in different environments. The genotype can be thought of in this regard as a "reaction norm" (Johannsen 1911). In other words, the action of the genotype is neither rigidly predetermined nor unrestrained; instead, the genotype can determine a series of phenotypic expressions within limits set by the genotype itself (Johannsen 1911).

The range of phenotypic variation in panicle length and flowering time, in diverse environments, was measured in different families of annual chess grass, *Bromus mollis* (Jain 1978). Large family-to-family differences were found in the amount of phenotypic plasticity. These differences indicate genetic control of phenotypic plasticity. Some genotypes engender highly modifiable phenotypes; other genotypes produce a narrower range of phenotypes.

A wide range of phenotypic expressions in a single individual plant can be demonstrated in plant groups capable of vegetative propagation. Subdivide an individual into clonal members and propagate these subdivisions under different environmental conditions in a controlled-environment greenhouse or growth chamber.

Good examples are furnished by Clausen, Keck, and Hiesey's (1948) experimental studies of environmental responses in the *Achillea millefolium* group (Compositae). These are perennial herbaceous plants that can be propagated from vegetative subdivisions. The subdivisions were grown in different chambers in a controlled-environment greenhouse in Pasadena, California, and also out of doors in the winter in Pasadena. The controlled environments differed in light duration, day temperature, and night temperature.

In a typical experiment, six subdivisions of one plant of *Achillea borealis*

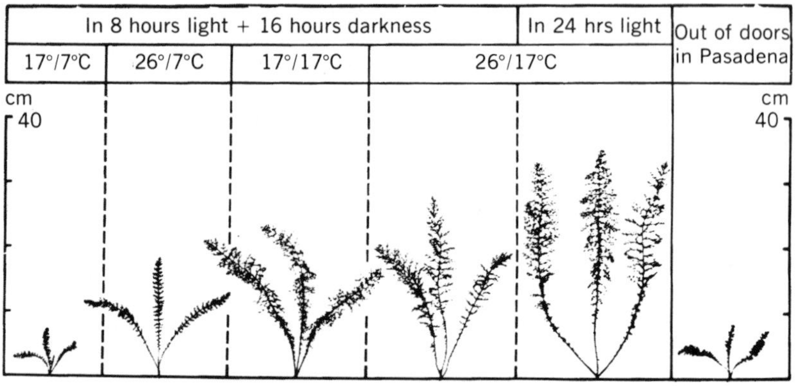

In 8 hours light + 16 hours darkness				In 24 hrs light	Out of doors in Pasadena
17°/7°C	26°/7°C	17°/17°C	26°/17°C		

Figure 18.1. Growth responses of clonal subdivisions of a single plant of *Achillea borealis* from Seward, Alaska, in six environments. The temperature figures are for days and nights, in that order. The growth period was 3.5 months. (Clausen, Keck, and Hiesey 1948)

from Seward, Alaska, were grown for 3.5 months in five controlled environments and in a sixth, out-of-doors, environment. The different growth responses at the end of the experimental period are shown in figure 18.1. The genotype of this plant produced good growth in some environments (long warm days) and slight growth in others (short days). Other genotypes of the same species responded in a different fashion to these environments (Clausen, Keck, and Hiesey 1948).

Adaptive Properties of Phenotypic Modifications

Phenotypic reactions to normal changes in the environment are usually adaptive. We see this in sun leaves and shade leaves of plants. Shade leaves have a larger surface area, which increases their photosynthetic capacity and compensates for the dimmer light, whereas sun leaves have a smaller surface area, thus cutting down on transpiration and water loss. The same plant genotype has the capacity to produce sun leaves in bright light and shade leaves in dim light. The genotypically determined capacity of a plant to respond phenotypically in this fashion enables that plant to adjust to varying light conditions.

The range of phenotypic modifiability shows wide differences in different major groups of higher organisms. Higher plants, and particularly herbaceous plants, are characterized by great phenotypic plasticity. Insects exemplify the opposite extreme of phenotypic inflexibility.

These differences in plasticity are correlated with differences in mode of development in the two groups. The plant body develops from growing points that are exposed to and strongly influenced by the environmental factors prevailing when the new parts are forming, so that sun leaves appear in bright light and shade leaves in dim light, and so on. The adult insect body, on the other hand, develops within a hard external skeleton that is formed in a preadult growth stage. The main external features of the insect body are laid down well before they can be functional, and have become set in their mold by the time they can be used.

Beyond these differences in mode of development lie more basic differences in the strategy of individual adjustment to fluctuating environmental conditions. Insects are motile, plants are sedentary. The adult insect can adjust to environmental variables, within limits, by moving to a warm spot in cold weather or to a moist spot in dry weather—in short, by habitat selection. A plant anchored to the ground by its roots does not have this option; for it, phenotypic modifiability is the main means of individual adjustment to variations in the environment.

Both the types of phenotypic modifications and the range of phenotypic modifiability are related to the requirements of the organisms. It is very probable, therefore, that genotypes have been selected for their capacity to make adaptive phenotypic responses to environmental variables.

The Role of Phenotypic Modifications in Evolution

Phenotypic modifications protect the individual organism against environmental stress, and, to the extent that they are successful in this, they constitute a buffer against environmental selection. Theoretically, a population composed of genotypes with a complete range of adaptive phenotypic reactions would not respond to the pressure of natural selection at all. In practice, phenotypic plasticity has a retarding effect on the action of environmental selection.

Phenotypic plasticity permits a population to persist in an environment that has changed in an unfavorable way. The postponement of selective elimination gives the population more chances of acquiring new genetic variations by mutation, gene flow, and recombination, from which it can construct a genotypic response to the new, unfavorable environment. This indirect connection between phenotypic modifications and evolution of new adaptive characters has been called the Baldwin effect (Simpson 1953b).

CHAPTER 19
Genetic Transformation

Transformation and Transduction in Bacteria
Experimental Transformation in Animals
Hybrid Sterility in *Drosophila paulistorum*
Experimental Transformation in Plants
Transformation in Nature
Bleaching in *Euglena*
Discussion

The induced hereditary changes of classical genetics—the mutations induced by high-frequency radiation and mutagenic chemicals—are "random," i.e., unoriented. The mutagenic agent induces a diverse array of mutant types at different loci. The induction of directed or oriented hereditary changes was a subject for speculation in classical genetics and early evolutionary genetics. Oriented hereditary change was seen in that era as an interesting possibility which depended upon the transfer of specific genetic material from one organism to another and its incorporation in the reproductive cell line of the receptor organism. In the third edition of *Genetics and The Origin of Species*, Dobzhansky included a section on "The quest for directed mutation" (Dobzhansky 1951:45 ff.).

The early work on induction of oriented hereditary change was done outside the field of genetics. The genetic significance of the early work on transformation in *Pneumococcus* was not appreciated at first. Bacterial transformation eventually became widely accepted, however, and workers turned their efforts to higher organisms with mixed results. In recent years, after many setbacks, directed hereditary change in higher organisms has become established on an experimental basis. Current work in the field may be developing in the direction of practical genetic engineering.

We will adopt the widely used term, genetic transformation, for oriented hereditary change resulting from the transfer of genetic material from a donor organism to a receptor organism. Other synonymous terms in common use are gene transfer and hereditary induction.

The evolutionary role of genetic transformation has been neglected by the synthetic theory of evolution. The starting point for evolution according to the synthetic theory is unoriented mutational variation. Directed hereditary changes do not fit into this picture. Such changes do occur, however, although we still do not know how important they are in evolution, and the synthetic theory will have to accomodate itself to them.

Transformation and Transduction in Bacteria

The first bona fide case of hereditary change stemming from the acquisition of foreign genetic material was discovered in the pneumonia bacterium, *Pneumococcus*. Transformation was discovered in *Pneumococcus* by Griffith in 1928, and was shown to take place in vitro by Avery, Macleod, and McCarty in 1944. The essential facts are now generally known and will be briefly summarized here for the record. Strains of *Pneumococcus* differ in the nature of the cell surface coat, which may be rough or smooth, and in correlated properties of virulence. These differences are hereditary. When one strain of the bacterium is grown in a sterile extract containing the DNA of another strain, some daughter cells of the former strain acquire the type of cell surface and virulence of the latter strain. The change is hereditary. The hereditary characters of a donor strain are transferred to a receptor strain by the DNA of the donor in a sterile growth medium (Avery et al. 1944).

Hereditary transformations brought about by sterile extracts have been demonstrated in other genera of bacteria, such as *Hemophilus* and *Streptococcus*, and for other characters, such as resistance to penicillin and streptomycin.[1]

Bacterial transformation is a laboratory phenomenon. It has a natural analogue in transduction. Under natural conditions, bacterial viruses or phage occasionally transfer genetic material from one bacterial cell to another. Transduction is this transfer of bacterial genetic material by phage.

The general sequence is as follows. The phage infects one bacterial cell, and then multiplies inside this cell at the expense of the host DNA. A daughter phage particle now infects a second bacterial cell, and, in so doing, intro-

1. For more information see Hayes (1968: ch. 20) and various recent genetic texts, e.g., Merrell (1975).

duces genetic material from the first host into the second host. Transduction occurs most successfully with temperate phages that do not completely destroy their bacterial hosts. The transduced bacteria thus live on and perpetuate their altered characteristics.

Transduction has been shown to occur in *Salmonella, Escherichia, Pseudomonas, Staphylococcus*, and other genera of bacteria. Among the characters transferred by transduction are synthetic ability, antibiotic resistance, and motility (see Hayes 1968: ch. 21). A case of conversion in *Clostridium* involves the type of toxin produced by the bacterium. One strain of *Clostridium botulinum* causes botulism in man and animals; but when it is infected by a certain bacteriophage it is converted to another strain of *Clostridium* that causes gas gangrene (Eklund et al. 1974).

Experimental Transformation in Animals

One of the early studies of transformation in a higher animal involved body mosaics in *Drosophila melanogaster*. Eggs containing young embryos of wild-type flies were treated with a solution containing DNA from flies with specific mutant characters. Mutant types of body pigmentation, bristle shape, and eye color inter alia were used as genetic markers. The treated embryos developed into adult flies showing the same phenotypic character as that in the mutant strain that furnished the transforming DNA. Furthermore, the transformed characters proved to be heritable. However, the transformed individuals were mosaics with patches of mutant and normal tissue; the body or body part did not become transformed as a whole (Fox and Yoon 1966, 1970).

Indirect evidence suggests that the transforming DNA did not become integrated into the chromosomes of the receptor flies in these experiments. The transforming DNA probably got into other parts of the cell in certain cell lines and cell sectors. This would account for the mosaicism of the phenotypic effects (Fox et al. 1970, 1971).

A recent experiment on transformation in *D. melanogaster* used the eye-color gene rosy as a marker and a transposable element *(P)* as the vector. The element *P* carries the chromosome segment that includes the rosy gene; in this experiment *P* carried the wild-type allele of rosy. The element *P* can become inserted at various sites on any chromosome of the complement. It can get into the germline as well as somatic tissues. Embryos of mutant flies with rosy eyes were injected with transposable elements carrying the wild-type allele. The injected flies produced progeny with normal eye color (Rubin and Spradling 1982).

Various laboratories are now inducing transformation in mice. Foreign DNA containing particular marker genes is injected into fertilized mouse eggs and the eggs are implanted in a foster mother. The first question of interest is whether the resulting mice will exhibit the foreign gene(s) phenotypically. Sometimes they do and sometimes they don't. The next question is whether the introduced gene(s) will get into the germline and be transmitted to the next generation. Here again the answer is sometimes yes and sometimes no (Gordon et al. 1980).[2]

In one very interesting experiment a gene in the rat for growth hormone was transferred to the mouse by splicing it onto a fragment of mouse DNA and injecting the fragment into mouse eggs. The rat gene became incorporated into the tissues of one-third of the mice that developed from the treated eggs. It was manifested phenotypically in large body size. The transformed mice had nearly twice the body weight of untreated litter mates (Palmiter et al. 1982).

Transformation has also been induced experimentally by a different method in chickens (Pandey and Patchell 1982).

Hybrid Sterility in Drosophila paulistorum

Drosophila paulistorum is a species group composed of a number of incipient species or semispecies in Central and South America. Hybrid sterility appears between some of the semispecies. Crosses between certain semispecies give sterile male hybrids. Other crosses produce all fertile progeny. The male sterility, where it exists, is determined in part by factors that are transmitted through the cytoplasm of the egg. In the final analysis, the hybrid male sterility is determined by the interaction between these cytoplasmic factors and the chromosomes.

The propensity for sterility is thus passed on by female flies with certain cytoplasmic constitutions to their hybrid male progeny. Consequently, it is possible for a hybrid cross made in one direction (e.g., Santa Marta ♀ × Mesitas ♂) to produce sterile males in F_1, whereas the reciprocal cross (Mesitas ♀ × Santa Marta ♂) yields fertile males in F_1 (Ehrman and Williamson 1965; Williamson and Ehrman 1967).

2. Brief reviews by Marx (1981, 1982a, 1982b).

Williamson and Ehrman (1967) made a homogenate of the cytoplasm of eggs of the Santa Marta strain, and injected this homogenate into females of the Mesitas strain. The treated Mesitas females were next crossed with Santa Marta males. The injected Mesitas females (in contrast to untreated Mesitas females) produced sterile male hybrids. Male sterility was induced by injection of the female parent with a particular type of cytoplasm.

What is the nature of the cytoplasmic sterility factors? Electron micrographs of reproductive tissues in sterile males reveal the presence in the cytoplasm of particles resembling *Mycoplasma*, a simple and primitive type of bacterium (Ehrman and Kernaghan 1971; Daniels and Ehrman 1974). More recently streptococcal L-forms have been correlated with the hybrid male sterility (Somerson et al. 1984).

A likely hypothesis to account for the various sets of facts is that *Mycoplasma* or similar organisms occur as symbionts in the cells of some semispecies of the *Drosophila paulistorum* group. in these semispecies there is a good adjustment between the chromosomal genes and the infected cytoplasm, so that normal fertility is not upset. However, crosses of these symbiont-carrying semispecies, used as female parents, with symbiont-free strains used as male parents produce combinations of cytoplasm and chromosomal genes that are not mutually adjusted. The phenotypic expression of this cytoplasm-nucleus interaction is hybrid male sterility (Williamson and Ehrman 1967; Ehrman and Kernaghan 1971).

Experimental Transformation in Plants

Attempts to effect genetic transformation experimentally in plants have so far ranged from unsuccessful to only partially successful or inconclusive (Kleinhofs and Behki 1977; Hess 1977; Coe and Neuffer 1977; Schiemann 1982; Barton and Brill 1983).

Unexpected and puzzling results emerged from early experiments with flax *(Linum usitatissimum)*. Different flax plants in the original generation were given different dosages of the nutrient elements, N, P, and K, and responded in the usual way by producing good growth and high yield. This standard phenotypic response is not expected to be repeated in later untreated generations. However, in the present case, the later-generation progeny of induced large plants tended to remain large even though the nutrient level was reduced; conversely, the progeny of induced small plants remained small despite increases in nutrient level. The induced size differences held up through ten generations. Crosses between derived large and small plants yielded intermediate F_1 hybrids, confirming the conclusion that the differences are hereditary. These hereditary differences do not appear to reside in the cytoplasm (Durrant 1962a, 1962b). Later studies revealed differences between the large and small lines in the DNA content of the nuclei, the derived large plants having the greater amount of nuclear DNA (Evans et al. 1966: Timmis and Ingle 1974). Apparently the nutrient treatment induced changes in DNA content.

Similar induced size changes correlated with nutrient treatment have been followed through three generations in the wild tobacco species, *Nicotiana rustica* (Hill 1967).

Recent experiments of Pandey (1976, 1978, 1980) in *Nicotiana* utilize a method of dual pollination. The female parent plant is pollinated with a mixture of live pollen of one genetic type and irradiated dead pollen of another type. The dead pollen is capable of producing pollen tubes and of discharging its DNA into the embryo sac but is not able to effect fertilization. Genetic markers in the dead pollen show up in some of the first-generation progeny and have been followed through two subsequent generations. Apparently chromosome fragments carrying the marker genes in the dead pollen get into the egg nucleus and become incorporated there.[3]

Transformation in Nature

Great interest attaches to the question whether genetic transformation takes place in nature and has had any important effects in evolution.

Although transduction is a firmly established process in bacteria, there does not seem to be any evidence that it has figured in bacterial evolution. On the other hand, the standard processes of mutation and selection are known to govern some microevolutionary changes in bacteria. The development of resistance to phage, for example, depends initially on phage-resistant mutations that appear spontaneously in bacterial colonies. These mutations arise at a given low rate that is not affected by the presence or absence of phage in the environment. The microevolutionary change thus follows the orthodox scheme. The negative evidence regarding the role of transduction in bacterial evolution should not be taken too seriously, however, since trans-

3. For a review of transformation and other unconventional methods of inducing hereditary change in plants see Pandey (1981). See also Caplan et al. (1983).

duction has been studied mainly from the molecular rather than the evolutionary standpoint.

The conditions for natural transformation in higher plants could exist where a good vector of DNA, such as the soil bacterium, *Agrobacterium tumefaciens*, transfers genes from one plant to another. This bacterium is known to carry, among other things, bacterial DNA fragments (plasmids) bearing bacterial genes for opine synthesis. Opines are amino acid derivatives which are not normally produced by plants, but which are used as nutrients by bacteria. *Agrobacterium*, on infecting a plant, inserts the plasmid containing the bacterial opine-synthesis genes into the nuclear DNA of the host plant where they are functional. The bacterium is transforming the host cells with genetic material of its own and in a way favorable to itself; but it is easy to imagine it carrying genetic material from other organisms to the plant hosts (Barton and Brill 1983).

A type of globin, leghemoglobin, occurs in the root nodules of some leguminous plants, where it helps in nitrogen fixation. Leghemoglobin is similar to the hemoglobins of vertebrates. One suggestion for the occurrence of a globin in plants is gene transfer from animals by an insect-born bacterium or virus (Lewin 1981). According to Dillon (1983:425–427), however, the actual homologies between the leghemoglobins and hemoglobins are too complex to be explained by simple gene transfer.

Genes of C-type viruses can get into the chromosomal DNA of mammals where they become incorporated and function to produce virus particles. They also get into the germline and are transmitted through the gametes. A C-type virus gene characterized by a particular nucleotide sequence occurs in old world monkeys and apes. Interestingly, a related gene with a similar sequence is found in the domestic cat and other cat species in the Mediterranean region, but not in cats native to other continental regions. The overlapping geographical distribution of the two related forms of the virus could be explained by gene transfer, by the virus and its vector, from monkeys to cats in the Mediterranean region (Benveniste and Todaro 1974).

According to the endosymbiont theory (Margulis 1970, 1981), the mitochondria of eukaryotes, and the chloroplasts of plants and green flagellates, originated as free-living prokaryotes which invaded primitive eukaryotic cells and became established as permanent self-reproducing symbionts in the cytoplasm. Mitochondria are considered to be derived from aerobic bacteria and chloroplasts from blue-green algae. The evidence consists of structural and biochemical similarities between these cell organelles and the corresponding types of modern prokaryotes. The DNA of mitochondria is double-stranded and circular, like that of bacteria. Chloroplasts and blue-green algal cells both contain a photosynthetic apparatus and bodies that look like ribosomes, in addition to DNA, and both are enveloped by a membrane. Furthermore, blue-green algae have a known tendency to form symbiotic associations with various eukaryotic organisms.

The evidence in favor of the endosymbiont theory of cell organelles is persuasive but not conclusive, and accordingly, opinion is divided on it. Proponents of the theory include Margulis (1970, 1981), Raven (1970), Ris and Plaut (1962), and Parthier (1982); criticisms of the theory are presented by Uzzell and Spolsky (1974) and Dillon (1981:504–508). The difficulty is one of deciding whether the similarities between the cell organelles and the prokaryotes in question are due to common ancestry or, alternatively, to convergence.

Intermediate positions between complete acceptance and complete rejection of the theory have to be considered. It may be that chloroplasts have had an endosymbiotic origin, but mitochondria have not originated from prokaryotes (Parthier 1982).

If the endosymbiont theory is true, even partly, it would add massive support to the idea that gene transfer outside of normal channels has had important evolutionary effects. The acquisition, by primitive eukaryotes, of endosymbiotic blue-green algae, would have been a critical step in evolution, making possible the development of the green flagellates and the plant kingdom.

Bleaching in Euglena

The phenomenon known as bleaching in the green flagellate, *Euglena*, is the permanent loss of chloroplasts in a cell line. This loss represents another type of induced and directed hereditary change.

The greenness of *Euglena* is due to the photosynthetic pigment chlorophyll in the chloroplasts. The chloroplasts are DNA-containing and ordinarily self-reproducing organelles in the cytoplasm; they constitute a type of cytoplasmic gene. Light and other factors are necessary for the proper development and functioning of the green chloroplasts.

Euglena gracilis, which manufacturers its own sugar in the light, can be maintained artificially in the dark on a growth medium containing sugar. In the dark the chlorophyll does not develop, and later the plastids themselves may fail to reproduce, so that they gradually dwindle in numbers and eventually become lost altogether in a cell line. The loss is permanent. A culture of *Euglena gracilis* that has lost its chloroplasts in a dark environment does not regain them when brought back into a light environment (see Lewin 1962).

Other environmental factors besides darkness that can cause bleaching in *Euglena gracilis* are heat (32–35°C), UV radiation, and antibiotics (streptomycin, etc.) (Lewin 1962).

These environmental factors can convert a green photosynthetic strain of *Euglena* into a colorless heterotrophic strain that grows only on a substrate containing sugar. Morphologically, the derived strain of *Euglena* resembles the colorless euglenoid, *Astasia longa*. One hypothesis of flagellate phylogeny holds that the colorless euglenoids are actually derived by bleaching from the green euglenas (Lewin 1962).

Discussion

The synthetic theory of evolution has followed classical genetics in emphasizing the role of chromosomal genes. The Mendelian behavior of chromosomal genes in sexual organisms is the basis of the Hardy-Weinberg law, the rules of recombinational variation, the genetical theory of selection, and other facets of this theory. The emphasis is not misplaced.

Let us make the starting assumption that the acquisition of foreign genetic material is sometimes beneficial to an organism or its descendants. Eukaryotic organisms have a regular mechanism for combining genetic material from different sources. This process is sexual reproduction, a precise and orderly process, taking place normally within a species population. The orderliness of the sexual process, it should be noted, has reference exclusively to the chromosomal genes.

Furthermore, the development of complex characters in eukaryotic organisms requires the action of well-integrated gene systems. Such gene systems can be assembled, maintained, and fine-tuned in the chromosome set. The assembling and fine-tuning of the chromosomal genotype are carried out by the orderly sexual process. Irregular, hit-and-miss methods of combining genetic material from unrelated sources will not do.

Hence the well-justified emphasis of the synthetic theory on chromosomal genes and their recombination according to precise rules by the sexual mechanism. When this is all said, however, the fact remains that irregular, hit-and-miss methods of genetic exchange also exist in the world. Foreign genetic material from an unrelated source occasionally becomes lodged, more or less permanently, in the cytoplasm of a eukaryotic host, or in its nucleus outside the chromosomes, or even at some site on a chromosome, and contributes to the hereditary variation of that host organism.

The irregular methods of trans-specific genetic exchange may represent ancient processes which worked well during the prokaryotic stage of evolutionary history, and continued to play a subservient role in eukaryotic evolution. Sexual reproduction, which evolved along with eukaryotic organisms, may have supplanted the irregular parasexual methods of genetic exchange in prokaryotes because of the inadequacy of the more primitive and inefficient methods once a certain stage of organic complexity was reached.

The emphasis in a broadened synthetic theory of evolution should fall on both sexual gene exchange and irregular trans-specific gene transfer.

PART V
Speciation

CHAPTER 20
Races and Species

In Parts 2 and 3 we considered evolutionary forces as they operate within a breeding population. The local population, however, is a component of a more extensive population system; it belongs to a race and to a species. If the single breeding population is a field in which evolutionary changes take place, so too is the interlocking system of populations, the race, and species. We turn now to a consideration of evolution at the level of these more inclusive population systems.

It is necessary to discuss the nature of races and species before attempting to describe evolutionary processes at this level. The present chapter gives a survey of the types of population systems and types of variation patterns in higher plants and animals. Isolating mechanisms and ecological interactions are considered in the next two chapters. With these aspects of population systems in mind, we can then proceed, in the second half of Part 5, to consider the processes involved in race and species formation.

Population Systems

The local population is a breeding unit in sexual organisms, but it does not exist in complete isolation. A certain amount of interbreeding occurs between local populations as a result of migration and gene flow. Numerous local populations are linked by their breeding relationships into races and biological species. These population systems represent breeding groups of a more inclusive scope than the local population.

We have seen that the members of a local population share in a common gene pool. The regional cohort of local populations—the race—also shares in a common gene pool, but in a more inclusive one. A still more inclusive and diversified gene pool is that of the biological species, which is the sum total of interbreeding races.

The visible result of interbreeding between populations and between races is intergradation in phenotypic characters that differ regionally within the species. It is by following the course of this intergradation in phenotypic characters over geographical areas that we get our first good indication of the occurrence and channels of interbreeding. When we follow out the inter-gradations in phenotypes, we come sooner or later to a prominent disconti-nuity in the variation pattern. This discontinuity marks the boundary line of the biological species. The interbreeding of populations and of races, in other words, is confined within limits, and these limits circumscribe the species.

We have stated that the boundary line of the biological species corre-sponds to the limits of normal interbreeding and integradation. With these outer limits set, we can go on to describe some types of intraspecific varia-tion patterns. The diverse variation patterns found in nature fall into three main classes, which will be discussed separately. The three modes are: (a) continuous geographical variation, (b) discontinuous geographical variation, and (c) ecological race differentiation. The modes are often mixed in actual cases. A given species may exhibit different variation patterns in different parts of the species area.

Allopatry and Sympatry

The spatial relations between local populations and between population systems are an important factor in determining the amount of interbreeding that takes place between them, and hence the type of variation pattern that develops. The terms allopatric and sympatric of Mayr (1942) are useful to describe these spatial relations.

1. Allopatry. Populations living in different areas are said to be allopatric. The amount of interbreeding between such populations is a function of the spatial distance between them. Geographical races are allopatric. If the geo-

graphical races are contiguous, the intergradation may be more or less continuous; if the races are disjunct, interbreeding is reduced between them and the intergradation is interrupted.

2. Sympatry. Two or more populations living in the same area are said to be sympatric. Non-interbreeding species coexisting in the same territory are sympatric. The term sympatry covers two different situations, which should be distinguished. In the first case, the populations coexist genetically but not ecologically, while in the second case they coexist both genetically and ecologically. The two situations are designated as neighboring sympatry and biotic sympatry, respectively (Grant 1963).

2a. Neighboring sympatry. One often finds two or more kinds of habitats occupying adjacent positions and having numerous interfaces over a large geographical area. In savanna country, for example, patches of woodland alternate with areas of open grassland in checkerboard fashion; in mountainous country, different altitudinal zones occur in the form of concentric rings. Populations inhabiting the adjacent but different habitats lie within the normal dispersal range of one another, so that interbreeding is possible from a strictly spatial standpoint, but they do not live or grow side by side. Such populations are neighboringly sympatric. Ecological races are usually neighboringly sympatric.

2b. Biotic sympatry. Now consider two or more populations living in the same habitat, e.g., in the woodland patches in savanna country. Not only do these populations have overlapping zones of dispersal of their gametes, they also come into contact in the non-reproductive or secular phases of their life cycles. Such populations are biotically sympatric.

3. Parapatry. This is a third and intermediate situation. Two populations are said to be parapatric when they occupy contiguous but non-overlapping areas without interbreeding.

Continuous Geographical Variation

This pattern is typical of outcrossing organisms that form large continuous populations. Examples are furnished by *Drosophila*, man, and many species of forest trees and plains grasses. The local populations are polymorphic. And each local population has its characteristic balance of polymorphic types resulting from gene flow and various modes of selection. Along a geographical transect there is a gradual shift in the frequencies of these polymorphic types.

Consider the ABO blood types in human populations ranging across Europe and Asia. Type B blood ranges from frequencies of nearly 40% in central Asia down to 5% and 6% in parts of western Europe. The gene frequency of the I^B allele, determining type B blood, decreases in parallel fashion from 36% in central Asia to 5% or less in parts of western Europe. The gradient in the allele frequency of I^B on a transect across Europe and Asia is shown in table 20.1 and figure 20.1. As I^B decreases in frequency from east to west, I^A and I^O increase (Mourant et al. 1976).

The gradient in type B blood and in the I^B allele has been explained as a result of repeated migrations of Mongolians and other Central Asian peoples into western Asia and Europe. Interbreeding of the Central Asians with the native races to the west, among whom the I^B allele is believed to have been rare originally, presumably raised the frequency of this allele in the resident populations. The numbers of Central Asian invaders reaching any region decreased with the distance from Central Asia. The gradient in the allele frequency of I^B across Eurasia thus reflects a gradual decrease in the amount of gene flow from Central Asia to successively more western parts of Asia and Europe (Candela 1942).

The inversion types in chromosome III of *Drosophila pseudoobscura* have been discussed previously as examples of polymorphism, microevolutionary changes and balancing selection (chapters 5 and 14). These inversions also exhibit continuous geographical variation in western North America.

Figure 20.2 shows the average annual frequencies of the ST and AR inversions at intervals on a transect through the central Sierra Nevada of Cal-

Table 20.1. Frequency of the blood-group allele I^B in a series of human populations on a west–east transect across Europe and Asia (Compiled from table 1.1 in Mourant et al. 1976)

Region	I^B allele (%)
Portugal: Lisbon	5.0–5.7
Spain: Cataluña	5.7–6.2
France: Paris	6.5–6.7
West Germany: Freiburg	7.7
Germany: Berlin	10.4–11.3
East Germany: Halle	11.1
Poland: Warsaw	14.2–15.3
USSR: Moscow	15.4–19.6
USSR: Urals	14.2–27.4
Mongolia	18.6–24.4
Afghanistan	28.1–36.2

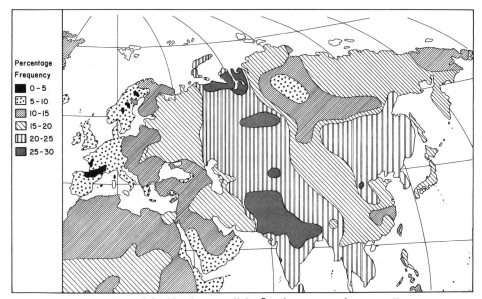

Figure 20.1. Frequency of the blood-group allele I^B in human populations in Europe, Asia, and northern Africa. (Drawn from data of Mourant et al. 1976)

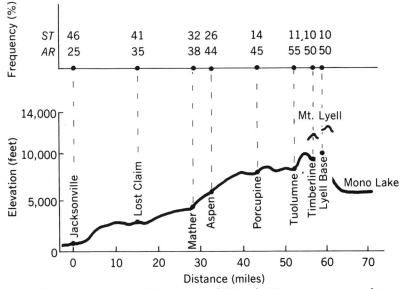

Figure 20.2. Average annual frequencies of *ST* and *AR* inversions in populations of *Drosophila pseudoobscura* on a transect through the Sierra Nevada, California. (Redrawn from Dobzhansky 1948)

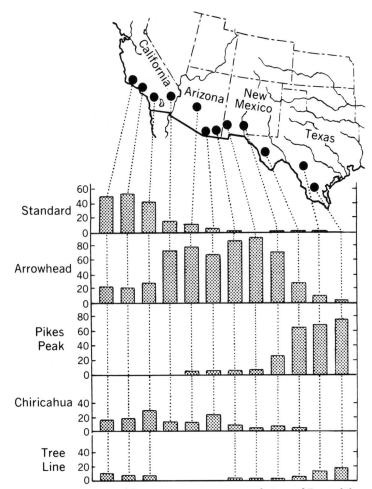

Figure 20.3. Frequencies of five inversion types in populations of *Drosophila pseudoobscura* on a transect from southern California to Texas. (Dobzhansky and Epling 1944)

ifornia. We see a gradual rise in frequency of AR and a decline in frequency of *ST* with rise in elevation in the mountains (Dobzhansky 1948). This transect is a segment of a much larger pattern of geographical variation, part of which is shown in figure 20.3. On a transect through the southwestern United States, *ST* has a high frequency in California but declines to the east; AR increases in frequency from California to Arizona and New Mexico, but then declines again in Texas. Pikes Peak *(PP)* on the other hand, has its maximum frequency in Texas, and declines to low levels to the west (figure 20.3). On a still larger scale, other trends occur to the north in the Rocky Mountain region, where AR is high, and to the south in Mexico, where CH is high (Dobzhansky and Epling 1944).

Geographical racial variation is real in the population systems of *Homo sapiens* in Eurasia and *Drosophila pseudoobscura* in North America. But the drawing of boundary lines between races is an arbitrary procedure, owing to the gradual character of the intergradation; any classification of the population system into races is therefore also arbitrary. How many races should we recognize in *Drosophila pseudoobscura?* One could with equal justification recognize four, or ten times that number.

In practice it is useful to distinguish between local races and geographical races. The former exhibit relatively minor shifts in the frequencies of polymorphic variants (e.g., Jacksonville and Lost Claim in *D. pseudoobscura*; see figure 20.2), whereas the latter are regional clusters of local races differing significantly from one another in their gene pools (e.g., the California, Arizona–New Mexico, and Texas population systems of *D. pseudoobscura*; see figure 20.3).

The change in character across a geographical transect may or may not take place at a uniform rate. Steep gradients may occur here and there in the variation pattern of the species.

An example is provided by the European house mouse, *Mus musculus.* The western European subspecies, *M. m. domesticus*, and the eastern European race, *M. m. musculus*, meet in a narrow zone of secondary intergradation running from north to south through Germany and Austria. In eastern Denmark this zone is 20 km. wide. The allele frequencies of various enzyme genes change abruptly in this zone from the condition characteristic of *M. m. domesticus* to that of *M. m. musculus* (Hunt and Selander 1973).

Areas of abrupt change in gene frequencies or phenotypic character states over short geographical distances are convenient taxonomically in those species which have them. They provide natural zones in which to draw racial boundary lines, thus reducing the arbitrariness involved in subdividing a continuous species population into geographical races.

Disjunct Geographical Races

An island-like population system develops where the inhabitable areas of the species are relatively small and widely spaced. The discontinuous distribution of a particular habitat is a common situation; examples are islands in an archipelago for terrestrial organisms, series of lakes for aquatic organisms, mountaintops for alpine organisms, rocky outcrops in a grassy plain, etc.

A characteristic variation pattern develops in this situation. The geographical variation exhibits discontinuities coinciding with the gaps in geographical variation. The colonies, if sufficiently isolated, tend to become adapted rather specifically to their local environmental conditions. The species then appears as a series of disjunct and distinct geographical races. This variation pattern is the expected result of the reduction of gene flow between colonies and of the action of selection or selection-drift within each colony.

The annual herbaceous plant *Gilia leptantha* (Polemoniaceae) occurs in openings in montane pine forests in southern California. The pine forests in this area occur at middle and high elevations in a series of isolated mountain ranges separated by many miles of unforested lowlands. The population system of *Gilia leptantha* accordingly has the disjunct distribution pattern shown in figure 20.4.

Furthermore, the populations in different mountain ranges group themselves into well-marked geographical races. Four such races, differentiated morphologically on floral characters, are recognized taxonomically as subspecies, as shown also in figure 20.4. Crossing experiments indicate that these four geographical races are highly interfertile. The floral characters of the races converge somewhat in their areas of closest proximity, but do not intergrade continuously (Grant and Grant 1956, 1960).

A closely related species, *Gilia latiflora*, occurs in arid lowland plains and valleys in the same general area in southern California. It forms continuous or semi-continuous populations over large areas. Its population system consists for the most part of continuously intergrading geographical races (Grant and Grant 1956, 1960). The contrasting patterns of geographical variation between *Gilia latiflora* and *G. leptantha* are correlated with their respective population structures.

The pattern of geographical variation in the *Nigella arvensis* group (Ranunculaceae) in the Aegean area of southeastern Europe is very interesting. This is a region of complex topography with numerous islands in the Aegean Sea bounded by the mainland areas of Greece and Turkey. *Nigella degenii* in this species group is entirely insular in distribution, occurring in the Cyclades Islands. This species consists of a series of more or less discrete local races on the different islands. The local races are grouped into four distinct geographical races with the distributions shown in figure 20.5. Three

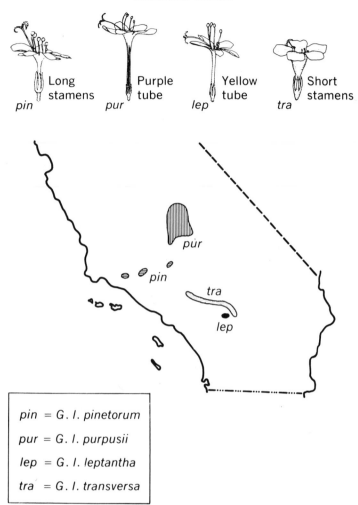

Figure 20.4. Disjunct geographical races of *Gilia leptantha* (Polemoniaceae) in separate mountain ranges of southern California. (Drawn from data of Grant and Grant 1956, 1960)

other islands on the periphery of the Cyclades archipelago are occupied by two closely related endemic species, *N. icarica* and *N. carpatha* (figure 20.5). *Nigella arvensis* proper, which is wide-ranging in Europe, shows continuous geographical variation on the mainland, but it too is represented by a distinct race on the islands of Crete and Rhodes (Strid 1970).

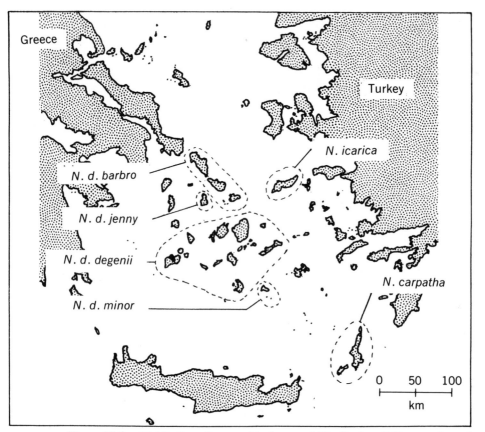

Figure 20.5. Distribution of geographical races of *Nigella degenii* (Ranunculaceae) and of two related endemic species in the Aegean Sea. (Redrawn from Strid 1970)

Ecological Races

The third main intraspecific variation pattern is that of ecological racial differentiation. Genetically and phenotypically different but interfertile races are adapted to different habitats in the same territory. These ecological races are consequently sympatric, usually neighboringly sympatric. The races interbreed and intergrade in numerous zones of contact throughout the area of the species. But they preserve their distinctive racial characters in their respective habitats.

Ecological races take a variety of forms. Among these are sun and shade races in savannah-inhabiting species, altitudinal races in montane species, host races in insects, and seasonal races in organisms with demarcated breeding systems.

Gilia achilleaefolia (Polemoniaceae) occurs in foothills of the Coast Ranges of California. It is a very variable species. The extreme forms are sun races and shade races with the following characteristics. The sun races, occurring in open grassy fields, have large blue-violet flowers clustered in small heads, and are cross-pollinated by bees. The shade races, by contrast, grow in live-oak woods, and have small, pale-colored, solitary, self-pollinating flowers. Shady woods and open fields have a mosaic pattern of distribution in the Coast Ranges, and so do the contrasting ecological races of *Gilia achilleaefolia*. The sun and shade races, which are interfertile, interbreed and inter-grade at numerous points of contact between their respective habitats (Grant 1954).

At high elevations in the Sierra Nevada of California several species of woody plants are represented by an erect arboreal race in the subalpine zone and by a low shrubby race in the alpine zone. Adjacent and interbreeding altitudinal races of this sort are found in *Pinus albicaulis*, *Pinus murrayana*, *Tsuga mertensiana* (hemlock), and two species of *Salix* (willows). Arboreal races of these species occur up to timberline, while shrubby races extend into the alpine zone above timberline (Clausen 1965).

Host races are common in the insects. In many groups of insects the species have narrow host ranges for feeding and breeding. The apple maggot (*Rhagoletis pomonella*, Tephritidae), for example, infests the fruits of two woody members of the Rosaceae in North America, the hawthorn *(Crataegus)* and cultivated apple *(Pyrus)*. The applies and hawthorns harbor different host races of the *Rhagoletis* fruit fly. Since apples and hawthorns grow close together in many places, the host races of *Rhagoletis pomonella* are neighboringly sympatric (Bush 1969a, 1969b; see also Jaenicke 1981).

The *Rhagoletis* fruit flies mate on their host plants, and this habit reinforces their fidelity to their particular type of host. Host recognition and host preference in *Rhagoletis* and the related genus *Procecidochares* are controlled by a single gene. Hence a mutation in this gene could initiate the formation of a new host race (Bush 1969a; Huettel and Bush 1972).

Biological Species

The biological species, the sum total of races, is a population system of central importance in nature and in evolutionary biology. It can be observed

on every hand—within biotic communities and within natural groups. Thus
the deer family (Cervidae) consists of the following well-marked species in
North America: whitetail deer, American elk, moose, and caribou *(Rangifer
tarandus)*; in Europe it consists of the fallow deer, roe deer, red deer, moose,
reindeer (another race of *Rangifer tarandus*), and, formerly, the Irish elk.

Biological species are kept separate by reproductive isolating mechanisms
that prevent or greatly reduce gene exchange between them. The varied forms
of reproductive isolation will be discussed later. The point here is that such
breeding barriers do operate and are responsible for the prominent discontinuities in the variation pattern that mark the boundary lines of species.

The presence of reproductive isolation between population systems is most
clearly evident where those population systems are sympatric and still remain distinct in their phenotypic characteristics. Sympatric population systems of sexually reproducing organisms are, ipso facto, separate biological
species. The maintenance of separate character combinations under conditions of sympatry is a natural test, and our best criterion, of the species status
of the population systems involved.

Because this situation of distinctness under sympatry is clear-cut, as regards the species status of the entities, it is useful to set it off terminologically
and to designate the entities as sympatric species. This case is thereby distinguished from the less clear one of allopatric species.

No one questions the fact that good biological species can live in different
territories. Yellow pines and white pines are specifically distinct whether they
are growing together in the same mixed-pine forest or in separate forests.
Distinct allopatric population systems are not necessarily good biological
species, however; they could also be disjunct geographical races that exhibit
similar features. The decision as to the rank of distinct allopatric populations
depends on the degree of distinctness, which is a relative matter, and on the
presence or absence of reproductive isolation, which is usually unknown in
this situation without further ad hoc study.

The amount of phenotypic difference between two population systems is
less significant for revealing their biological species rank than is the presence
of reproductive isolation. In fact, pairs or sets of morphologically very similar but reproductively isolated species are known in many genera of insects,
flowering plants, protozoans, and other groups. Such morphologically very
similar species are known as sibling species.

For example, the malaria mosquito of Europe, known in the older literature as *Anopheles maculipennis,* turned out on fine analysis to be a complex of six sibling species (segregated and named as A. *maculipennis sens.*

str., A. *melanoon*, A. *messeae*, and others). These sibling species, though reproductively isolated, are virtually indistinguishable morphologically. Yet they differ in egg color, chromosome morphology, certain physiological traits, behavior, and the medically important character of being or not being a vector of malaria (Mayr 1963:35–37, 41).

Semispecies

We sometimes find population systems in nature that are neither good races nor good species. There are intermediate stages of differentiation between disjunct geographical races and allopatric species, or between ecological races and sympatric species. Problematical population systems of this sort are connected by levels of gene flow that are intermediate between the levels typical of races on the one hand and species on the other. Their variation pattern is consequently intermediate between intergradation and discontinuity.

In some cases a combination of race-like and species-like features may be mingled in the same set of population systems. Thus, a pair of sympatric population systems may remain isolated in most places but hybridize in one or several localities. Or a population system may consist of integrading allopatric races throughout most of its area; but its extreme or terminal races, in one part of the total area, coexist sympatrically without interbreeding.

The traditional categories of race and species are not sufficient to cover the situation in nature. We need an intermediate category for the intermediate condition. The term semispecies is used to designate population systems that are intermediate between races and species in interbreeding, intergradation, and reproductive isolation.

A set of semispecies constitutes a populational unit of more inclusive scope than the species. A set of allopatric semispecies is known as a superspecies. A set of sympatric or at least marginally sympatric semispecies is a syngameon. A series of allopatric races that achieve sympatry in their extreme members is known as an overlapping ring of races (see chapter 23).[1]

The definitions of the biological species given earlier in this chapter are applicable primarily to sets of races, and require some qualifications when the component units are not races, but semispecies. The biological species

1. For more on semispecies see Mayr (1963) and Grant (1981: chs. 6 and 18). Concerning syngameons in plants see Grant (1981: ch. 18). Regarding overlapping rings of races in animals see Mayr (1963:507–512).

Table 20.2. Classification of population systems in sexual plants and animals

Level of Divergence	Type of Population System	Relationships Between 2 or More Population Systems of Same Type		
		Geographical relationship	Breeding relationship	Phenotypic relationships
1	Local races	Microgeographical in extent	Interbreeding; intergrading continuously or semicontinuously	Relatively slight differentiation
2	Contiguous geographical races Disjunct geographical races Ecological races	Allopatric Allopatric Neighboringly sympatric	As in Level 1	More differentiation
3	Allopatric semispecies Sympatric semispecies	Allopatric Sympatric	Intermediate between levels 1 and 4	More differentiation
4	Allopatric species Sympatric species	Allopatric Sympatric	Reproductively isolated; not intergrading	Relatively great differentiation (except in sibling species)
5	Superspecies Overlapping ring of races Syngameon	Varies Often sympatric	As in level 4	Great differentiation

is not the most inclusive breeding group in all cases. It is the most inclusive *normally* interbreeding group. A more inclusive population system in special cases is the syngameon.

The diverse array of population systems in nature requires a more sophisticated classification than the simple and traditional dichotomy of race and species. Such a classification, based on the preceding discussion in this chapter, is presented in table 20.2.

Types of Species

A distinction must be made between biological species and three other types of biological entities also commonly referred to as species. The types of species in biology are: (a) biological species; (b) successional species (paleospecies, chronospecies); (c) microspecies; and (d) taxonomic species (morphological species). Types b–d will be discussed in this and the following sections.

An additional species concept exists in traditional philosophy, where the species is regarded as a "natural kind." Natural kinds are characterized by being distinctive, immutable, and eternal (Hull 1983). Pre-evolutionary versions of the taxonomic species were influenced by this philosophical concept.

Let us compare successional species with biological species. A set of biological species is a group of sexually reproducing, non-interbreeding lineages seen at one time level. Consider one of these lineages through geological time. If it undergoes substantial phenotypic changes in time, as it well may, the paleontologist will assign different species names to the phenotypically different stages. Such species are successional species.

Successional species are nodal points in a phyletic evolutionary trend that differ enough to warrant the assignment of different species names. They are a product of phyletic evolution rather than of evolutionary divergence. Therefore they are not coordinate with biological species.

Biological species, when sympatric, interact ecologically and genetically. Such interactions do not occur between successional species which live in different time levels.

Species in Uniparental Organisms

Races, semispecies, and biological species are population systems in sexual organisms. Uniparental organisms, whether they are asexual or parasexual, fall outside this class of entities and require separate consideration. The biological species concept does not cover uniparental organisms.

However, uniparental organisms do form populations. Their colonies consist of genotypically identical or similar individuals, occupying a specific habitat and making particular ecological demands on the environment (Simpson 1961). They are populations but not breeding populations. And the uniparental populations do not link up into discrete biological species.

The variation pattern in uniparental organisms is quite different from the combination of intergradation and discontinuity that characterizes biparental organisms. What one observes is an array of uniform local (but not breeding) populations built up by the multiplication of one or a few adaptive genotypes. From locality to locality the population genotype is likely to change slightly. If the group occurs in a wide range of environments over a large area, the number of genotypically different kinds of populations may have to be reckoned in the hundreds or thousands, and the extreme forms may differ greatly. The total range of variation within the population system may be comparable to that in a species group in sexual organisms. But the variations are not grouped naturally into race and species units.

The biological species concept obviously breaks down in such cases. And so too, for that matter, does any practical taxonomic species concept, for the variations form a more or less continuous network that is not subdivided into distinct taxonomic units.

Botanists, who encounter this situation in many plant groups, have attempted to deal with it by using a special category, microspecies. A microspecies is a genotypically uniform population or population system with recognizable phenotypic traits of its own in a uniparental group. An array of related microspecies then makes up a particular uniparental complex.[2]

The taxonomic species *Crepis occidentalis* (Compositae) is a very variable group of perennial herbs in western North America. The taxonomic species includes a diploid sexual population system—a biological species—with a localized distribution in the mountains of eastern and northern California. Throughout the remainder of the area of *Crepis occidentalis sens. lat.*, from British Columbia to southern California and from the Pacific Coast to the Great Plains, the populations consist of derived agamospermous types, i.e., plants that reproduce by seeds formed by asexual means (Babcock and Stebbins 1938).

The agamospermous and hence uniparental populations are uniform. But

2. See Grant (1981: ch. 5).

morphological characters of the leaves, flowering heads, fruits, and other plant parts show recognizable differences from one agamospermous type to the next. The morphological variability in *Crepis occidentalis sens. lat.* is clustered into some 27 recognizable asexual microspecies with their respective geographical ranges (Babcock and Stebbins 1938).

The situation in autogamous plants calls for special comment. Many groups of annual herbaceous flowering plants and some perennial plant groups are autogamous, or automatically self-pollinating and self-fertilizing, and are thus sexual but uniparental. However, the autogamy and inbreeding are usually incomplete and are supplemented by small amounts of cross-pollination. The outcrossing, though constituting only a small fraction of the total reproduction of the autogamous plant, is biologically significant and is often sufficient to link the individuals together into partially breeding populations, and the populations together into species.

The Species in Taxonomy

The species is a basic unit in taxonomy as well as in population biology. The objectives of taxonomy and of population biology are similar but different. Taxonomy is concerned with the formal classification of organisms, and minor systematics, with formal classification at the race and species level. The species in taxonomy is therefore primarily a unit of classification. And the main criterion for blocking out species units in taxonomy is convenience and workability in practical classification, identification, and museum filing.

Now the discontinuities in variation patterns that delimit biological species also serve as convenient boundary markers for the recognition of taxonomic species. It is convenient taxonomically as well as significant biologically to subdivide the deer family in North America into the species: whitetail deer, mule deer, elk, moose, and caribou. In cases such as this, which are widespread and common in eukaryotic organisms, the species of taxonomy is synonymous with the biological species.

But situations often arise in which the species criterion of taxonomy does not coincide with that of population biology. In some groups it is not convenient, or feasible, or even possible, to recognize biological species in taxonomic practice. There are three fairly common situations in which the taxonomic species necessarily diverges from the biological species.

In the first place, the population systems of sexual organisms do not fall into just two mutually exclusive categories, race and species, but also include intermediate categories, or semispecies. The taxonomic system, however, contains a fixed, artificial hierarchy of categories, i.e., genus, subgenus, section, species, subspecies, and forma. There is no place in the hierarchy for intermediates between species and subspecies; the semispecies of population biology has no counterpart in the taxonomic system. In taxonomy, in order to meet the requirements of a binomial or trinomial nomenclature, a population system at the semispecies level of divergence must be treated as either a species in itself or a subspecies (race) of some other species.

A second set of discrepancies is posed by sibling species. The species units of taxonomy should be identifiable on external morphological characters as preserved in museum specimens. This criterion is not fulfilled by sibling species. Yet sibling species are real biological species, on their breeding relationships, whether they are identifiable in ordinary taxonomic practice or not.

A third area of non-correspondence between the two species concepts occurs in asexual and parasexual organisms. It is a traditional practice in taxonomy, and a requirement of the rules of nomenclature, to classify all organisms into species, whether they are sexual and biparental or not. But biological species simply do not exist in asexual groups.

Conflicts between taxonomic and population-biological systems of classification thus arise in several problem areas. In these situations each field must be true to its own objectives. The resolution is to recognize the legitimacy of both species concepts, that of taxonomy and that of population biology. Semantic confusion can be avoided by designating the two types of species as taxonomic and biological species, respectively, as has been done in the preceding discussion.[3]

3. For a more detailed recent discussion of the species problem in biology, with references, see Grant (1981: ch. 7).

CHAPTER 21
Isolating Mechanisms

Classification
Ecological and Temporal Isolation
Mechanical, Ethological, and Gametic Isolation
Incompatibility Barriers and Hybrid Inviability
Hybrid Sterility
Hybrid Breakdown
Combinations of Isolating Mechanisms

Classification

Gene exchange between different populations or population systems is reduced or blocked by barriers of various types, known collectively as isolating mechanisms. The varied array of isolating mechanisms is classified in different ways by different authors.[1] Our classification is presented in table 21.1.

A fundamental distinction is drawn between spatial and reproductive isolation in table 21.1. In this we follow Dobzhansky (1937a, 1951a) rather than Mayr (1942, 1963, 1970) who does not include geographical isolation in his concept of isolating mechanisms. Unlike other authors we separate ecological isolation from reproductive isolation and elevate it in rank to a third main category, environmental isolation. The rationale for doing so is given below. Table 21.1 subdivides reproductive isolation into premating and postmating barriers following various other authors (e.g., Mayr 1970; Dobzhansky 1970).

In spatial isolation, the gametes of different populations do not get together because the populations live in areas separated by distances that are large relative to the dispersal potential of the organisms. In short, spatial isolation is isolation by geographical distance. In reproductive isolation, on the

1. Good recent reviews of isolating mechanisms are given by Dobzhansky (1970), Mayr (1970), and Levin (1978). Levin's review of isolating mechanisms in plants is especially rich in details.

Table 21.1. Classification of isolating mechanisms

I. Spatial
    1. Geographical isolation
II. Environmental
    2. Ecological isolation
III. Reproductive
    A. Premating barriers
        3. Temporal isolation
            a. Seasonal
            b. Diurnal
        4. Ethological isolation
        5. Mechanical isolation
        6. Gametic isolation
    B. Postmating barriers
        7. Incompatibility barriers
            a. Prefertilization
            b. Postfertilization
        8. Hybrid inviability
        9. Hybrid sterility
            a. Genic
            b. Chromosomal
            c. Cytoplasmic
      10. Hybrid breakdown

other hand, the barriers to interbreeding stem from inherent characteristics of the organisms themselves. Reproductive isolation permits populations to live in the same territory with little or no gene exchange.

These two modes of isolation enter into the definitions of races and species. Spatial isolation is a characteristic of local populations, local races, and geographical races. Well-developed reproductive isolation is a distinguishing characteristic of biological species. Partial or incomplete reproductive isolation is a characteristic of semispecies.

Environmental or ecological isolation forms are a third and somewhat intermediate category. Here the populations differ genetically in their ecological requirements and preferences. The ability of the populations to live in the same territory is determined by the availability of the appropriate habitats or niches and by the force of interspecific competition. If the populations do coexist, hybridization is then controlled by the availability of habitats suitable for their hybrid progeny. The barriers to gene exchange are secular-ecological in nature.

Ecological isolation is a universal feature of species, but it is not a distinguishing feature of species. Ecological isolation is also found between ecological races and between sympatric semispecies. Some ecological differentiation is probably generally present between geographical races.

Reproductive isolation is subdivided into premating and postmating barriers as already mentioned. Premating barriers are those that prevent the gametes or, in plants, the gametophytes from coming together, whereas postmating barriers begin to operate after the gametes or gametophytes do get together.

Ecological and Temporal Isolation

Ecological isolation is a consequence of the secular ecological differentiation between sympatric species. This latter condition is general and takes numerous forms. *Drosophila* species living in the same area in California or Brazil have different nutritional preferences and feed on different kinds of yeasts. Certain oak species in Texas grow on different soil types, one species *(Quercus mohriana)* occurring on limestone, another *(Q. havardi)* on sand, and still another *(Q. grisea)* on igneous outcrops.

Ecological differentiation in the secular or non-reproductive stages of life reduces the chances of successful hybridization between sympatric species, to one degree or another, and therefore contributes to their isolation. We will discuss ecological differentiation between species more fully in the next chapter.

Most animals and plants have demarcated breeding or flowering seasons. Mating or cross-pollination takes place during a particular season of the year (e.g., summer or fall), and often at a particular time of day (sometimes by day, sometimes by night). Related species may differ as to the seasonal or diurnal timing of their sexual reproductive periods. Such interspecific differences lead to temporal isolation. Temporal isolation includes seasonal isolation and isolation due to different diurnal periodicities.

Two related species of pines, *Pinus radiata* and *P. attenuata*, come into sympatric contact on the central California coast. Here *Pinus radiata* sheds its pollen early, in February, whereas *P. attenuata* sheds its pollen about six weeks later, in April, and the two species are thus seasonally isolated (Stebbins 1950:209–210). *Drosophila pseudoobscura* and *D. persimilis*, which are sympatric over a large area in western North America, breed during the same season of the year but at different times of day, *D. pseudoobscura* being sexually active in the evening and *D. persimilis* in the morning (Dobzhansky 1951*b*).

Temporary isolation, like any other form of reproductive isolation, may be complete or incomplete. Related species of plants in the same territory often have flowering seasons with different peaks but overlapping ranges, so that the seasonal isolation is incomplete.

Mechanical, Ethological, and Gametic Isolation

Higher animals and plants generally have complex genitalia or flowers consisting of female and male organs that are coadapted structurally so as to facilitate copulation, insemination, or pollination in normal intraspecific matings. Let two such species differ in the configuration of their genitalia or flowers. Then interspecific copulation, insemination, or pollination may be impeded. Such impediments constitute mechanical isolation.

Mechanical isolation has been described in various species pairs in flowering plants with complex floral mechanisms. *Polygala vauthieri* and *P. monticola brizoides* (Polygalaceae) coexist sympatrically without hybridizing in Brazil. The flowers of both species are visited and pollinated by bees. The pollen presentation mechanism of the flowers deposits and glues the pollen to the bees' heads. Each *Polygala* species deposits the pollen on a different part of the head. The positioning of the pollen is appropriate for transfer to the stigma of the same species, but not to the other species, and thus interspecific pollination is avoided (Brantjes 1982).

In eastern North America, *Impatiens capensis* and *I. pallida* (Balsaminaceae) often grow together but do not hybridize. The floral mechanisms of the two species are adapted for different pollinating animals. *Impatiens capensis* is normally visited and pollinated by ruby-throated hummingbirds (*Archilochus colubris*), and *I. pallida* by bumblebees, and these pollinator differences account for much of the isolation between the species (Wood 1975). A parallel situation exists in *Monarda* (Labiatae) in the same area, where *M. didyma* is pollinated by hummingbirds and *M. clinopodia* by bumblebees (Whitten 1981).

In higher animals copulation is ordinarily preceded by courtship. Courtship comprises a series of stimuli and responses—dances, displays, songs, pheromones, and the like—that prepare the females and males for copulation. The courtship behavior varies from species to species in a group, and the signals are often species-specific. Thus the pheromones in Lepidoptera, which play an important role in mating, are often species-specific, and so are the calls and songs in amphibians and birds. The result is failure of attraction between females and males of different species. The behavioral inhibitions on interspecific mating constitute ethological isolation. Ethological isolation plays a key role in the prevention of interspecific hybridization in many groups of animals, both vertebrate and invertebrate.

Many aquatic organisms release their gametes into the water. External fertilization then depends on the free-living eggs and sperms coming together, and their migration and union may be guided in turn by biochemical substances. These biochemical attractants may be species-specific. Mutual attraction and fertilization then occur between eggs and sperms of the same species, but not between the eggs of one species and the sperms of another in the same water body. This failure of attraction between foreign gametes is gametic isolation.

A good example of gametic isolation is found in the sea urchin, *Strongylocentrotus*, in which external fertilization in seawater is the norm. Under controlled experimental conditions, with mixtures of gametes contained in vessels of water, intraspecific fertilizations take place freely in either *S. franciscanus* or *S. purpuratus*, but the interspecific fertilizations (*S. f.* ♀ × *S. p.* and *S. p.* ♀ × *S. f.*) are strongly inhibited (Lillie 1921; Dobzhansky 1951a).

Incompatibility Barriers and Hybrid Inviability

Let us consider the next stage of isolation in higher animals and plants with internal fertilization and gestation or seed-ripening. Assume that pre-mating barriers have failed and interspecific copulation or pollination has occurred. A new series of internal barriers to hybrid formation may now come into play in either the parental or F_1 generation.

Many developmental steps occur between pollination and seedling growth in an angiospermous plant. Pre-fertilization steps are: (a) germination of the pollen grains on the stigma, (b) growth of the pollen tube in the style, (c) growth of the pollen tube to the embryo sac in the ovule, (d) release of the sperm nuclei, and (e) their attraction to the female gametes. Fertilization is a step in itself, and in angiosperms it is a twofold process; (f) fertilization of the egg and (g) fertilization of the endosperm nucleus. The early post-fertilization steps follow: (h) first divisions of the zygote, (i) endosperm development, (j) embryo development, (k) seed formation, and (l) seed germination. And the subsequent stages of the new generation continue: (m) establishment of the young seedling, (n) seedling growth, and (o) development of the adult plant.

Blocks to gene exchange following interspecific crossing can occur at any one of these steps.

A parallel series of stages occurs in mammals: insemination, migration of spermatozoa, conception, implantation, embryo development, birth, and growth and development of the young. And here too, blocks to successful hybridization can occur at any point in the long developmental process.

The blocks are known collectively as incompatibility barriers and hybrid inviability. Where do we draw the line between these two? Let us consider two possibilities, each of which has its advantages and disadvantages. The main point to bear in mind, however we choose to draw the dividing line, is that development is a continuous process that can be blocked at any one of numerous stages.

We can arbitrarily assign the pre-fertilization blocks to incompatibility, and the post-fertilization blocks to hybrid inviability. This is embryologically sound but not always practical. A plant or animal breeder may make a particular cross but not get a viable young F_1 individual, for unknown reasons. The stage of the block could of course be determined by further study, but is somewhat irrelevant as far as the net result of the artificial (or natural) crossing is concerned. It may be more convenient, therefore, to draw the dividing line between incompatibility barriers and hybrid inviability at a later stage.

The second possibility, then, is to define incompatibility so as to include the various blocks after insemination or pollination and up to birth, egg-laying, or seed-ripening. Hybrid inviability would then denote the visible depressions in vigor and malformations in the F_1 individuals after birth, hatching, or germination. Incompatibility blocks in the early stages following insemination or pollination are analogous to gametic isolation in organisms with external fertilization.

Hybrid Sterility

Interspecific crosses in many groups of animals and plants yield F_1 hybrids that are vigorous but sterile, the mule (horse ♀ × donkey ♂) being a well-known example. The sterility phenomena, though confined by definition to the reproductive stage in the F_1's, are still quite heterogeneous. There is variation from case to case in the exact developmental stage affected and in the underlying genetic causes of the infertility.

The development of sex organs and the course of meiosis are complex processes that can easily be upset by disharmonious gene interactions in hybrids. Failure of development of the sex organs can be illustrated by some species hybrids in plants that produce flowers with abortive anthers. In species hybrids in animals the course of cell division in the germ line frequently breaks down owing to genic disturbances. Spermatogenesis may stop before it reaches meiosis, or meiosis may be aberrant, and in either case spermatozoa are not formed. Failure of spermatogenesis in the pre-meiotic stage is the main immediate cause of hybrid sterility in male mules; meiotic disturbances are the cause of sterility in male hybrids of some *Drosophila* crosses (e.g., *D. pseudoobscura* × *persimilis*).

A generalization known as Haldane's rule applies to the sex-limited distribution of hybrid sterility and inviability in dioecious animals. The F_1 hybrids of interspecific crosses in dioecious animals will consist, at least potentially, of a heterogametic sex (carrying the XY chromosome pair) and a homogametic sex (XX). Haldane's rule is that, where sex differences exist in hybrid sterility or inviability, the heterogametic sex is more likely to exhibit these conditions than is the homogametic sex. The heterogametic sex in most animals, including mammals and Diptera, is the male; we have just seen examples of sterile male hybrids in horses and *Drosophila*. There are, however, numerous exceptions to Haldane's rule.[2]

A third developmental stage in which hybrid sterility can be expressed is the gametophyte generation in plants. In flowering plants the products of meiosis lead directly to gametophytes—pollen grains and embryo sacs—which contain two to several nuclei and house the gametes. Inviability of the gametophytes is a common cause of hybrid sterility in flowering plants. Meiosis is completed but the pollen and embryo sacs fail to develop and function properly.

The causes of hybrid sterility at the genetic level of determination are threefold: genic, chromosomal, and cytoplasmic.

Genic sterility is a widespread and basic condition. Unfavorable combinations of the nuclear genes of parental types differentiated at the species level can and do lead to developmental and cytological aberrations in the F_1 hybrids that prevent the formation of gametes. Genetic analysis of genic sterility in *Drosophila* hybrids (*D. pseudoobscura* × *persimilis*, *D. melanogaster* × *simulans*, etc.) shows that sterility genes are located on all or nearly all of the chromosomes of the parental species (see Dobzhansky 1951a: ch. 8; 1970: ch. 10).

Unfavorable interactions between cytoplasmic genes and nuclear genes also

2. See White (1973:569 ff.).

lead to sterility in species hybrids in various plant and animal groups.[3] We reviewed a case of cytoplasmic hybrid sterility in the *Drosophila paulistorum* group in chapter 19.

Species of plants and animals often differ by translocations, inversions, and other rearrangements that, in the heterozygous condition, cause semi-sterility or sterility. The degree of sterility is proportional to the number of independent rearrangements; thus heterozygosity for one translocation gives 50% sterility, for two independent translocations, 75% sterility, and so on. The sterility, in plants, is of the gametophytic type. Meiosis in the chromosome-structural heterozygotes yields daughter nuclei carrying deficiencies and duplications for particular segments; these deficiency-duplication products are unable to form functioning pollen and ovules. Chromosomal sterility of this sort is very common in species hybrids in flowering plants.

The course of meiosis in a hybrid can be disturbed by either genic factors or chromosome-structural differences. Either genic or chromosomal sterility can be expressed in aberrations of meiosis. But the types of meiotic aberration are different. Genic sterility is common in animal hybrids, and chromosomal sterility in plant hybrids. Genetic analysis of some plant species hybrids indicates that chromosomal and genic sterility are often combined in the same hybrid.

Hybrid Breakdown

Let us suppose that a species hybrid is viable enough and fertile enough to reproduce. Its F_2, B_1, F_3, and other later-generation progeny will then generally contain a substantial proportion of inviable, subvital, sterile, and semisterile individuals. These types are the unfavorable recombination products of interspecific hybridization. The depression of vigor and fertility in hybrid progeny is hybrid breakdown. Hybrid breakdown is the last in the sequence of barriers to interspecific gene exchange.

Hybrid breakdown is invariably found in the progeny of species hybrids in plants, where it can be observed more readily than in most animal crosses. The F_1 hybrid of *Zauschneria cana* × *septentrionalis* (Onagraceae) is vigorous and semifertile, but most F_2 plants are dwarfish, slow-growing, rust-susceptible, or sterile. In one cross between these species, 2133 F_2's were grown, but not one of these was vigorous (Clausen, Keck, and Hiesey 1940).

3. This subject is treated by Grun (1976).

Eighty percent of the F_2 generation of *Layia gaillardioides* × *hieracioides* (Compositae) are inviable or subvital (Clausen 1951:108–111). An estimated 75% of the zygotes in the F_2 to F_6 generations of *Gilia malior* × *modocensis* (Polemoniaceae) were inviable or subvital, and a high percentage (70% or more) of the vigorous fraction were sterile or semisterile (Grant 1966a).

Combinations of Isolating Mechanisms

The isolation of species is rarely dependent on a single isolating mechanism acting alone. It is usual to find several different isolating mechanisms acting in concert.

Consider the species pair *Drosophila pseudoobscura* and *D. persimilis*. The combinations of isolating mechanisms between them include ecological isolation, temporal isolation, ethological isolation, hybrid sterility, and hybrid breakdown. No one of these mechanisms is sufficient in itself to block hybridization. Working in combination, however, they bring about a complete isolation of the two sympatric species in nature (Dobzhansky 1951b, 1955).

CHAPTER 22
Ecological Relationships

Interspecific Competition
Competitive Exclusion
Species Replacement
Species Coexistence
Selection for Ecological Differentiation
Character Displacement
Ecological Niche
The Influence of Ecological Demands

Sympatric species interact in their secular or vegetative phase as well as in the reproductive phase. The ecological interactions take different forms. The following modes of interaction are recognized (Boucher et al. 1982):

1. Predation, parasitism, grazing; benefit for species A, disadvantage for species B.
2. Commensalism; benefit for species A, neutral for species B.
3. Competition; disadvantage for both A and B.
4. Mutualism; benefit for both A and B.

These modes of interaction have different evolutionary effects.

Sympatric species that belong to different major groups, have very different ecological requirements, and perform different roles in the biotic community, do not enter into relations of competition. They may form specific mutualistic associations.[1] Or, they may function as the more or less interdependent components—interdependent in a generalized way—of a biotic community. A biotic community is an ecological unit that is self-sustaining

1. For reviews of coevolution, involving both mutualism and predation sensu lato, see Gilbert and Raven (1975) and Futuyma and Slatkin (1983).

in energetics, and hence an association of sympatric species playing complementary roles in nutrition and energetics.

Interspecific competition comes into play when the sympatric species have similar ecological requirements in general, and utilize some particular resource that is present in a limiting quantity. The species involved in the competition are often more or less closely related, but may also be distantly related but convergent.

The subject of population and community ecology is currently very active and sometimes controversial. The literature is voluminous. Although ecology overlaps broadly with evolution, it is essentially a different field, and therefore beyond the scope of this book.[2] Our objective here is the more modest one of singling out certain aspects that have an important bearing on speciation.

Interspecific Competition

Interspecific competition can be demonstrated in mixed laboratory populations or natural communities. Artificially remove one species and observe the change in abundance, if any, in another sympatric species with similar ecological requirements. If the second species shows an increase in numbers after removal of the first, it can be concluded that it was being held down previously by interspecific competition.

This result has been obtained in mixed laboratory populations of *Paramecium aurelia* and *P. caudatum* (Gause 1934), and in natural intertidal communities of barnacles *(Chthamalus* and *Balanus)* (Connell 1961), to name two classical examples. And it has been obtained in a number of more recent studies, e.g., in heteromyid rodents and plethodont salamanders (Lemon and Freeman 1983; Hairston 1983).

Interspecific competition has two broad aspects, consumptive competition and interference competition. The first is passive use by different species of the same resource.

For example, different species of shrubs in a desert community are likely to compete passively or non-aggressively for a limited amount of soil mois-

2. For reviews of modern ecology with an evolutionary slant see inter alia Hutchinson (1965, 1978), Odum (1971), Krebs (1972), Emlen (1973), Pianka (1978), Roughgarden (1979), and Tilman (1982). Recent reviews of interspecific competition are given by Schoener (1982, 1983) and Connell (1983).

ture. The species of *Geospiza* and other ground-finches on the Galapagos Islands compete for food, and this competition is an important factor determining their ecological and geographical distribution on the several islands (Lack 1947; B. R. Grant and P. R. Grant 1982).

The second aspect, often superimposed on the first, is direct inhibition of one species by a competitor species.

A number of plant species produce chemical substances in the leaves that leach out into the soil and inhibit the germination and growth of other plants in the vicinity (Muller 1966, 1970; Whittaker and Feeny 1971). In animals, inhibition may be achieved by aggressive behavior or by the exercise of dominance based on the threat of aggression. Native bighorn sheep *(Ovis canadensis)* and feral burros *(Equus asinus)* compete for water and forage in the Mojave Desert of California and Nevada. Burros are dominant over the bighorn sheep in direct confrontations; when the burros move in on water holes occupied by bighorns, the latter move out and in some cases abandon the area (Laycock 1974; see also Monson and Sumner 1980).

Consumptive competition has been given most of the attention in ecological theory, but, as Hairston (1983) points out, interference competition would be the more favorable mode for any given species. Hairston (1983) analyzed a large number of field experiments on interspecific competition and found the proportions of the two modes to be: 67 cases of interference competition, and 54 cases of consumptive competition.

The terrestrial salamander, *Plethodon jordani*, exhibits racial variation in competitive ability with a foreign species, *P. glutinosus*. These two species are sympatric and competitive in some areas in the southern Appalachian region of eastern North America; in other areas there is little competitive interaction between them. Populations of *P. jordani* from the latter areas show only slight competitive ability when brought into contact with *P. glutinosus*, whereas populations from competitive areas show strong competitive ability. Selection apparently builds up interference mechanisms in races of *P. jordani* occupying areas of strong competition with a foreign species (Hairston 1983).

Competitive Exclusion

The experiments of Gause, the observations of field naturalists (especially Grinnell and Lack), and mathematical equations (of Volterra and Lotka) have led to a generalization known as the competitive exclusion principle; and also known as Gause's law and the Lotka-Volterra principle (see Hardin 1960).

The competitive exclusion principle has served as a focal point in species ecology since the 1930s and has been extensively discussed by many authors. Different formulations of the principle are given by different authors. It would take us too far afield to trace the various formulations here; instead we will quote one passage from Lack (1947), which expresses the essential idea, and then give a general statement of the principle. Lack states (1947:62):

> My views have now completely changed, through appreciating the force of Gause's contention that two species with similar ecology cannot live in the same region (Gause 1934). This is a simple consequence of natural selection. If two species of birds occur together in the same habitat in the same region, eat the same types of food and have the same other ecological requirements, then they should compete with each other, and since the chance of their being equally well adapted is negligible, one of them should eliminate the other completely. Nevertheless, three species of ground-finch live together in the same habitat on the same Galápagos islands, and this also applies to two species of insectivorous tree-finch. There must be some factor which prevents these species from effectively competing.
>
> The above considerations led me to make a general survey of the ecology of passerine birds (Lack 1944a). This has shown that, while most closely related species occupy separate habitats or regions, those that occur together in the same habitat tend to differ from each other in feeding habits and frequently also in size, including size of beak. In a number of the latter cases it is known that the beak difference is associated with a difference in diet, and this correlation seems likely to be general, since it is difficult to see how otherwise such species could avoid competing.

The competitive exclusion principle makes two general statements about sympatric species. If two species occupy the same ecological niche, one species is practically certain to be superior to the other in this niche, and will eventually replace the inferior species. Or, more briefly, "complete competitors cannot coexist" (Hardin 1960). The second statement is a corollary of the first: If two species coexist in a stable equilibrium, they must be ecologically differentiated so that they can occupy different niches.

There are two ways of looking at the competitive exclusion principle: as an axiom and as an empirical generalization. Considered as an axiom, it is logical, coherent, and has been very heuristic. Considered as an empirical generalization, it is widely but not universally valid, as will be shown later.

Species Replacement

Interspecific competition can have a variety of end results. One of these is species replacement.

Interspecific selection (or species selection) is the increase in numbers and ecological dominance of one species relative to another ecologically similar species. Interspecific competition leads to interspecific selection where one species has some inherent competitive advantage over another sympatric species. Thus a great increase in numbers of feral burros has been observed in the Death Valley area of California since 1930, and along with this there has been a marked decline in numbers of the ecologically similar bighorn sheep (Laycock 1974).

We may note in passing that sympatry and direct competition are not necessary conditions for interspecific selection in general. Climatic changes could affect two allopatric species in such a way as to favor the spread of one and reduce the other. This would be a case of interspecific selection in an allopatric field, comparable to interdeme selection as discussed in chapter 15.

The interspecific selection process can go all the way to species replacement. Species A is likely to replace species B completely in a territory if the environmental conditions in which A has the advantage remain constant. The dingo *(Canis familiaris dingo)* thus replaced the Tasmanian wolf *(Thylacinus)* in most or all of Australia during historical times.

Species replacement has been studied in laboratory experiments with *Paramecium, Tribolium,* and other organisms. In his classical experiments with *Paramecium,* Gause (1934) put *P. aurelia* and *P. caudatum* together in a medium of water, salts, and food bacteria *(Bacillus pyocyaneus)* in glass containers, and kept the temperature and composition of the medium constant. *Paramecium aurelia* replaced *P. caudatum* completely in a few weeks. Different replications of this experiment carried out under identical environmental conditions always gave the same result. When the conditions were changed, by using a different strain of *Bacillus* as a food organism, *P. caudatum* replaced *P. aurelia.* In general, whenever two species of *Paramecium* were forced to compete in a homogeneous culture medium, one species always replaced the other eventually.

Species replacement undoubtedly occurs commonly in nature. The process is usually difficult to observe and interpret correctly in the modern world of nature, apart from a few clear-cut cases like that of the Tasmanian wolf and dingo in Australia. But the indirect evidence for frequent species replacement is inescapable when we consider evolutionary changes on a geological time scale. An estimated 98% of the living families of vertebrate animals are descended from about eight species in the early Mesozoic (Wright 1956). Those eight species obviously comprise only a very small fraction of the thousands of vertebrate species living at that time.

Species Coexistence

Species replacement is not the only outcome of interspecific competition, as is evident from the fact that related species with similar ecological requirements frequently coexist in nature. Complete species replacement can be circumvented in any one of several ways. Four such ways are listed below.

1. Species replacement is a time-consuming process. It is to be expected that observations made at any given moment of time will find some pairs of competing species in an uncompleted stage of replacement.

The number of species of phytoplankton in some lakes is greater than the estimated number of niches for such plankton. This greater than expected species diversity could be due to interspecific competition which has not yet run its course (Richerson et al. 1970).

2. Ecologically similar species may coexist without ever reaching a stage of direct interspecific competition. This would be the case if their numbers were kept in check by some factor other than direct competition.

Populations of herbivores in natural terrestrial communities are often held down by predators. Predation then prevents interspecific competition between herbivores for food from becoming an important factor affecting herbivore coexistence (Hairston et al. 1960).

3. The environmental conditions may change reversibly during the course of interspecific selection so as to give a selective advantage to species A in one phase and to species B in another phase. The two species would then coexist in a cyclical equilibrium.

Some ephemeral ponds harbor mixtures of green algae in the genera *Haematococcus*, *Chlamydomonas*, *Scenedesmus*, and *Chlorella*. *Haematococcus* is favored when the pond dries up, the other algae when there is standing water. Where the pond alternately dries up and refills on a frequent cycle, the opposing types of green algae can coexist indefinitely (Hutchinson 1957).

4. A situation which is very important in nature is that where the environment is heterogeneous and contains a range of conditions for some critical factor. Species A may be superior to species B in one facet of the envi-

ronment while species B is superior in another facet. Species A and B can coexist in this situation by living partly or mainly in their preferred respective facets of the environment.

Laboratory experiments with *Paramecium* are relevant in this connection. It will be recalled that *P. aurelia* replaces *P. caudatum*, or vice versa, in a homogeneous culture medium (Gause 1934). If, however, the laboratory environment is heterogeneous, species replacement may not occur. Combinations of *P. aurelia* and *P. bursaria* were placed in tubes containing vertically stratified suspensions of food yeasts. Here *P. aurelia* feeds mainly in the upper layers and *P. bursaria* mainly on the bottom. Given different food niches of this sort, and different feeding preferences of the two species, the two species can continue to exist in equilibrium indefinitely (Gause 1935).

Natural habitats, are, of course, always heterogeneous, and are often heterogeneous in ways that permit some ecological segregation of sympatric species, and hence a continuous coexistence.

Selection for Ecological Differentiation

We find that interspecific selection can lead to complete species replacement in some situations and to ecological divergence and coexistence in others. A model of the latter process, where ecological differentiation is the outcome of interspecific selection, is presented here.

The model contains the following assumptions: There are two ecologically similar and competitive species, A and B. Their environment is heterogeneous, with facies E_1, E_2, and E_3. Each competing species contains polymorphic types (A_1, A_2, A_3; B_1, B_2, B_3) fitted to the three environmental facies (to E_1, E_2, and E_3, respectively). Polymorphic type A_1 is superior in E_1, while B_3 is superior in E_3.

The setup at the beginning of the interspecific selection process is as follows:

Environment	E_1	E_2	E_3
Species A	A_1 (superior)	A_2	A_3
SpeciesB	B_1	B_2	B_3 (superior)

Letting interspecific selection run its course, and ignoring various complicating factors such as interbreeding between polymorphic types in each species, the final result will be:

Environment	E_1	E_2	E_3
Species A	A_1	A_2	
Species B		B_2	B_3

We see that ecological divergence develops between competing species A and B. Each species becomes more narrowly specialized for the aspects of the common environment in which it has the selective advantage. Furthermore, the process of specialization is accompanied by a loss of genetic variability in each species.

Considerable evidence from natural populations is in agreement with this model.

Lack (1947) points out that the white-eye bird (*Zosterops palpebrosa*) in southeast Asia has a different altitudinal range in areas where it occurs alone from that in areas where it is sympatric with other species of *Zosterops*. In Burma, where *Z. palpebrosa* is the sole member of its species group, it breeds over a wide altitudinal range from sea level to the high mountains. But in Malaya and Borneo, where a related species of *Zosterops* occupies the middle and higher zones, *Z. palpebrosa* is restricted to the lowlands. And in Java, Bali, and Flores, where two other related species occur in the higher and coastal zones, respectively, *Z. palpebrosa* is restricted to the middle altitudinal zone.

A second line of evidence comes from the phenomenon of character displacement, to be described next.

Character Displacement

Character displacement can be observed in some pairs of species that have overlapping areas, being sympatric in one part of their range and allopatric elsewhere. Let allopatric and sympatric races of the two species be compared with respect to their morphological, ecological, or behavioral characteristics. It frequently turns out that the sympatric races of two such species are more distinct than the allopatric races. The enhanced degree of differentiation in the sympatric races of overlapping sympatric species is known as character displacement (Brown and Wilson 1956).

Character displacement is observed in a number of animal groups and in a few plant groups. Examples are found in nuthatches (*Sitta*) in Eurasia for bill size and body size, in ground-finches (*Geospiza*) in the Galapagos Islands for bill size and body size, in *Lasius* ants in North America for various characters, and in other groups of vertebrates and insects (Brown and Wilson 1956).

Characters exhibiting character displacement may owe their differentiation to different modes of selection. Sympatric divergence of characters affecting success in the secular-ecological phase of life, such as bill size and body size in birds, is likely to be a result of interspecific selection for ecological differentiation, as suggested by Brown and Wilson (1956). Some cases of character displacement in reproductive characteristics may also be products of selection for ecological differentiation (Whalen 1978). Other cases of reproductive character divergence, however, are due to selection for reproductive isolation (see chapter 26).

Ecological Niche

"Ecological niche" is a key term in the competitive exclusion principle. How do we define this term? And can we define it in a way that is independent of the competitive exclusion principle?

Much effort has been devoted to the search for a satisfactory formal definition of niche in either verbal or mathematical terms.[3] The subject is an elusive one, and as far as I can see, the search has not been entirely successful.

Hutchinson (1965:32 ff.) distinguished between the so-called fundamental niche and the realized niche. The fundamental niche, according to Hutchinson, is the volume in multiple dimensions occupied by a species, where each dimension corresponds to one variable factor necessary for the life of the species. The multidimensional volume of the fundamental niche is considered to be undistorted by the presence of competitor species. The realized niche is then the fundamental niche as restricted by the presence of competitor species. However, Hutchinson's definition of the fundamental niche seems to be a definition of what plant ecologists and plant geographers have long called the "tolerance range" of a species. It specifies the potential species area, not the niche. Hutchinson's "realized niche" *is* the ecological niche. The inclusion of the effects of interspecific competition in the definition of the realized niche is important.

The field naturalist knows what an ecological niche is in practice, whether

3. See Connell and Orias (1964), Hutchinson (1965, 1978), Levins (1968), Krebs (1973), Whittaker et al. (1973), Emlen (1973), Rejmanek and Jenik (1975), and Pianka (1978).

he can formally define it or not. Deep groundwater and topsoil water represent different ecological niches for the root systems of plants. Large seeds and small seeds are different food niches for seed-eating birds.

A good example of niche diversity is provided by the feeding zones of warblers in spruce forests in the northeastern United States (MacArthur 1958). Five sympatric species of *Dendroica* feed on insects in three different parts of the crowns. Their characteristic feeding zones are:

Upper crown: Cape May warbler *(Dendroica tigrina)* and blackburnian warbler *(D. fusca)*

Mid-part of crown: Black-throated green warbler *(D. virens)* and bay-breasted warbler *(D. castanea)*.

Lower part of tree: Myrtle warbler *(D. coronata)*.

There are finer degrees of ecological differentiation between warbler species that feed in the same zones. Thus in the middle zone *D. castanea* gets much of its insect food by hawking, whereas *D. virens* engages only rarely in hawking (MacArthur 1958).

Examples such as the foregoing, which could be multiplied endlessly, suggest that the niche can be characterized in terms of two sets of factors: (a) the habitat or broad environmental field for which a species is adapted; and (b) the restrictions on exploitation of the habitat imposed by interspecific competition and interspecific selection. Hence the niche can be regarded as a facet of the habitat for which a species is particularly specialized.

Following logically from this is a recognition of differences in niche breadth. Ecological niches may be relatively narrow or relatively broad. And the niche breadth is expected to correlate with the degree of specialization of the species occupying the niche in question. Various workers have pointed out that niche breadth diminishes with increase in interspecific competition. The presence of a number of ecologically similar species in a community tends to promote narrow specialization and reduced niche breadth (Lack 1945; Dobzhansky 1950b; Connell and Orias 1964; Grant and Grant 1965: 164–165; Levins 1968).

The Influence of Ecological Demands

Interspecific competition for essential environmental resources sets in motion the process of interspecific selection, which can result eventually in either ecological divergence or complete replacement. It is logical to assume that interspecific selection will run its course more rapidly where the interspecific competition is intense. A general rule, subject to some exceptions, is that large-bodied animals and plants make heavier demands on the environment, and more quickly exhaust the resources present in limiting amounts, than do small-bodied organisms. These differences between large and small organisms can be expected to lead to correlated differences in the intensity of interspecific competition and the rate of interspecific selection.

Ross studied the geographical and ecological distribution of six related species of leafhoppers in Illinois (*Erythroneura lawsoni* group, Jassidae, Hemiptera). These small insects feed and breed on sycamore leaves (*Platanus occidentalis*) and thus have a narrow ecological niche. Nevertheless, biotic sympatry is common in the *E. lawsoni* group, and several species are frequently found coexisting in the same niche. Signs of interspecific competition were looked for in such sympatric associations but were not found (Ross 1957; see also McClure and Price 1975).

Ross (1957) attributes the apparent absence or weakness of interspecific competition to the small size of the leafhoppers, which makes it possible for them to maintain their populations in a restricted niche without depleting the food supply. He remarks that "It is difficult to visualize half a dozen species of elephants all maintaining reproductive units if restricted to sycamores for food."

He goes on to state the general conclusion that "The number of species which [can] occupy the same niche [is] inversely proportional to the food requirements, hence usually to the absolute size, of the organism."

Distribution patterns in a number of animal groups are in agreement with Ross's generalization. It is significant that related species of large mammals are nearly always allopatric, e.g., the African and Asiatic elephants, the species of big cats, and bears in most species combinations. The Black rhinoceros and White rhinoceros occur together in the same general area in South Africa, but they are neighboringly sympatric there and do not occupy the same niche; furthermore, one species is rare.

The opposite extreme among higher animals is found in small insects such as *Erythroneura*, *Drosophila*, and *Anopheles*, in which biotically sympatric assemblages composed of several or many species are normal.

Parallel trends are seen in higher plants in a temperate-zone flora. Related species of dominant trees in genera such as *Quercus* and *Pinus*, and of perennial herbs in genera like *Iris* and *Polemonium*, mostly have allopatric or neighboringly sympatric distributions. But small annual herbs commonly form sympatric assemblages of several species.

The larvae of butterflies are heavy feeders on plant herbage, whereas the adult butterflies are light-to-moderate flower feeders. It is significant that a given species of butterfly often has a broad food niche in the adult stage but a narrow and specialized food niche in the larval stage. The adult butterflies feed on a fairly wide range of flower species, but lay their eggs on specific host plants, where the larvae later complete their development.

Now, different groups of butterflies have different larval host plants. Thus for the pipevine swallowtail *(Battus philenor)*, the larval food plant is *Aristolochia*; for the monarch butterfly *(Danaus plexippus)*, milkweed *(Asclepias)*; and for the alfalfa butterfly *(Colias eurytheme)*, alfalfa, clover, and vetch (Fabaceae). But the adults of these same species of butterflies feed on broadly overlapping arrays of flower species. Parallel differences between the food-niche breadths of larval and adult stages occur in various groups of moths. Ecological differentiation has thus progressed further in the heavy-feeding larvae of many butterflies and moths than in the moderate- or light-feeding adults of the same species.

CHAPTER 23

Race Formation and Gradual Speciation

Evolutionary Divergence

Divergence between related evolutionary lines, if gradual and long-continued, goes through a series of stages with time. A common ancestral population gives rise, in succession, to two or more local races, geographical races, semispecies, biological species, and species groups (figure 23.1). The divergence can continue to higher levels from genus to class and phylum.

The populational units and taxonomic categories are stages in evolutionary divergence. The higher taxonomic categories are products of macroevolution. Our concern here is with the process of divergence in the lower middle range involving races and species. Divergence at these levels constitutes problems of speciation.

Divergence at the race and species levels is manifested in three main sets of characteristics. Progression from lower to higher levels on the branching phylogeny is accompanied by (a) increasing differentiation in genotype; (b) increasing differentiation in morphology, physiology, ecology, and behavior; and (c) stronger isolation (figure 23.1).

The increase in genetic differentiation with increasing level of divergence is illustrated by the *Drosophila willistoni* group (Ayala et al. 1975). Numer-

DEGREE OF ISOLATION

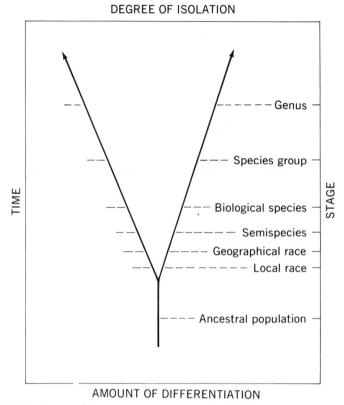

AMOUNT OF DIFFERENTIATION

Figure 23.1. Stages of Divergence.

ous populations belonging to 14 taxa were assayed for 36 enzyme loci. The similarities or dissimilarities between populations were expressed in terms of the genetic similarity index (I). The data were next grouped by level of divergence, and the average similarity index (Ī) determined for each level, with the results shown in table 23.1. Genetic similarity decreases at three of the four stepwise advances in level of divergence.

It should be recognized that the three sets of characteristics listed above may evolve at different rates in different groups of organisms. In some groups a relatively small amount of genotypic differentiation is amplified into striking phenotypic differentiation, while other groups have extensive genotypic but small phenotyic differentiation. Reproductive isolation is only loosely correlated with degree of phenotyic differentiation when a broad sample of plant and animal groups is considered.

Table 23.1. Average genetic
similarity (Ī) between populations,
races, and species of the
Drosophila willistoni group
(Ayala et al. 1975)

Level of Divergence	Ī
Local populations	0.970
Subspecies	0.795
Semispecies	0.798
Sibling species	0.563
Non-sibling species	0.352

Variation also exists in the sequence of stages. Some of the middle stages of speciation may be bypassed in some cases. Reversals in the process of divergence are also possible. Such variant modes can be considered later.

Factors in Race Formation

The formation of races—be they local or geographical, contiguous or disjunct—is brought about by the same evolutionary forces as those that control microevolutionary changes within populations. It will be recalled that the basic variation-producing processes are mutation, gene flow, and recombination, and that the variation-sorting processes are selection and drift. Race formation takes place when these processes follow different courses, and consequently produce, different gene pools in different geographical subdivisions of a species area.

The Role of Selection in Race Formation

The chief effective force involved in race formation is undoubtedly natural selection. Two groups of phenomena furnish evidence for the role of selection in race formation. First, racial traits are commonly seen to be adaptive for the environment inhabited by the race in question. Second, parallel racial variation is often found in different species that occur throughout a similar range of environments.

The generalization in animal systematics known as Bergmann's rule applies to species of warm-blooded vertebrates that occur in both warm and cold areas. Bergmann's rule states that such species tend to be represented by large-bodied races in areas with cold climates and by small-bodied races in warm areas. This rule holds true for about 75–90% of the bird species and 60–80% of the mammal species in various faunas. It is thus a valid generalization, subject to some exceptions (see Mayr 1963:319–321).

The adaptive value of the body-size differences described by Bergmann's rule is a function of the thermodynamic properties of warm bodies. In a cold climate a large warm body has an advantage in respect to heat conservation because of the low ratio of surface to volume. This advantage tends to disappear in a moderately warm climate. And in a hot climate a small body may have a positive advantage for temperature regulation because of a high surface-to-volume ratio, which promotes heat dissipation.

If racial variation in body size along the above lines were known in only a few isolated cases, the interpretation that such variation is adaptive, and hence controlled by selection, would rest on relatively weak foundations. But the fact that the same body-size trend occurs in many distinct species in different faunas, so as to warrant recognition as Bergmann's rule, makes the case for a controlling role of natural selection compelling.

The related ecogeographic rule, Allen's rule, pertains to the length of protruding body parts—ears, nose, bill, tail—in species of warm-blooded animals ranging through different climatic zones. According to Allen's rule, the races of a species of bird or mammal inhabiting a cold area generally have short protruding body parts, while the races in warm areas have long extremities.

This general trend in racial variation, like that in body size, is related to the problems of temperature regulation in warm-blooded vertebrates. Long protruding parts give off body heat. Therefore, for a warm-blooded animal, long extremities are usually selectively disadvantageous in a cold climate but advantageous in a warm climate.

Many other regular patterns of racial variation which have been recognized in various groups of animals are probably also adaptive.[1]

Parallel racial variation correlated with environmental features is also found in many wide-ranging plant species. Here the subject is subsumed under the heading of ecotypes rather than codified as ecogeographic rules. Plant species ranging along a coastline may be represented by prostrate ecotypes on sand dunes and by bushy ecotypes on cliffs. Species of woody plants in high mountains may be represented by arboreal ecotypes in the subalpine zone and by shrubby elfinwood ecotypes in the alpine zone (see discussion of al-

1. For reviews see Rensch (1960a, 1960b:43–46), and Mayr (1963:318 ff.).

titudinal races in chapter 20). Such ecotypes are undoubtedly produced by environmental selection. Good experimental studies of ecotypes have been carried out in numerous plant species.[2]

The Role of Drift

Haphazard racial variation is found in some species that have colonial population systems. By "haphazard" we mean that the racial differences are not obviously correlated with environmental differences in transects through the species area. An illustration is found in the New Guinea kingfisher (*Tanysiptera galatea*) in the New Guinea territory (Mayr 1942:152–153).

New Guinea proper is a very large island, 1,500 miles long and 100 to 500 miles wide. It spans a wide range of climates and vegetation types, from evergreen rainforest at one end to seasonally dry monsoon forest at the other. New Guinea is surrounded by numerous medium-sized and small islands with climates and vegetation similar to those of the neighboring parts of the main island.

The mainland of New Guinea proper is occupied by three races of *Tanysiptera galatea*, namely, *T. g. galatea*, *T. g. meyeri*, and *T. g. minor*. These races extend, with some gaps, from one end of the large main island to the other. They are very similar. Five additional races occur on some of the small neighboring islands. These insular races are quite distinct from one another and from the adjacent mainland races from which they probably descended. A closely related species, *T. hydrocharis*, occurs on two islands as well as on part of the mainland (figure 23.2).

Mayr's (1942, 1954) interpretation of this pattern of racial variation is as follows. The orthodox racial variation on the New Guinea mainland is a product of selection in a large and widely interbreeding population system. The island populations were founded by small numbers of colonizing individuals migrating from the mainland. Drift (or founder effect in Mayr's terminology) occurred during the early generations of colonization, and its effects were perpetuated by spatial isolation from the large mainland population. These factors enabled the small-island races to diverge markedly from the neighboring and probably ancestral mainland population. One insular population system, *T. hydrocharis*, has diverged to the species level.

It is not necessary to assume that selection is not involved in race forma-

2. See Turesson (1922, 1925), and Clausen et al. (1940, 1948).

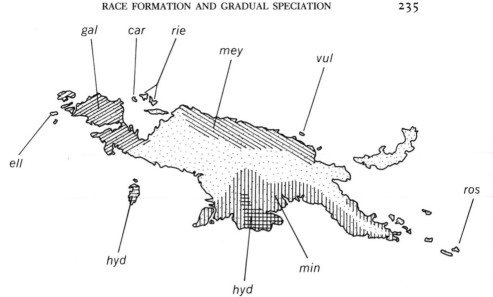

Mainland
gal	T. g. galatea
mey	T. g. meyeri
min	T. g. minor

Islands
ell	T. g. ellioti
car	T. g. carolinae
rie	T. g. riedelii
vul	T. g. vulcani
ros	T. g. rosseliana

Both
| hyd | T. hydrocharis |

Figure 23.2. Distribution of races of the kingfisher *(Tanysiptera galatea* group) in New Guinea. (Redrawn from Mayr 1942)

tion in the *Tanysiptera galatea* group. Drift and selection usually operate hand in hand in natural populations, as pointed out in chapter 16. The biologically realistic distinction regarding factors responsible for race formation is that between selection in large populations, on the one hand, and the selection-drift interaction, on the other.

Many cases of racial variation are known that conform to the *Tanysiptera* pattern. An example in the *Nigella arvensis* group (Ranunculaceae) in the Aegean Islands and surrounding mainland was described earlier (figure 20.5).

The Role of Introgression

Introgression is a special mode of natural hybridization known particularly in plants. Introgression consists of the formation of natural F_1 hybrids and their backcrossing to one or both parental species or semispecies. The backcrossing often occurs repeatedly in successive generations. The result is a flow of certain genes from the donor species or semispecies into the recipient population system. Introgression is an extension of gene flow and recombination to the species level.

Where two species or semispecies, A and B, are marginally sympatric and engage in introgression, races develop in the zone of overlap, and such races are convergent in various morphological characters and ecological preferences. Species A gives rise, by introgression, to a race in the area of species B that approaches B in various characteristics, and vice vesa. Examples have been studied in *Iris, Tradescantia, Helianthus, Gilia*, and many other plant groups.[3]

Here again, as in the case of drift, we should not assume that selection is not involved. Introgression generates the variations, but natural selection sorts them out. Introgressive races are products of both introgression and selection.

The Role of Gene Flow

Extensive dispersal and gene flow would be expected to retard race formation. A limited body of evidence is in agreement with this expectation.

Some highly migratory butterflies and moths have wide ranges with only a slight development of geographical races. The painted-lady butterfly (*Vanessa cardui*), which is highly migratory and nearly worldwide in distribution, has no geographical race formation throughout most of its range. There is one geographical race in Australia and New Zealand (Williams 1958). The white-lined sphinx moth (*Hyles lineata*), which is also highly migratory and widespread, occurring throughout North and South America, except in the far north, is likewise represented by only one subspecies H. l. lineata)

3. For reviews see Heiser (1973b), and Grant (1981: ch. 17).

Table 23.2. Correlation between migratory habits and geographical race formation in North American species of hummingbirds (Based on data of Johnsgard 1983)

Geographical Variation	Migratory Habits		
	Strongly migratory	Migratory but with residential races	Residential or mainly so
Monotypic	7 species	—	5 species
2–4 subspecies	—	2 species	9 species

throughout most of this area. Two additional subspecies occur in Uruguay and the Galapagos Islands respectively (Schreiber 1978). These are by no means isolated examples in the Lepidoptera.

A contrasting pattern is found in non-migratory Lepidoptera, which often have a more "typical" amount of geographical race formation. A series of 9 species of non-migratory Lepidoptera in Scandinavia all show definite geographical variation, in contrast to three migratory species in the same area with little or no geographical variation (Williams 1958).

The Hawaiian honeycreepers (Drepanididae) present a parallel case. The strong fliers in this group, *Himatione sanguinea*, *Vestiaria coccinea*, and *Psittirostra psittacea*, show no differentiation into subspecies on the different islands, with the exception of one subspecies of *H. sanguinea* on Laysan Island. On the other hand, weak fliers such as *Loxops virens* and *L. maculata* often have a different subspecific form on each island or small island group (Baldwin 1953).

The 23 North American species of hummingbirds (Trochilidae) include both migratory and residential forms. A survey of the distributional and taxonomic data given by Johnsgard (1983) in his treatment of this group reveals the pattern shown in table 23.2. Migratory habits tend to be correlated with monotypy, and residential habits with a differentiation into geographical subspecies. Two predominantly migratory species *(Selasphorus sasin* and *S. platycercus)* have residential races which are distinguished as subspecies.

Incipient Reproductive Isolation

Reproductive and environmental isolation can and does develop as a by-product of evolutionary divergence, as illustrated by the following experiment.

A long-term experiment with *Drosophila melanogaster* was started with a single population. Two replicate populations derived from this foundation population were maintained in separate population cages and exposed to different temperatures and humidities for six years. Strong reproductive isolation, involving ethological discrimination and hybrid sterility, developed between the two populations. The ethological isolation was determined by genes distributed over all chromosomes, and the hybrid sterility had a cytoplasmic basis (Kilias and Alahiotis 1982).

Different geographical races of the same species are often partially isolated reproductively. Races living in different geographical areas are likely to have different blooming seasons or breeding seasons. They are likewise apt to have different ecological requirements. Partial internal barriers may also exist between geographical races. Artificial interracial F_1 hybrids in herbaceous flowering plants are usually semisterile, and their F_2 progeny usually show some hybrid breakdown. All of the known reproductive isolating mechanisms surveyed in chapter 21 can be found, in partial or incipient form at least, between certain pairs of geographical races.

Consider the case of two geographical races of a plant species living in different areas, one area being cool and moist and the other warm and dry. The two races will be exposed to selection for different climatic and perhaps different edaphic conditions; this is the first stage of ecological isolation. They are likely to come into flower at different season; partial seasonal isolation develops. Different groups of bees are apt to be on the wing in the cool moist and warm dry areas, and as the indigenous plant races become adapted to their normal bee visitors and pollinators, their floral characteristics undergo changes. We now have the beginnings of mechanical or ethological isolation or both. Some changes in flower structure, such as style length, affect the ease of crossing, and can initiate an incompatibility barrier. Finally, the large number of gene differences accumulated by the two races in the course of their differentiation may well result in the formation of some disharmonious recombination products in their hybrids. Partial hybrid inviability, semisterility, and partial hybrid breakdown come into the picture.

Divergence at the racial level thus sets the stage for possible future speciation.

Geographical Speciation

The geographical theory of speciation developed by K. Jordan, D. S. Jordan, Rensch, Mayr, and others in the first half of this century, holds that biological species are normally derived from geographical races (see Mayr 1942). Spatial isolation is a normal prelude to the development of reproductive isolation.

The sequence of stages in species formation as envisioned in this theory is shown in figure 23.3. A wide-ranging uniform species (23.3 A) becomes differentiated into contiguous geographical races (B) which later break up into disjunct races (C). Reproductive isolation develops as a byproduct of divergence in the disjunct races. Migration then reestablishes contact between the races. If this contact comes too early the result is interbreeding and secondary intergradation (D). But if the contact comes later, after reproductive isolation has reached a certain critical threshold, the divergent populations can live in the same territory without interbreeding, and have thus attained the stage of young biological species (E).

One common mode of species formation, known as geographical speciation, does follow this sequence from ancestral population through geographical races and allopatric semispecies to biological species. All of the postulated stages are actually found in various species groups in both animals and plants.

A set of examples from a single natural group, the cowbwebby gilias (Polemoniaceae) is shown in figure 23.4. These are diploid, annual, predomi-

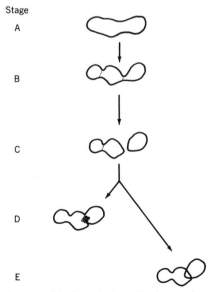

Figure 23.3. Stages of geographical speciation. Further explanation in text. (Rearranged from Mayr 1942)

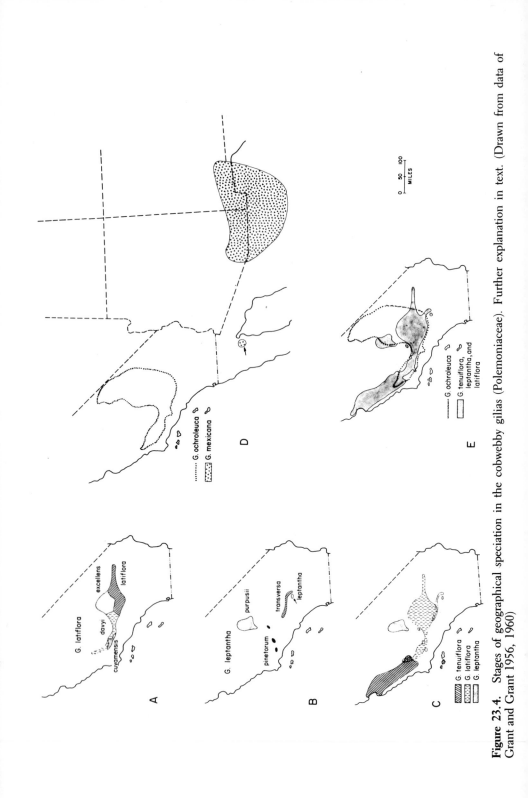

Figure 23.4. Stages of geographical speciation in the cobwebby gilias (Polemoniaceae). Further explanation in text. (Drawn from data of Grant and Grant 1956, 1960)

nantly outcrossing plants of desert and mountain regions in the American Southwest. In this group we find the following series of stages (Grant and Grant 1956, 1960). A species composed of contiguous geographical races (figure 23.4A). Disjunct and distinct geographical races (B). A syngameon composed of three marginally sympatric and sporadically hybridizing semispecies (C). A pair of intersterile allopatric species (D). A pair of intersterile sympatric species (E).

Borderline cases between geographical races and full-fledged species are particularly significant in this connection. Such borderline cases, or semispecies, are not uncommon as every animal and plant systematist knows. Mayr (1942:165) estimates that 12.5% of the taxonomic species of birds of North America are in a borderline condition between species and subspecies, so that ornithologists sometimes treat them in one taxonomic rank and sometimes in the other. Examples are the red-shafted flicker and yellow-shafted flicker *(Colaptes)* and the Audubon warbler and myrtle warbler *(Dendroica)*.

Comparable examples are found in many other animal and plant groups. Take the red deer *(Cervus elaphus)* of Eurasia and the wapiti *(C. canadensis)* of North America and eastern Asia; are they geographical races of one species or separate allopatric species? The same question can be raised with regard to the eastern North American white pine *(Pinus strobus)* and the western white pine *(P. monticola)*.

Overlapping Rings of Races

An especially interesting type of borderline case is the overlapping ring of races. Here we have a chain of intergrading races—A, B, C, D, E—stretched out in a great circle with overlap between A and E. Races A and E represent the morphological and ecological extremes in the population system. They intergrade in the series A–B–C–D–E. Considering this pattern alone, the population system is a normal species composed of geographical races. But the terminal and extreme members of the series, A and E, coexist sympatrically, and the relationship between A and E, therefore, is that of sympatric species.

An example is furnished by the salamander *Ensatina eschscholtzii* of the Pacific slope of North America (Stebbins 1949, 1957). Geographical variation for color pattern exists in this species. One series of races is reddish, another black and yellow, and another black and white; and the yellow or

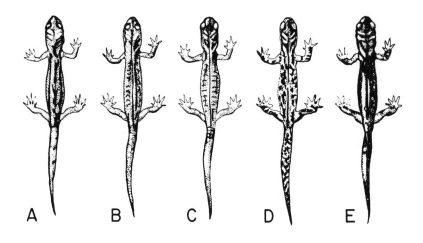

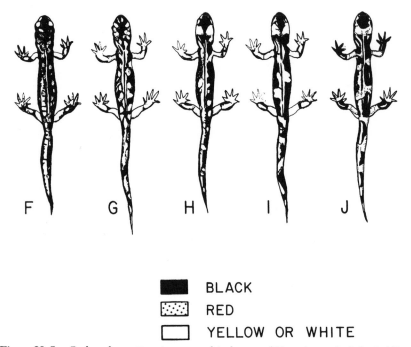

BLACK

RED

YELLOW OR WHITE

Figure 23.5. Body color pattern in geographical races of *Ensatina eschscholtzii*. (A) *E.e. eschscholtzii*. (B) *E.e. xanthoptica*. (C) *E.e. oregonensis*. (D) *E.e. picta*. (E) Intergrade *picta-platensis*. (F) *E.e. platensis*. (G) Intergrade *platensis-croceator*. (H) *E.e. croceator*. (I) Intergrade *croceator-klauberi*. (J) *E.e. klauberi*. (Rearranged and redrawn from Stebbins 1949)

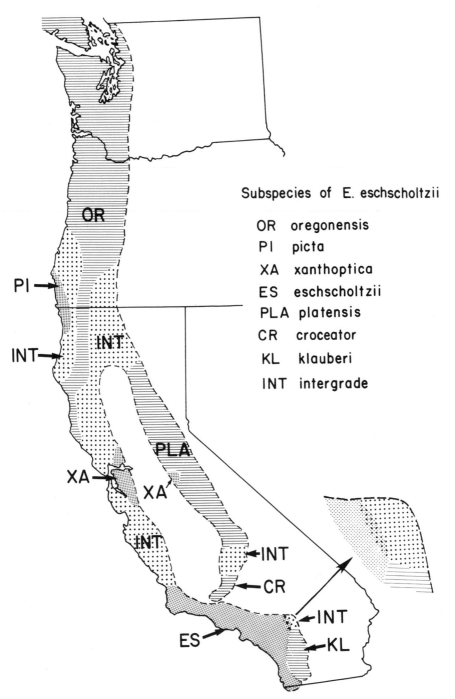

Figure 23.6. Distribution of geographical races of *Ensatina eschscholtzii* in California, Oregon, and Washington. Further explanation in text. (Redrawn from Stebbins 1949)

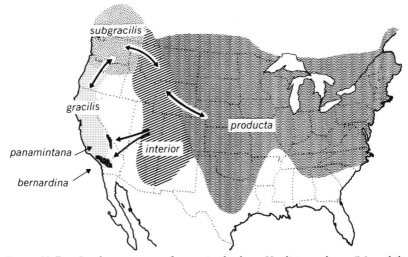

Figure 23.7. Overlapping ring of races in the bee, *Hoplitis producta* (Megachilidae). Intergradation between geographical races indicated by double-pointed arrows. (Modified from Michener 1947)

white spots may be small or large (figure 23.5). Complete intergradation occurs between the various racial color patterns (figures 23.5 and 23.6). In the San Bernardino Mountains of southern California, however, two terminal races coexist sympatrically (figure 23.6). One of these is the red-bodied *E. e. eschscholtzii* and the other the black-and-white *E. e. croceator-klauberi* intergrade (figure 23.5., A and I).

In the North American bee, *Hoplitis producta* (Megachilidae), integradation between geographical races occurs in the pathway: *gracilis — subgracilis — interior — producta* (figure 23.7). In addition, *H. p. interior* has given rise to two derivative races, one of which, *bernardina*, overlaps in range with *gracilis* (figure 23.7) (Michener 1947).

Other cases of circular overlap have been described inter alia in gulls (*Larus argentatus* group), deer mice (*Peromyscus maniculatus*), land snails (*Partula olympia* group), and *Drosophila paulistorum*. The titmouse (*Parus major* group) in Eurasia is often cited as an example of an overlapping ring of races; but since the zones of overlap are marked by hybridization it is probably better categorized as a syngameon.[4]

4. For review see Mayr (1963:507–512).

Reversals

Evolutionary divergence at the race and species level is not an inexorable process. Divergence from geographical races to biological species continues only so long as the evolutionary forces producing that divergence continue to act. The process of divergence may be arrested at any stage. Geographical races in some species are very ancient. Divergence may be not only arrested, but reversed at any stage up to the formation of complete sterility barriers between the related species. Natural hybridization, which is common in higher plants and well known in many groups of animals, represents a reversal in the process of divergence.

CHAPTER 24
Modes of Speciation

Geographical speciation as described in the preceding chapter is one of several known modes of speciation in higher organisms. Some other modes are theoretically possible but unverified and controversial at present. In this chapter, we will review in synoptical form the various modes of speciation, both known and controversial.

There is some redundancy in the terminology of speciation now. This is due to the fact that different students often apply different terms to what is essentially the same process or mode. The duplication of terminology obscures the unity of the processes in such cases. For our discussion we will adopt the terms for the various speciation modes that we consider preferable, and will indicate the synonyms in passing.

Quantum Speciation

"Quantum speciation" is the budding off of a new daughter species from a small peripheral isolate of a large polymorphic ancestral species. The process was first described by Mayr (1954) and has since been elaborated on by Mayr (1963) and other workers.[1]

The term, quantum speciation, was proposed for this process by Grant (1963:459) for reasons which will become apparent later. Mayr himself (1954) employed the term, speciation by genetic revolution. Other synonymous terms are: speciation by catastrophic selection (Lewis 1962), and speciation by population flush-crash cycles (Carson 1971). Mayr has recently (1982b) proposed an additional synonym, peripatric speciation, without referring to the older and now widely used terms.

In geographical speciation the pathway is local race—geographical race—allopatric semispecies—biological species (as in figures 23.1 and 23.3). In quantum speciation, by contrast, the pathway runs directly from local race to new species. Quantum speciation thus represents a shortcut method of species formation.

The starting point for quantum speciation, as mentioned earlier, is a large, polymorphic, outcrossing, ancestral population. Let a small daughter colony be founded by one or a few migrants somewhere beyond the ancestral species border. The daughter colony will be spatially isolated in its peripheral location. The founding individuals bring only a small and non-random sample of the gene pool of the ancestral population to this daughter colony. Drift therefore takes place during the founding of the new colony. Furthermore, since the founder individuals will have to interbreed with one another at first, inbreeding and more drift occur during the early generations in the new colony.

Inbreeding in the small isolated daughter population leads to the formation and establishment of new homozygous gene combinations that would ordinarily be swamped out by outcrossing in the ancestral population. This is a source of phenotypic novelty in itself. But there is an additional source of novelty.

Habitually outcrossing populations tend to develop homeostatic buffering systems that ensure that normal phenotypes develop from highly heterozygous genotypes. To put it in other words, selection in outcrossing populations favors alleles that interact with one another in heterozygous combinations so as to produce normal phenotypes.

The corollary is that enforced inbreeding and homozygosity remove these

1. See Mayr (1963: chs. 15–18; 1970: chs. 15–18); Lewis (1962); Grant (1963: ch. 16; 1981: chs. 13, 14); Carson (1971, 1975); Stanley (1979: ch. 2).

homeostatic buffers and permit the formation of novel phenotypes, often radically different from the norm. Such novel phenotypes resulting from enforced inbreeding in normally outcrossing populations are referred to as phenodeviants. Phenodeviants are well known from experimental studies in *Drosophila*, chickens, primroses, *Linanthus*, and other animals and plants (Lerner 1954). They are predictably formed in small isolated daughter colonies derived from large outcrossing ancestral populations.

We have, then, two levels of novel phenotypes to distinguish: first, simple homozygous segregates, which are swamped out in the ancestral population but can be fixed in the small isolated daughter colony; and second, phenodeviants.

Now there is no assurance that the novel phenotype in the daughter colony, whether it is a simple homozygous segregate or a phenodeviant, will be adaptively valuable in its environment. If it is not, the colony will soon become extinct. In fact, most new phenotypes will probably be misfits and the colonies harboring them will probably not persist. But in a large series of trials, as where numerous daughter colonies are founded by a large ancestral population, over periods of time and in different parts of the species border, some adaptively valuable genotypes and phenotypes are likely to arise once or a few times.

Such adaptively valuable genotypes, when they do arise, are quickly fixed in the derivative colony by the combination of inbreeding, drift, and selection. The colony, with its novel phenotypic characters and new adaptive mode can then go on to increase and spread as a divergent species.

The evolutionary force controlling quantum speciation is considered to be the selection-drift combination. Other determinants have indeed been suggested, namely, founder effect (Mayr 1954, 1963) and catastrophic selection (Lewis 1962), but these are preferably viewed as special cases of selection-drift as I pointed out in an early discussion of the problem (Grant 1963). Wright's (1931) concept of the selection-drift interaction in small populations, though not originally related to speciation modes, is thus fundamental for quantum speciation. So too is Simpson's (1944) concept of quantum adaptive shifts in small populations.

Geographical and Quantum Speciation Compared

Geographical speciation and quantum speciation both start with large sexually reproducing populations. Both modes involve passage of the divergent lines through a period of spatial isolation on their way to attainment of species status. Here the similarities cease.

In geographical speciation, the allopatric precursor of the new species is a geographical race; in quantum speciation it is a local race. In the former, the forces involved are regional environmental selection in large outcrossing populations; in the latter, they are selection combined with inbreeding and drift in small daughter colonies. In geographical speciation the changes are gradual, slow, and conservative, whereas quantum speciation occurs at a rapid rate and may bring about drastic changes.

The difference in controlling forces between the two modes of speciation has some ramifications. Directional selection for new gene combinations, as involved in geographical speciation, entails a high substitutional genetic load. Such selection is costly in numbers of genetic deaths, and places a burden on the reproductive potential of the population. This burden, in turn, restricts the evolutionary rate to levels of selective mortality that can be tolerated by the population (see chapter 17).

If, however, the parental population breaks away from reliance on directional selection in large populations, and instead exposes some small populations to the joint action of selection and drift, it can avoid some of the restrictive cost of selection (see again chapter 17). And therefore it can change more rapidly. The cost-of-selection factor thus operates in a restrictive manner on evolutionary rates during geographical speciation; but quantum speciation is a way out of this same cost-of-selection restriction, and, in theory at least, can proceed more rapidly than geographical speciation.

Examples of Quantum Speciation

Mayr (1954) described the variation pattern in the New Guinea kingfisher, the *Tanysiptera galatea* group, in relation to what we call quantum speciation. This variation pattern was reviewed in the preceding chapter. It will be recalled that the insular races of *T. galatea*, derived probably from small numbers of migrants from the mainland population, are markedly different. Their degree of differentiation stands in contrast to the conservative racial variation of the intergrading mainland populations of the same species. One insular population has diverged to the species level (*T. hydrocharis*) (see figure 23.2).

Some species of *Drosophila* in the Hawaiian Islands appear to have arisen by quantum speciation. Probable phylogenies in the Hawaiian drosophilas can be traced by similarities and differences in the inversion types as seen in salivary-gland chromosomes. Also, the known ages of the islands inhabited by the flies help to establish the relative ages of the species.

The *Drosophila planitibia* group contains three closely related species (*D. planitibia*, *D. heteroneura*, and *D. silvestris*) with the same chromosomal formula; that is, they carry the same set of inversion types in homozygous condition. The three species occur on two islands with the following distribution: *D. planitibia* on Maui, and *D. heteroneura* and *D. silvestris* on Hawaii. Now the island of Hawaii is only 700,000 years old, whereas Maui is older. This fact, combined with the chromosomal homologies, suggests that *D. planitibia* is ancestral and *D. heteroneura* and *D. silvestris* are derived (Carson 1970; Carson and Kaneshiro 1976).

The sequence of events as regards one speciational change, e.g., *D. planitibia-D. heteroneura*, is probably as follows: (a) colonization of Hawaii by one or a few migrant individuals of *D. planitibia* from Maui, perhaps by a single fertilized female; (b) fixation of formerly polymorphic genes in homozygous condition in the new daughter colony on Hawaii; (c) this fixation is accompanied by rapid divergence of the Hawaiian colony to the level of a new derivative species, *D. heteroneura* (Carson 1970).

The second daughter species on Hawaii, *D. silvestris*, could have arisen by an independent event of colonization from the same ancestral population on Maui, or, alternatively, by divergence from the same original foundation colony on Hawaii that produced *D. heteroneura* (Carson 1970).

It is very interesting that parallel cases of quantum speciation occur in other species groups of *Drosophila* in the same archipelago (Carson 1970, 1981; Carson and Kaneshiro 1976; Carson et al. 1970).

Quantum Speciation Involving Chromosomal Rearrangements

In many groups of plants and animals the related species are differentiated with respect to translocations, inversions, and other segmental rearrangements. The segmental rearrangements may or may not be associated with aneuploid differences in chromosome number. The cytogenetics of interspecific karyotype differences has been extensively investigated. Indeed it is fair to say that the cytogenetic aspects of this problem are better understood than the evolutionary aspects. The problem clearly involves speciation as well as cytogenetics.

Quantum speciation is probably an important mode of formation of species-specific karyotypes. An ancestral population is assumed to be poly-morphic for translocations, inversions, and other rearrangements. Its daugh-ter species, derived by quantum speciation, and passing through a bottleneck of small size and inbreeding, will represent an array of chromosome-seg-mental homozygotes, differing karyotypically from one another and from the ancestral population, and separated inter se by chromosomal sterility bar-riers.[2]

A good example is provided by the species pair, *Clarkia biloba* and *C. lingulata* (Onagraceae), of annual herbs in the Sierra Nevada of California. The wide-ranging *C. biloba* ($2n = 16$) is clearly ancestral on chromosomal and ecological evidence, and the narrowly endemic *C. lingulata* ($2n = 18$), which occurs on the southern periphery of the area of *C. biloba* is just as clearly derived (Lewis and Roberts 1956).

These species differ karyotypically by two independent translocations, an inversion, and in basic number. The differences produce a chromosomal sterility barrier. One of the translocations is associated with the extra ninth chromosome in *C. lingulata*. The distinctive floral characteristics of *C. lingulata* are also associated with this ninth chromosome (Lewis and Roberts 1956).

The origin of *C. lingulata* is best explained as a product of quantum spe-ciation. A set of segmental rearrangements in a chromosomally polymorphic population of *C biloba* on the southern edge of its species area probably be-came fixed by drift and selection to give rise to the new deviant species *C. lingulata* (Lewis and Roberts 1956; Lewis and Raven 1958; Lewis 1962; Gottlieb 1974).

The adaptive role of species-specific karyotypes in animals and plants has been explained in two ways. One hypothesis holds that the species-specific segmental arrangement determines a particular set of developmental and physiological responses as a result of a given gene arrangement. Numerous position effects add up to an overall "pattern effect" (Goldschmidt 1940, 1955; King and Wilson 1975; King 1975). The second suggestion is that the chro-mosome rearrangements bring about the linkage of adaptive gene combina-tions in both the ancestral and the daughter species, though in different ways in the two stages (Darlington and Mather 1949; Grant 1956, 1964, 1971).[3]

There is much merit in the pattern effect hypothesis. However, it will ex-plain only a part of the interspecific chromosome-structural differentiation

2. For more on this process with reference to plants see Grant (1981: chs. 13, 14).
3. See Grant (1981: ch. 14).

in plants at least. In plants a correlation exists between structural differentiation and a life-history strategy calling for restrictions on recombination. This correlation can be accounted for by the gene linkage hypothesis. The latter hypothesis has not yet been seriously considered or tested in insects or other animal groups with interspecific chromosomal rearrangements; this is a desirable task to carry out.

Allopolyploid Speciation

The formation of new species by allopolyploidy has long been known in angiosperms and other land plants. This is indeed the mode of speciation that is best documented experimentally in any major group of organisms. Polyploidy is the occurrence of three or more chromosome sets (or genomes) in a cell, individual, or population; we are of course interested here in polyploidy in populations. Allopolyploids (or amphiploids) are polyploids derived from interspecific hybrids.[4]

Assume that the chromosome sets or genomes of two diploid species differ by several or many chromosomal rearrangements. Their genome constitutions can be written as AA and BB, respectively, where A and B represent the structurally differentiated haploid sets. The interspecific F_1 hybrid AB will have aberrations of meiosis and consequently will be chromosomally sterile.

Now suppose that the sterile hybrid AB undergoes chromosome-number doubling. This happens spontaneously in plants. Doubling can occur in somatic cells to give rise to a tetraploid branch and, eventually, tetraploid flowers with the genomic constitution $AABB$. Or, the diploid hybrid AB produces unreduced gametes (AB) as an end result of the aberrant meiosis, and union of two such unreduced diploid gametes yields tetraploid zygotes in F_2 with the constitution $AABB$.

The tetraploid plant $(AABB)$, having two sets of A chromosomes and two sets of B chromosomes, has normal chromosome pairing in bivalents at meiosis in the combinations A/A and B/B, and normal separation of chromosomes to the poles, so that it forms chromosomally and genically balanced gametes (AB). The allotetraploid is fertile. It has its own hybrid combination of morphological, physiological, and ecological characteristics, a combination that is different from that of either parental diploid species. And it breeds true for

4. For review see Grant (1981: chs. 19, 22–25).

its intermediate hybrid constitution because of the intragenomic chromosome pairing and separation *(A/A* and *B/B)* at meiosis.

There is a chromosomal sterility barrier between the allotetraploid and its diploid parental species. Their F_1 hybrid, if it should be formed at all, would be triploid, and triploids are sterile in plants, owing to irregularities of meiosis. Our new allotetraploid *AABB* is fertile, possesses its own particular character combination, for which it breeds true, and is intersterile with its most closely related diploid species. It is therefore a new biological species of hybrid origin.

Allopolyploidy is a very common mode of speciation in angiosperms, ferns, and some other groups of plants. Recent estimates indicate that 47% of the species of angiosperms and 95% of the species of pteridophytes (ferns and fern allies) are polyploids, and most of these polyploids are allopolyploids.

Some common economic plants are allopolyploids. Well-known examples are bread wheat *(Triticum aestivum)*, New World cottons *(Gossypium hirsutum* and *G. barbadense)*, and tobacco *Nicotiana tabacum)*. *Nicotiana tabacum* ($2n = 48$), with the genome constitution *SSTT* is an allotetraploid derivative of two South American diploids ($2n = 24$) close to the present species *N. silvestris (SS)* and *N. tomentosiformis (TT)* (Clausen 1941).

In a number of instances a natural-occurring allopolyploid species has been resynthesized artificially from its diploid ancestors. Thus *Galeopsis tetrahit* ($2n = 32$) (Labiatae) is a tetraploid species derived from the diploids *G. pubescens* and *G. speciosa* ($2n = 16$). Müntzing (1930, 1932) produced the diploid hybrid *G. pubescens* × speciosa, and from it obtained the synthetic allotetraploid *G. pubescens-speciosa* ($2n = 32$), which possessed the characters of *G. tetrahit*. He next outcrossed *G. pubescens-speciosa* to natural *G. tetrahit*. The F_1 hybrids derived from this cross were fertile with regular meiosis, proving that *G. pubescens-speciosa* was indeed a synthetic form of *G. tetrahit* (Müntzing 1930, 1932). Similar resyntheses of natural allopolyploid species, and successful outcrosses of the synthetic to the natural forms, have been carried out in the wheat genus *Triticum* and other genera.

Hybrid Speciation in Plants

Allopolyploidy is one of several modes of hybrid speciation in plants. By hybrid speciation we mean the formation, in the progeny of a natural hybrid, of a new, true-breeding line that is isolated from the parental species and from its siblings in the hybrid population. The new line must circumvent the obstacles of hybrid sterility and hybrid breakdown. Allopolyploidy accomplishes this feat. Some other mechanisms achieve a similar end result without a change in the number of chromosome sets. Two of these will be described briefly here.[5]

The first mode is recombinational speciation. This can be defined as the formation, in the progeny of a chromosomally sterile species hybrid, of a new, structurally homozygous recombination type that is fertile within the line but isolated from other lines and from the parental species by a chromosomal sterility barrier.

Assume that two parental species differ by two or more independent translocations $(P, Q, \ldots)$. The chromosomal arrangement will be $PPQQ$ in one parental species and $ppqq$ in the other. Their F_1 hybrid is a double (or multiple) translocation heterozygote $P/p \ Q/q$), and is consequently chromosomally sterile in some degree. It can, however, segregate several classes of structurally homozygous, fertile types in F_2 or later generations.

Two such classes of fertile progeny are the parental types, but they do not represent new lines. In addition, the hybrid can produce the homozygous recombination types $PP \ qq$ and $pp \ QQ$. These are fertile and new. Furthermore, they are separated from the parental species and from one another by a chromosomal sterility barrier, albeit a weak one in this case.

With larger numbers of independent translocations the chromosomal sterility barriers surrounding the new homozygous recombination types become stronger and the new line accordingly becomes better isolated. The model has been presented here in terms of translocations. Sets of independent inversions or transpositions have similar effects on sterility and, with appropriate modifications, can be used in the model.

The process of recombinational speciation has been followed in experimental hybrid progenies in several plant groups including *Nicotiana* (Solanaceae), *Elymus* (Gramineae), *Oryza* (Gramineae), and *Gilia* (Polemoniaceae). We still don't know how large a role it plays in nature. It probably occurs occasionally but less frequently than allopolyploidy.

Internal isolating mechanisms are determined by genic factors—sterility genes, gamete lethals, incompatibility genes, etc.—as well as by segmental rearrangements involved in chromosomal sterility. There is no apparent theoretical reason why a process parallel to recombinational speciation could not be carried out with sets of genic sterility factors rather than with chro-

5. A more detailed treatment is given in Grant (1981: chs. 19, 20).

mosomal rearrangements. However, this possible mode of hybrid speciation has not yet been investigated thoroughly.

The second mode to be discussed here is hybrid speciation involving external barriers.

In some plant groups the species are interfertile and are isolated mainly by external barriers. Ecological, seasonal, and floral isolation are the main species-separating barriers. The morphological, physiological, and behavioral differences between the species that lead to these barriers are, of course, gene-controlled. The progeny of natural hybrids between the species, if such arise, will segregate for the gene differences and corresponding character differences determining the external isolation. This opens the possibility of the formation of interspecific recombination products with new character combinations so as to set up new, externally isolated subpopulations. If the external isolation persists, these could develop into new species of hybrid origin.

Probable examples of hybrid speciation by segregation and recombination of external barriers have been described in several plant groups. The California foothill species of larkspur, *Delphinium gypsophilum* is probably derived from *D. recurvatum* × *D. hesperium* (Ranunculaceae) in this manner (Lewis and Epling, 1959). Parallel cases are found in *Amaranthus* (Amaranthaceae), *Epilobium* (Onagraceae), the tree-ferns *Alsophila* and *Nephelea* (Cyatheaceae), and other genera.

Whether hybrid speciation plays a significant role in animal evolution or not remains an open question. Most animal evolutionists consider that it does not (e.g., Mayr 1963; White 1973). However, reports are occasionally published of the probable hybrid origin of new species in various animal groups, e.g., woodpeckers (Miller 1954) and leafhoppers (Ross 1958). Hybrid speciation in animals requires further study.

The Problem of Sympatric Speciation

Sympatric speciation has been a controversial topic for many years. Unfortunately the issue is usually posed as sympatric speciation in general, without any further qualification, and arguments are then presented pro or con. This approach is bound to lead to conflicting results. If, however, we bring the issue into clearer focus, by specifying the conditions in which sympatric speciation is supposedly taking place, then we can investigate the feasibility of the process under these particular restrictive conditions.

Four sets of conditions are relevant to the discussion and should be specified.

1. What spatial field do we have in mind when we use the term "sympatric"? Sympatry may be either neighboring or biotic (see chapter 20). The type of sympatry should be distinguished since sympatric speciation is much more feasible in a neighboringly sympatric field than in a biotically sympatric one.

2. Are we talking about sympatric speciation as a product of primary divergence or as a product of hybridization? Allopolyploid speciation is sympatric and is common in the plant kingdom. Other modes of hybrid speciation in plants are also sympatric and occur occasionally, at least, indicating that allopolyploidy is not an exceptional case. The real problem, then, centers on the possibility of sympatric speciation during primary divergence.

3. What is the breeding system of the organism involved in sympatric speciation? If it is self-fertilizing there is no particular problem. The problematical situation is sympatric speciation in outcrossing organisms, because of the swamping effect of cross-fertilization.

Let us say that we have narrowed the problem down to primary sympatric speciation in a biotically sympatric field and with obligate outcrossing.

4. Then are we seeking to find out whether the process, under such conditions, is a theoretical possibility or a real process in nature? It may well be a theoretical possibility, but one which requires unrealistic assumptions, so that the theoretical models have little relevance for the real world.

Neighboringly Sympatric Speciation

We are concerned here with primary divergence in an outcrossing population where the divergent lines occupy neighboringly sympatric zones. The divergence is brought about by disruptive selection in the neighboring zones, which differ environmentally, and is protected from swamping to some extent by the partial spatial separation of the zones.

A number of actual cases are known, in both plants and animals, of divergence under these conditions to the *race* level. In plants, one good example, among others, is the outcrossing grass. *Agrostis tenuis*, in Great Britain, which has developed different edaphic races on different adjacent soil types (Antonovics 1971). An interesting case in insects involves the fruit fly *Rhagoletis* (Tephritidae).

The larvae of the apple maggot, *Rhagoletis pomonella*, live in and feed on the fruits of cultivated apples *(Pyrus)* and native hawthorns *(Crataegus)* in North America. The populations on apples and on hawthorns are racially different. The flies mate on the host plant, where the females then deposit the eggs in the fruits. The larvae develop in the fruits and the next generation of adults emerge and mate again on the host plant, completing the cycle (Bush 1969a).

Hawthorns and apples grow close together in some apple-growing districts of the northern United States. And here the host races of *Rhagoletis pomonella* also maintain themselves in close proximity. Their habit of mating on their respective host plants helps them to preserve their racial differentiation in their neighboring habitats (Bush 1969a).

Now hawthorns are native in North America, whereas apples are introduced. The original host plant of *Rhagoletis pomonella* in North America was therefore probably *Crataegus*. The fruit fly is known to have infested apples in North America since 1866. It probably switched over from hawthorns to apples sometime in the nineteenth century (Bush 1969a).

The formation of the new host race was probably an event taking place in a neighboringly sympatric field. Host-plant recognition and preference in a related genus, *Procecidochares*, and probably also in *Rhagoletis*, are controlled by a single gene. It will be recalled that mating takes place on the host plant. Therefore a mutation in the gene governing host recognition could initiate a process of neighboringly sympatric race formation (Bush 1969a; Huettel and Bush 1972).

This brings the divergence to the stage of host races on the basis of plausible circumstantial evidence. But the problem before us is sympatric species formation.

Rhagoletis pomonella belongs to a sibling species group containing also *R. mendax*, *R. cornivora*, and *R. zephyria*. All four species are sympatric in eastern North America. They infest fruits of different plant families. Their plant hosts are as follows.

R. *pomonella*	*Crataegus, Pyrus, Cotoneaster, Prunus* (Rosaceae)
R. *mendax*	*Vaccinium, Gaylussacia* (Ericaceae)
R. *cornivora*	*Cornus* (Cornaceae)
R. *zephyria*	*Symphoricarpos* (Caprifoliaceae)

This host distribution could have arisen by shifts from one plant family to another in a sympatric field (Bush 1969b). The case for neighboringly sympatric speciation in fruit flies rests on an extrapolation from race formation to species formation. Such extrapolations are hazardous. Nevertheless, the pattern of distribution in the fruit flies is suggestive of neighboringly sympatric divergence at different taxonomic levels.

Narrow, species-specific food niches are common in many groups of insects. It is tempting to think that abrupt shifts from one food niche to another, like those postulated for fruit flies, have constituted a common pattern of speciation in the insects.

Biotically Sympatric Speciation

The formation of a genetically different population within a biotically sympatric field in an outcrossing organism is beset with serious difficulties. The divergence requires intense disruptive selection, which is accompanied by a heavy selection cost that the population may not be able to bear. And the new divergent population is vulnerable to swamping and reabsorption in the parental population whenever the disruptive selective pressures are relaxed.

Strong disruptive selection has been carried out in experimental populations of *Drosophila melanogaster* and *Musca domestica*. Divergences developed between the high selection lines and the low lines for particular characters in spite of continual intercrossing (Thoday and Gibson 1962, 1970; Streams and Pimentel 1961; Pimentel, Smith, and Soans 1967; Soans, Pimentel, and Soans 1974). These results have been considered promising for the thesis of biotically sympatric speciation. However, we should note that the extracted lines in the disruptive selection experiments fall far short of species status.

The disruptive selection experiments tested the ability of populations to undergo sympatric divergence for alternative character states of a single character. Species, however, differ in character combinations, which are determined by complex gene combinations. Sympatric divergence of this magnitude by disruptive selection would entail a selection cost enormously greater than that associated with single-character divergences.

The prospects for biotically sympatric speciation would be improved if disruptive selection were combined with assortative mating. This combination of factors would come into play if at least one of the characters undergoing disruptive selection were to induce positive assortative mating. The disruptive selection could then bring about a given amount of divergence with a reduced selection cost.

In summary, theoretical difficulties of speciation by disruptive selection in an outcrossing population, with or without assortative mating, are very great. It seems unlikely that this process can occur with any significant frequency in nature. There is no valid biogeographical evidence as yet to indicate that this mode of speciation occurs at all in nature.

Stasipatric Speciation

White (1978 and earlier) has proposed a mode of sympatric speciation which he terms stasipatric speciation. The starting point is a chromosomally polymorphic ancestral species. A new chromosome-segmental homozygote becomes established within the ancestral area. It is isolated from the ancestral species by a partial chromosomal sterility barrier, and forms the nucleus for a new daughter species.

This hypothetical mode is based mainly though not exclusively on a group of wingless grasshoppers, the *Vandiemenella viatica* group (Eumastacidae), in southern Australia. White (1978) recognizes seven, karyotypically different species in this group, of which three are ancestral and four derived. The ancestral species are V. *viatica*, P-45C, and P50; the derived ones are V. *pichirichi*, P45B, P24, and P25. The derived species are supposed to have originated from the ancestral ones by stasipatric speciation.

However, there are some difficulties with this interpretation of this example. First of all, are the seven taxa in question really species? They are similar morphologically and are mostly allopatric. Three zones of marginal sympatry are shown (White 1978:179). No data are given regarding reproductive isolation between the seven taxa. Thus it is possible that some of the taxa are geographical races; others could be marginally sympatric semispecies or species.

Nor does the geographical distribution of the seven taxa suggest a sympatric origin of the derived forms. The derived taxa are almost entirely allopatric with respect to the ancestral taxa and occur to the north of the latter. The simplest interpretation of the evidence is that the four derived taxa have diverged from the ancestral populations in separate allopatric areas and have subsequently attained marginal sympatry in certain localities.

Space does not permit a detailed analysis of the other examples cited by

White as support for the hypothesis of stasipatric speciation.[6] Suffice it to say that other examples, like the *Vandiemenella* grasshoppers, can be interpreted in terms of allopatric divergence. The case for stasipatric speciation seems unconvincing to me.

6. For the full exposition of White's case see White (1978:177–216).

CHAPTER 25
General Theory of Speciation

The Goal
The Steps
The Pathways
The Modes

We have considered several modes of speciation in the preceding two chapters. Let us now look for the common denominators in the various modes. We will follow an earlier treatment (Grant 1963:448–454).

The speciation process can be broken down into four basic components: (a) the goal, (b) the steps for reaching the goal, (c) the pathways composed of different series of steps, and (d) the fields. We can then consider the various modes of speciation as different combinations of these basic components.

The Goal

Sympatric species are adapted for different niches or habitats in a common territory. The adaptations of each species for its niche or habitat are based on character combinations and on the underlying gene combinations. Conversely, each species represents a unique set of adaptive gene combinations. The species is a field for the process of gene recombination, or, more particularly, a field for the generation of adaptively useful gene recombinations. This is the Dobzhanskian or gene-combination concept of species (Dobzhansky 1937a, 1937b, 1941, 1951a, 1970).[1]

Not all the recombination products that can be generated by the sexual process are adaptively useful. Many recombination products are inviable or

1. See Grant (1981b) for review.

subvital and represent a wastage of the reproductive potential of the population. Reproductive isolating mechanisms function to keep the interbreeding and the generation of recombinations within adaptively useful limits. Those limits are normally the boundaries of biological species.

Speciation is therefore essentially the formation of different reproductively isolated sets of adaptive gene combinations.

Speciation is indeed the most rapid and efficient means of fixing a new adaptive gene combination. It is far more efficient than mass directional selection in a large polymorphic breeding population, where the favored type is continually broken up by the sexual process; and it is definitely more efficient than selection-drift in a subdivided species population, where the new type can still interbreed with the older types. Viewed from the standpoint of the strategy of the genes, the best way for a new gene combination to become established is to form a new species of its own (Grant 1981b).

The Steps

The process of speciation requires a series of consecutive steps. The steps are: (1) production of multiple-gene variation, (2) formation of the new allele combination, (3) fixation of the new allele combination in a derivative population, and (4) protection of the new allele combination by reproductive isolating mechanisms.

Let us consider each step in turn. It will be helpful to think in terms of a specific though oversimplified genotype change. An ancestral species with a genotype AA BB gives rise to a daughter species *aa bb*.

1. Production of multiple-gene variation. The ancestral population AA BB must acquire some *a* alleles and *b* alleles. Three main sources of such variation in natural populations are relevant to this discussion: first, normal existing variability in polymorphic populations, variability accumulated over long periods of time; second, special new variability resulting from events of hybridization; and third, special new variability resulting from sporadic bursts of mutation, either genic or chromosomal.

2. Formation of the new allele combination. The sexual process now produces some *aa bb* genotypes in the ancestral AA BB population. But these are rare at first, and furthermore, they are broken up again by the same sexual process.

3. Fixation of the new allele combination in a derivative population. The

fixation of the new adaptive gene combination *aa bb* in a daughter population is a critical step in the speciation process. There are two main ways.

The first is selection under wide outcrossing in a geographical race of the ancestral species. Directional selection for the genotype *aa bb* in a large outcrossing population containing predominantly A and B alleles in its gene pool gradually increases the frequency of the *a* and *b* alleles, and eventually brings about the substitution of the *aa bb* genotype. The process has a high selection cost and is slow. But a population that is able to afford numerous selective deaths and has a long time to effect the substitution will eventually reach the new adaptive goal.

The other way is inbreeding. If the rare *aa bb* individuals can cross inter se or self-fertilize they can multiply quickly in numbers. The fixation of gene combinations can be more rapid and more economical of selective deaths in the case of inbreeding than it is with selection and outcrossing.

Two modes of inbreeding are known to be significant in natural populations, namely, crossing in small populations, which amounts to crossing between sibs and cousins in the main, and self-fertilization. A third mode, assortative mating, or preferential mating of genetically similar individuals, could turn out to be significant also, though its role in speciation must be considered uncertain in the present state of knowledge.

4. Protection of the new allele combination by reproductive isolation. The reproductive isolating mechanisms arise in two ways: as by-products of the divergence in genotype (see chapter 23), and as products of selection for reproductive isolation per se (see chapter 26).

The Pathways

Steps 1 and 3 have different variant forms. These forms can be combined in different ways to give a variety of alternative pathways.

Let us assign letter symbols to the forms of variation production and variation fixation for convenience of reference.

Step 1
Normal existing variability *(v)*
Hybridization *(h)*
Novel mutations *(u)*

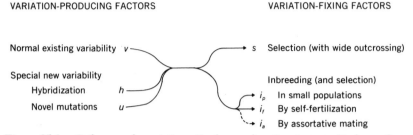

Figure 25.1. Pathways of speciation. Further explanation in text. (Redrawn from V. Grant, *The Origin of Adaptations,* © 1963 Columbia University Press, New York; by permission)

Step 3
Selection (with wide outcrossing) *(s)*
Inbreeding (and selection) *(i)*
In small populations (i_p)
By self-fertilization (i_f)
By assortative mating (i_a)

We can now link the variant forms of steps 1 and 3 together into alternative pathways as in figure 25.1. Thus there is pathway *v-s*, pathway *v-i_p*, and so on. The number of possible pathways shown in figure 25.1 is nine or twelve, depending on whether assortative mating (i_a) is counted or not.

The Modes

The speciation process can take place in three types of spatial fields, namely, allopatric, neighboringly sympatric, and biotically sympatric fields. The three types of fields can be combined with the nine or twelve pathways to produce a classification system of speciation modes. Most of the possible modes are, to be sure, either unlikely or unknown, but a fair number of modes remain and warrant consideration.

Several modes of speciation discussed earlier in chapters 23 and 24 are listed again here and fitted into the classification system based on pathway and field. The first set listed below consists of modes which are either known or highly probable.

v-s, allopatric.   Geographical speciation.
v-i_p, allopatric.   Quantum speciation.

$h\text{-}i$, biotically sympatric.	Recombinational speciation, and hybrid speciation with segregation of external barriers, in plants.
$h\text{-}u\text{-}i$, biotically sympatric.	Allopolyploid speciation.
$v\text{-}i_p$, neighboringly sympatric.	Neighboringly sympatric speciation in outcrossing insects and other invertebrates with narrow ecological niches. (Supported by some good circumstantial evidence.)

The second set consists of modes which are not known to occur but are possible and discussed in the literature.

$v\text{-}i_f$, biotically sympatric.	Biotically sympatric speciation in self-fertilizing plants, invertebrates, and protistans. (A definite possibility.)
$v\text{-}s$, biotically sympatric.	Biotically sympatric speciation in outcrossing organisms under disruptive selection. (Regarded as a real possibility by some workers, but as unlikely by others including myself.)
$v\text{-}i_a$, biotically sympatric.	Biotically sympatric speciation in outcrossing organisms with assortative mating. (A possibility.)

CHAPTER 26
Selection for Reproductive Isolation

Reproductive isolation is believed, on good evidence, to develop in two ways. First, reproductive isolating mechanisms develop as by-products of evolutionary divergence from the race to the species level, as noted in preceding chapters. This enables the divergent species to coexist sympatrically.

But the primary, by-product isolating mechanisms may not prevent hybridization completely. Some hybrids may be produced. These are likely to be inviable, sterile, weak, or poorly adapted. Their production accordingly represents a wastage of the reproductive potential of the parental species.

Under these conditions, it may be selectively advantageous for the parental species to reinforce the existing primary isolating mechanisms with new barriers that do effectively block hybridization. Selection can act to build up special reproductive isolating mechanisms for the sake of reproductive isolation itself, if and when such isolation is selectively advantageous.

This second mode of formation of reproductive isolation will be described in the present chapter.

Selection for reproductive isolation per se had also been referred to in the literature as the Wallace effect or Wallace process, and as reproductive character displacement.

The Process

Selection for reproductive isolation per se comes into operation when all of the following conditions are fulfilled. (a) The species populations are in sympatric contact. (b) They engage in hybridization; (c) The hybrids and hybrid progeny are inviable or sterile or otherwise adaptively inferior. (d) The loss of reproductive potential in the parental species due to the hybridization is selectively disadvantageous; in other words, prevention of hybridization has a definite selective advantage over the status quo. (e) Individual variation exists within one or both parental species with respect to characteristics affecting reproductive isolation.

The last two conditions go to the heart of the selective process itself. It is necessary to postulate that the species population is polymorphic for ease of hybridization with a foreign species and that the hybridization has significant deleterious effects. Under such conditions, those individuals in the species population that possess barriers to hybridization contribute more viable and fertile progeny to future generations of their population than do sister individuals that hybridize freely. And the genes of the former will therefore increase in frequency and spread throughout the population.

The conditions listed above vary greatly from one species pair to another and from one group of organisms to another. Biotic sympatry (condition a), for example, is generally more common among insects and annual plants than among large land vertebrates and large woody plants. And the selective disadvantage of a loss of reproductive potential (condition d) is much greater in ephemeral organisms, such as annual plants and small insects, than in long-lived perennial plants with a high excess fecundity.

The conditions promoting selection for isolation are sometimes realized and sometimes wanting in an array of species groups. Selection for isolation is not expected to be a universal process.

When the process of selection for isolation does come into play, it has reference primarily to isolating mechanisms that operate in the parental generation. The adaptive goal is the prevention of hybridization. And this goal is achieved most efficiently by the building up of mechanisms, such as ethological isolation or incompatibility, that act prior to fertilization.

This rule can be stretched in the case of organisms where the maternal parent carries and nurtures the embryo, as in mammals and most seed plants. Here it is theoretically possible for selection to build up barriers that operate in the early stages of the F_1 generation. The blocking of effective hybridization by the formation of abortive young hybrid embryos may not be as economical reproductively as ethological isolation or failure of fertilization, but it nevertheless represents an improvement in reproductive efficiency over the production of sterile or inviable adult hybrids.

Evidence

Selection for reproductive isolation has been shown to occur in both experimental and natural populations. Experiments have been carried out in mixed populations of *Drosophila* (*D. pseudoobscura* and *D. persimilis*) and *Zea mays* (white flint and yellow sweet varieties) (Koopman 1950; Paterniani 1969). The related species or varieties hybridize spontaneously. Artificial selection against the hybrid types led to a marked increase in reproductive isolation between the species or varieties in just a few generations.

The first attempt to demonstrate the efficacy of selection for isolation in natural populations made use of single species pairs with partial sympatry. The setup is that shown in figure 26.1, top. Two species A and B have races that are allopatric with respect to one another (A_a and B_a) and other races that are sympatric (A_s and B_s). One can compare the strength of ethological or other pre-fertilization barriers in the allopatric combination of races with that in the sympatric combination.

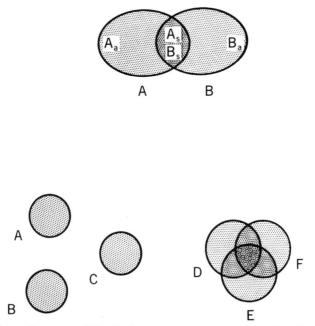

Figure 26.1. Two types of distribution patterns in species groups in which it is possible to compare allopatric and sympatric populations with respect to parental-generation isolation. *(Top)* A partially sympatric species pair. *(Bottom)* A species group containing allopatric species and sympatric species.

This comparison has been made for partially sympatric species pairs of *Drosophila* (*D. pseudoobscura* and *D. miranda*), tree frogs (*Microhyla carolinensis* and *M. olivacea, Hyla ewingi* and *H. verreauxi*), phlox (*Phlox pilosa* and *P. glaberrima*), and some other organisms. In each case, ethological isolation itself or some measurable component of ethological isolation is more strongly developed in the sympatric areas of the two species than in the allopatric areas (Dobzhansky 1951*a*; Blair 1955; Littlejohn 1965; Levin and Kerster 1967; Levin and Schaal 1970).

Thus in the southern United States, the mating calls of *Microhyla carolinensis* and *M. olivacea*, which function in ethological isolation, are very different in their zone of sympatric overlap but are not well differentiated when allopatric races of the two species are compared (Blair 1955). The same pattern of racial variation for call notes is found in the partially sympatric species pair *Hyla ewingi* and *H. verreauxi* in southeastern Australia (Littlejohn 1965).

More extensive comparisons of the same sort were made in the superspecies *Drosophila paulistorum*. This tropical American superspecies consists of six semispecies (Amazonian, Andean, etc.), which are partially sympatric in various combinations (Dobzhansky et al. 1964). The strength of ethological isolation between the semispecies can be measured in mixed laboratory populations and expressed quantitatively as an isolation coefficient (I). This coefficient varies from 0 to 1, where 0 is random mating and 1 is complete isolation (Dobzhansky et al. 1964; Ehrman 1965).

The isolation coefficient was computed for allopatric race combinations and for sympatric race combinations representing eight semispecies pairs (e.g., Amazonian × Andean, Amazonian × Guianan, etc.) in the *D. paulistorum* superspecies. The values of I for the race combinations in the different semi-species pairs were as follows:

	I, range	I, average
Allopatric race combinations	0.46–0.76	0.67
Sympatric race combinations	0.68–0.96	0.85

The isolation coefficient is consistently higher between sympatric races of a semispecies pair than between allopatric races of the same semispecies pair throughout the large and complex *D. paulistorum* superspecies (with one exception) (Ehrman 1965).

Cross-Incompatibility in Gilia

Another setup in which it is possible to detect the occurrence of selection for isolation is that shown in figure 26.1, bottom. A species group contains some allopatric biological species (A, B, C) and some sympatric species (D, E, F). The strength of parental-generation isolating mechanisms is then compared in two classes of interspecific combinations, allopatric × allopatric and sympatric × sympatric.

A favorable group of organisms in which to make this comparison is the leafy-stemmed gilias, a section of the genus *Gilia* (Polemoniaceae). This natural group of annual herbs consists of nine biological species in Pacific North America and South America. The species fall into two classes on geographical distribution. Five species in the foothills and valleys of California are extensively sympatric. Four other species on the coastline and offshore islands are entirely allopatric with respect to one another and only rarely sympatric with outlying populations of the foothill-and-valley species. The artificial F_1 hybrids between the nine species are highly sterile, with chromosomal sterility, in all combinations (Grant 1965, 1966b).

Since the plants are annual herbs, and flower during only one season, seed set is critical for them. By the same token, the production of seeds that develop into sterile hybrids represents a serious loss of reproductive potential.

An effective block to hybrid formation in *Gilia* is cross-incompatibility. The strength of incompatibility barriers between species can be determined quantitatively in artificial interspecific crosses. A good measure is the average number of hybrid seeds produced per flower cross-pollinated for a large number of flowers in a given species cross (*S/Fl*) (Grant 1965, 1966b).

The value of *S/Fl* was determined for 20 interspecific combinations of the nine species of leafy-stemmed gilias. Incompatibility was found to differ greatly between the two geographical classes of hybrid combinations, as indicated by the data in table 26.1. It is apparent that very strong incompatibility barriers are developed between the sympatric species but not between the allopatric species in this group.

Ethological Isolation

Species-specific recognition marks and courtship behavior are well developed in most groups of animals. One can think of courtship plumage, dis-

Table 26.1. Comparative strength of incompatibility barriers between species of leafy-stemmed *Gilia* with different geographical relationships (Grant 1965, 1966b)

Geographical Relationship of Parental Species	No. Interspecific Combinations	S/Fl,* Range	S/Fl, Mean of Means
Allopatric species inter se	5	7.7–24.8	18.1
Sympatric species inter se	9	0.0–1.2	0.2
Allopatric × sympatric species	6	0.0–6.8	3.2

* S/Fl = average number of hybrid seeds per flower pollinated for each species cross.

plays, and songs in birds, of courtship dances and olfactory signals in insects, and so on. The varied types of recognition signals usually differ from species to species. They function to promote intraspecific matings and to discourage interspecific matings.

The origin of species-specific courtship characteristics can probably be attributed mainly to selection for reproductive isolation, although other modes of selection may also be involved. Courtship characteristics act as stimuli and releasers of sexual behavior, and could be products of selection for reproductive output per se, to some extent at least; but this mode of selection does not provide an explanation of the species specificity of the courtship characteristics. Again, *some* differences in courtship characters and behavior may develop as by-products of divergence; but it is very difficult to explain the more elaborate species-specific paraphernalia of courtship as side effects of differentiation in respect to food-getting and other secular aspects of life. The best hypothesis is that these courtship characteristics, which are so widespread in the animal kingdom, are mainly products of selection for ethological isolation.

Experimental and field evidence for a direct selective origin of ethological barriers in *Drosophila* and frogs has been presented above.

Historical Background

In the late nineteenth century A. R. Wallace (1889: chs. 7, 11) suggested that the prevention of hybridization between species would be selectively advantageous to such species. The selective advantage would be due to the "inferiority of the hybrid offspring" as compared with the intraspecific progeny, and it would lead to the spread of the non-hybridizing types within each species population. Wallace believed that this selection mode provided an explanation for the development of two types of isolating mechanisms: internal barriers (Wallace 1889: ch. 7), and species-specific recognition marks involved in ethological isolation (1889: ch. 11).

Wallace, Darwin, and their contemporaries used the terms sterility or infertility in a collective sense to include what we now distinguish as cross-incompatibility and hybrid sterility. Wallace (1889: ch. 7), in his discussion of a selective origin of sterility barriers, explicitly included both cross-incompatibility ("the infertility of crosses between distinct species") and hybrid sterility ("the usual sterility of their hybrid offspring"). He was on the right track as regards cross-incompatibility and can perhaps be excused for including hybrid sterility too.

Romanes (1890) also expressed the idea that parental-generation isolation could be built up by selection as a means of preventing the formation of inferior hybrid offspring.[1] The whole question then dropped out of sight for several decades after 1890.

In the modern era, the hypothesis of a direct selective origin of reproductive isolation was independently suggested again by Fisher in 1930. Fisher's discussion (1930:130 ff.; 1958:143 ff.) was very brief and general. So also was an early discussion by Huxley (1942:287 ff.).

The hypothesis was taken up and developed further by Dobzhansky (1940, 1941:285 ff.; 1951:209 ff.). The first good experimental evidence was provided by Dobzhansky's student, Koopman (1950).

In the 1940s and 1950s, attempts were made to explain the distribution of reproductive isolating mechanisms in various species groups in terms of selection for isolation. The evidence was not extensive at that time, and the hypothesis was greeted with skepticism in many quarters. In the 1960s, however, more convincing evidence was obtained in *Drosophila* and *Gilia* that met some of the earlier objections.

1. This reference of Romanes was discovered by M. J. Kottler, who called it to my attention.

PART VI
Macroevolution: Basic Processes

CHAPTER 27

Geological Time

Main Stages of Earth History
Origin of Life
Primitive Evolution
The State of Complex Multicellular Organisms
Discussion

Main Stages of Earth History

The origin of the solid planet earth occurred 4.6 billion years ago. Earlier time in the solar system can be extended back to 5 billion years B.P. (before present). One suggestion is to assign this early time, from 5–4 billion years B.P., to the Priscoan eon (Harland et al. 1982); an alternative procedure would be to submerge the Priscoan into the Archean eon, which would then run from 5.0–2.5 billion years B.P. (p. 9). The Archean is followed by the Proterozoic eon (2.5–0.59 billion years B.P.) and Phanerozoic eon. The latter represents the last 590 million years of earth history. The chronology is summarized in table 27.1.

Calvin (1956) proposed a subdivision of earth and solar system history into four main stages according to the dominant mode of cosmic developmental process at each stage: (a) atomic evolution, in which nuclear reactions occurred to form hydrogen atoms, and, from hydrogen, to build up the other elements; (b) chemical evolution, in which the atoms combined into chemical compounds of various degrees of complexity, including non-living organic molecules; (c) organic evolution, encompassing the developments from the origin of life through the formation of higher animals; (d) cultural evolution, the accumulation and transmission of a cultural heritage that commenced with the rise of mankind above the animal level.

Atomic evolution prevailed in the formative period of the earth, and graded into chemical evolution later. Chemical evolution eventually reached a stage

Table 27.1. Main subdivisions of geological time (Harland et al. 1982)

Eon	Era or event	Million years B.P.
Phanerozoic	Cenozoic	65
	Mesozoic	248
	Paleozoic	590
Proterozoic		2,500
Archean		4,000
	Origin of Earth	4,600
Priscoan		5,000

in which complex organic molecules were formed. Such non-living organic compounds paved the way for the origin of life during the Archean eon. The stage of organic evolution has been in progress for some 3 billion years or more of earth history. The stage of cultural evolution belongs to the latter phase of human evolution and is thus a very recent phenomenon, on the planet Earth at least.

Origin of Life

The essential step for the origin of a simple form of life was the formation of DNA or DNA-like macromolecules possessing the properties now found in genes. These properties are the ability to direct the synthesis of proteins and other substances that help them maintain themselves as entities, for awhile at least, and the ability to replicate themselves.

The carrying out of these processes requires food as an energy source. The accumulation of energy-rich carbon compounds during the previous period of chemical evolution provided a store of potential food.

The accumulation of organic compounds in the course of chemical evolution depended on two favorable conditions that existed then but ceased to exist later. The two conditions were the absence of living organisms and the absence of free oxygen in the atmosphere. These are factors that bring about the breakdown of organic molecules in the modern world. In the pre-biotic world the absence of these factors meant that such molecules, once formed by spontaneous chemical reactions, would persist and accumulate.

The chemical reactions that produce organic molecules do take place spontaneously, as has been confirmed in experiments. Set up an atmosphere composed of hydrogen, water, ammonia, and methane. Expose it to lightning discharges, natural radioactive emissions, cosmic rays, or ultraviolet rays. Various organic compounds, including amino acids, are then formed spontaneously.

Therefore it is logical to infer that an "organic soup" containing stored energy would form and accumulate during the stage of chemical evolution. And at some point the processes of chemical evolution must have produced nucleic acid molecules with genetical activity. These first living particles would be "naked genes" living on the stored energy in the organic soup.[1]

Genes have the capacity to replicate, multiply, and mutate. Replication spins off a fraction of mutant types. Multiplication goes on to the physical limits of the environment. Consequently a competition sets in sooner or later among the genes or their carriers for limiting environmental resources, and in this competition some types will be more successful reproductively than others. Thus the tendency to evolve is one of the basic attributes of life in a world of limited resources. With the origin of life, organic evolution was on its way.

Primitive Evolution

Several major advances in nutrition and structural organization took place during the first 2 billion years, more or less, of organic evolution, advances that were essential for the later evolution of multicellular organisms. The important developments were: autotrophic nutrition, particularly photosynthesis; aerobic respiration; eukaryotic cellular organization; and sexual reproduction.

The continued growth of the primitive heterotrophic particles necessarily led to the gradual depletion of the original organic soup. The depletion of the food reserves would have put a premium on the ability of organisms to synthesize their own foods out of inorganic raw materials such as water and carbon dioxide. The development of photosynthesis and chemosynthesis in primitive prokaryotes was a major advance in early evolution. It paved the way for a second important advance, aerobic respiration.

1. These ideas have been developed by Haldane (1933), Oparin (1938, 1964), Urey (1952), Miller (1953), Blum (1955), Wald (1955), Calvin (1956), and others. Good reviews are given by Keosian (1964), Dickerson (1978), and Dillon (1978).

The primitive atmosphere contained hydrogen but not free oxygen; it was a reducing rather than an oxidizing atmosphere. Under these conditions the primitive heterotrophic particles and cells would have had to get the energy out of the organic soup by fermentation. Now fermentation is metabolically inefficient in that it leaves most of the energy of the carbon compounds untapped. Aerobic respiration, involving a more complete breakdown of the carbon compounds, releases far more energy. Organisms obtaining their energy by cellular respiration can operate at a much higher metabolic rate than organisms relying on fermentation (Oparin 1938, 1964; Wald 1955).

Aerobic respiration requires an oxygen-containing atmosphere. Oxygen is a by-product of photosynthesis. Students of early earth history believe that the change from the original oxygen-free atmosphere to an oxygen-containing atmosphere was a product of the activity of primitive photosynthetic organisms. The transition to an oxidizing atmosphere permitting respiration must have been gradual and very slow (Oparin 1938, 1964; Wald 1955).

Its slowness may account for the long lag of about 2 billion years between the first appearance of primitive photosynthetic organisms and the appearance of eukaryotes.

The eukaryotic cellular organization, with its grouping of numerous gene centers onto true chromosomes, its division of labor between nucleus and cytoplasm, and its capacity to harbor organelles such as chloroplasts and mitochondria, represented another big step forward in complexity and in the ability to carry out diverse self-sustaining life processes.

The two basic components of sexual reproduction—fertilization and meiosis—were made possible by the grouping of genes onto true chromosomes and of chromosomes into a nucleus in eukaryotes. Sexual reproduction, as an orderly and symmetrical method of generating recombinational variability, enters the picture at the level of unicellular eukaryotes.

The above stages are exemplified in the living biota by (a) simple heterotrophic bacteria; (b) photosynthetic or chemosynthetic prokaryotes, e.g., blue-green algae and autotrophic bacteria; and (c) simple aerobic eukaryotes, e.g., unicellar green algae and plant flagellates. Stage c passes into a fourth primitive stage (d) of simple multicellular organisms, e.g., filamentous algae and lower fungi.

The fossil record of primitive evolution is very fragmentary, as would be expected, although significant finds have been made in recent years. A list of the early fossil floras known to me is given below (compiled from Cloud et al. 1969; Licari and Cloud 1972; Cloud 1974; Barghoorn 1971; and Horodyski and Bloeser 1978).

1. Figtree formation, South Africa. 3.2 billion years old.
   bacteria
   unicellular blue-green algae
2. Gunflint chert, Ontario, Canada. 1.6–1.9 billion years old.
   bacteria
   blue-green algae
3. Paradise Creek, Queensland, Australia. 1.6 billion years old.
   blue-green algae
4. Little Belt Mts., Montana. 1.4 billion years old.
   blue-green algae
   green algae?
5. Beck Spring, California. 1.2–1.4 billion years old.
   blue-green algae
   green algae
   yellow-brown algae
6. Bitter Springs formation, Australia. ca. 0.85 billion years old.
   bacteria
   blue-green algae
   green algae
   fungi?

On the basis of these records, the origin of primitive heterotrophic forms, and the beginning of stage a, must have occurred sometime before 3.2 billion years B.P. but after chemical evolution had reached an advanced phase. The stage in which blue-green algae were the main photosynthetic organisms (stage b) ran from 3.2 to 1.6 billion years B.P. in the Archean and Proterozoic eons. Simple green algae appear about 1.4 billion years B.P., marking the beginning of stage c. Simple multicellular eukaryotes representative of stage d enter the scene about 0.85 billion years ago in the Proterozoic.

The Stage of Complex Multicellular Organisms

The stage of simple multicellular organisms grades into a stage of complex organisms as exemplified by large-bodied plants and metazoan animals. This latter stage begins in the Late Proterozoic, picks up momentum in the Cambrian, and continues through the Phanerozoic. The chronology of this era is summarized in tables 27.2 and 27.3.

Table 27.2. Age of periods of Phanerozoic and late
Pre-Cambrian (Harland et al. 1982)

Era	Period	Million years B.P.
		0
Cenozoic	Quaternary	
		2
	Tertiary	
		65
	Cretaceous	
		144
Mesozoic	Jurassic	
		213
	Triassic	
		248
	Permian	
		286
	Carboniferous	
		360
Paleozoic	Devonian	
		408
	Silurian	
		438
	Ordovician	
		505
	Cambrian	
		590
Sinian (Proterozoic)	Vendian	
		670

A significant late pre-Cambrian fauna is the Ediacara fauna of Australia, containing coelenterates and annelids (Glaessner 1961). Its age is now dated as about 625 million years old, in the middle of the Vendian period of the Proterozoic (see table 27.2) (Harland et al. 1982).

The major groups of aquatic metazoan animals became abundant later in the Cambrian. Animals and plants were aquatic through the Cambrian and

Table 27.3. Age of epochs of Cenozoic
(Harland et al. 1982)

Period	Epoch	Million years B.P.
Quaternary	Holocene	
		0.01
	Pleistocene	
		2.0
	Pliocene	
		5.1
	Miocene	
		24.6
Tertiary	Oligocene	
		38.0
	Eocene	
		54.9
	Paleocene	
		65.0

Table 27.4. First appearances of various major groups in the fossil record (Compiled from Harland et al. 1982)

Group	Million years B.P.	Epoch
Hominids	15	Miocene
Rodents	45	Eocene
Grasses	56	Paleocene
Primates	61	Paleocene
Angiosperms	120	Early Cretaceous
Birds	150	Late Jurassic
Dinosaurs	245	Early Triassic
Therapsids	265	Early Permian
Winged insects	310	Pennsylvanian
Gymnosperms	370	Late Devonian
Wingless insects	390	Early Devonian
Land plants	420	Silurian
Jawed fishes	425	Silurian
Jawless fishes	510	Late Cambrian
Foraminiferans, mollusks	575	Early Cambrian
Echinoderms, trilobites, brachiopods	580	Early Cambrian
Coelenterates, annelids	625	Mid Vendian

Ordovician. Some groups emerged onto land in the Silurian and Devonian, and became better adapted for terrestrial and aerial life during and since the Carboniferous. Animals and plants of a modern cast belong to the Cenozoic era of the last 65 million years.

A series of first appearances of various major groups of both animals and plants is presented in table 27.4. Changes in diversity of metazoan animals through Phanerozoic time are portrayed in figure 27.1.

Different major groups of animals achieved dominance at different times during the Late Proterozoic and Phanerozoic. The periods of expansion and dominance of the various groups are as follows.

Soft-bodied invertebrates, Vendian
Hard-shelled marine invertebrates, Cambrian–Silurian
Placoderms, Devonian
Cartilaginous fishes, Carboniferous
Insects, Late Carboniferous—Cenozoic
Bony fishes, Triassic—Cenozoic
Amphibians, Mid Carboniferous—Triassic
Reptiles, Permian—Cretaceous
Mammals and birds—Cenozoic.

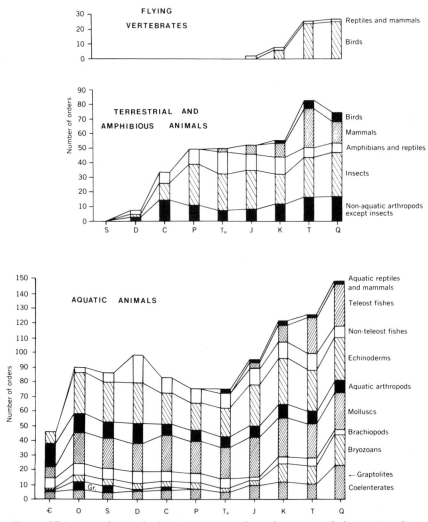

Figure 27.1. Fossil record of metazoan animals in three major habitats. Number of orders known for various major groups in each period of Phanerozoic. (Redrawn from Simpson 1969)

Discussion

It has often been stated that the course of organic evolution, with its general increase in the complexity of biological organization through time, runs counter to the second law of thermodynamics, which would predict changes in the reverse direction. It has also been pointed out repeatedly that the second law of thermodynamics is not entirely applicable to organic evolution.

In the first place, the second law applies to a closed system, whereas organic evolution takes place in an open system. Furthermore, organisms take in and use energy to maintain the biological organization of the cell and body. Food provides the energy for the maintenance of organization by the individual organisms, and ultimately and indirectly, for the increase in complexity during the course of organic evolution.

CHAPTER 28
Evolutionary Trends

The fossil record of many animal groups and some plant groups reveals a series of morphological changes in a given direction over a long period of time. Such long-term oriented changes within a natural group are evolutionary trends. They are a fact of paleontology. They can also be inferred with reasonable confidence from the evidence of comparative morphology in many contemporaneous plant groups that lack a fossil record but have preserved an array of primitive, intermediate, and advanced forms.

The purpose of this chapter is to describe one well documented example of an evolutionary trend in moderate detail, and to call attention to some others. Our approach to evolutionary trends here is descriptive rather than analytical; we will examine the nature of evolutionary trends in chapter 31.

Examples

A classical example of an evolutionary trend involves the coiling and sculpturing of the shell in the freshwater snail *Paludina* (= *Vivipara*, Viviparidae) in southeastern Europe during the Late Pliocene. Series of fossil snails of different ages were found in deposits in former lakes in Yugoslavia and Aegean Islands. A series from Yugoslavia is shown in figure 28.1. It runs from an ancestral form *P. neumayri* (figure 28.1, neu) through intermediate stages to a derived form *P. hoernesi* (figure 28.1, hoer). The shells

hoer

neu

Figure 28.1. Evolutionary trend in the snail *Paludina* in Late Pliocene time. The series runs from left to right and from bottom to top. *(neu)* ancestral form *P. neumayri*. *(hoer)* derived form *P. hoernesi*. (Rearranged from Abel 1929, after Neumayr)

gradually become more tightly coiled and more elaborately sculptured with the advance of time (Abel 1929).

The Pliocene *P. neumayri* is close to a living species, *P. unicolor*, which still occurs in Yugoslavia and neighboring areas. The derived form, *P. hoernesi*, is obviously very different from the ancestral form and is sometimes placed in a separate genus, *Tulotoma*. Its relatives no longer occur in Yugoslavia, but do have living representatives in North America and China (Abel 1929).

Another classical example is the increase in size of the horn in titanotheres (Brontotheriidae, Perissodactyla) from Eocene to Oligocene (see chapter 33 and figure 33.2).

Some trends occur independently and in parallel in a number of separate groups. Thus a trend toward increased body size occurs in many groups of animals. This trend is found in many groups of land mammals during the Tertiary period, e.g., in titanotheres, horses, and proboscidians; in dinosaurs during the Mesozoic; and in some marine animals such as sea snails in the Paleozoic. So widespread among animals is the trend to increased body size that it has become codified as Cope's rule.

Evolutionary trends in flowering plants are usually recognized on comparative morphological rather than paleontological evidence, because of the ready availability of the former and difficulties with the latter.[1]

In plants as in animals, independent groups often exhibit parallel trends. Trends toward reduced size and simpler morphology, known collectively as reduction series, are common. One type of reduction series, exhibited independently in numerous groups, runs from flowers with numerous petals, stamens, and/or carpels to flowers with only a few of these organs. Another common reduction series runs from large perennial herbs to small annual herbs. This trend often continues within annual herbs from large-flowered cross-pollinating forms to small-flowered self-pollinating forms.

Trends in the Equidae

The most famous and best documented evolutionary series is that involving characters of the teeth, feet, and other body parts in the Equidae during the Tertiary. The very good fossil record of the horse family spans a time period of about 60 million years since the Paleocene. The extreme members of the family are the small four-toed *Hyracotherium* (or *Eohippus*) of Late Paleocene and Eocene age and the modern genus *Equus*, consisting of horses, burros, onagers, and zebras. A nearly continuous series of gradations is found between these extreme forms in the fossil record.

The Equidae developed out of a still earlier Paleocene stock, the condylarths, which were small dog-like animals with five-toed padded feet. The fossil series extends downward from *Hyracotherium* to the condylarths.

The phylogeny of the horse family has been worked out from fossil evidence and is well known down to the level of genera. A simplified version of the phylogeny is shown in figure 28.2. The geological ages and geographical distribution of the principal equid genera are also listed for reference purposes in table 28.1. The main theater of equid evolution was North America, but branch lines also evolved in Eurasia, Africa, and South America at various times during the Tertiary and Quaternary.

The main trends in horse evolution involve: (a) body size, (b) tail length, (c) foot mechanism, (d) leg length, (e) head shape, (f) brain size and complexity, and (g) cheek teeth. The account given here is based primarily on Simpson (1951, 1953), which should be consulted for additional details.

---

1. Many such trends are described by Cronquist (1968), Takhtajan (1969), Stebbins (1974), and Carlquist (1975).

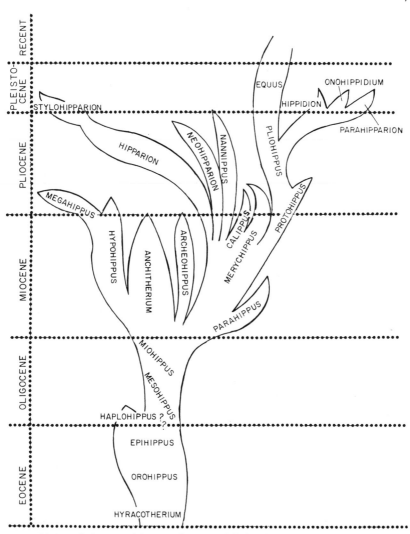

Figure 28.2. Phylogeny of the Equidae. Simplified; only generic branches are shown. (Redrawn from G. G. Simpson, *The Major Features of Evolution*, copyright 1953, Columbia University Press, New York; by permission)

As regards body size, *Hyracotherium* was small, like the earlier condylarths, standing 25–50 cm tall in different species, in contrast to the well-known large size of *Equus*. A general increase in body size occurred in many lines including the equine horses from Oligocene to Pliocene time. Milestones in this trend are indicated by the scaled drawings of feet and heads in

Table 28.1. Ages and geographical distribution of genera of the Equidae (Simpson 1951, and pers. comm.)

Epoch	North America	Old World	South America
Recent	*Equus*	*Equus*	*Equus*
Pleistocene	*Equus*	*Equus* *Stylohipparion*	*Equus* *Hippidion* *Onohippidium* *Parahipparion*
Pliocene	*Equus?* *Pliohippus* *Calippus* *Nannippus* *Neohipparion* *Hipparion* *Megahippus* *Hypohippus*	*Stylohipparion* *Hipparion* *Hypohippus*	
Miocene	*Merychippus* *Archaeohippus* *Megahippus* *Hypohippus* *Parahippus* *Anchitherium*	*Anchitherium*	
Oligocene	*Miohippus* *Mesohippus* *Haplohippus*		
Eocene	*Epihippus* *Orohippus* *Hyracotherium*	*Hyracotherium*	
Late Paleocene	*Hyracotherium* (= *Eohippus*)		

figures 28.3 and 28.5. It should also be noted that the size-increase trend, though real enough, did not occur at an even rate through time, was not universal throughout the horse family, and was subject to reversals in various horse genera, including *Equus* itself.

The tail changed from long relative to body size in *Hyracotherium* to relatively short with long hairs in *Equus*.

The most conspicuous trend in horse feet is reduction in the number of toes. The starting point of this trend is in the ancestral condylarths, which were five-toed. *Hyracotherium* was four-toed and three-toed in the forefeet and hindfeet, respectively (figure 28.4A). Oligocene, Miocene, and most

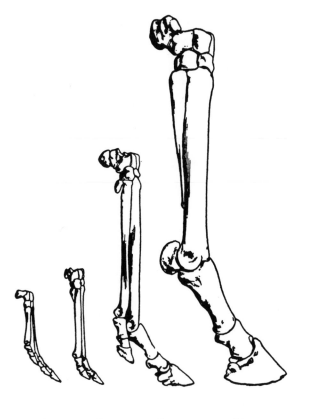

A   B   C   D

Figure 28.3.   Forefeet of horses. (A) *Hyracotherium.* (B) *Mesohippus.* (C) *Merychippus.* (D) *Equus.* (Rearranged from G. G. Simpson, *Horses,* copyright 1951, © 1979, Oxford University Press, New York; by permission)

Pliocene horses were three-toed (figure 28.4B, C). Among three-toed horses a difference is apparent between earlier genera, in which the two lateral toes are well developed (figure 28.4B), and later forms, in which they are shortened (figure 28.4C). The trend culminates in one-toed feet in the Pliocene genus *Pliohippus* and its later descendants, including *Hippidion* and *Equus* (figure 28.4D).

The reduction in toe number accompanies a change in the foot mechanism. The feet of condylarths and *Hyracotherium* were padded, and the animals walked on these pads, as do dogs. The earliest three-toed horses of the Oligocene and some of their direct descendants in the Miocene were also

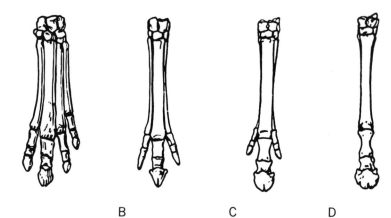

A   B   C   D

Figure 28.4. Toes on forefeet of horses. Not shown to same scale. (A) *Hyracotherium*. (B) *Mesohippus*. (C) *Merychippus*. (D) *Pliohippus*. (Redrawn from Simpson 1951, and Romer 1966)

pad-footed (figure 28.4B). In another branch of three-toed Miocene horses, however, and in the later one-toed horses, the foot has a spring action. The body weight is borne not on pads, which are now absent, but on the hoof of the central toe (Figures 28.3C, 3D; 28.4C, 4D).

In the three-toed spring-footed horses the two short side toes may have functioned as shock absorbers, according to Simpson, to relieve some of the strain on the main central toe (figures 28.3C, 28.4C). These side toes eventually disappeared in the one-toed horses. Here the body weight rests entirely on the hooves of the central toes. Powerful ligaments attached to the bones of these toes produce a spring action in running.

The changes in foot structure were episodic rather than continuously slow; these changes came in certain lineages only and at relatively rapid rates.

Increase in leg length in equid evolution is partly a function of increase in body size. But the change in foot mechanism from padded feet to hooved spring-feet added to the effective length of the legs, for spring-footed horses walk and run on their tiptoes.

Inspection of figure 28.5 reveals numerous and complex changes in proportions in the skull as well as the change in size. The muzzle becomes longer and the jaw bones wider. These alterations in the head are related to changes in dentition, to be considered next.

Other changes in the equid head are correlated with a trend in brain capacity. *Hyracotherium* probably had a small brain. Oligocene *Mesohippus*

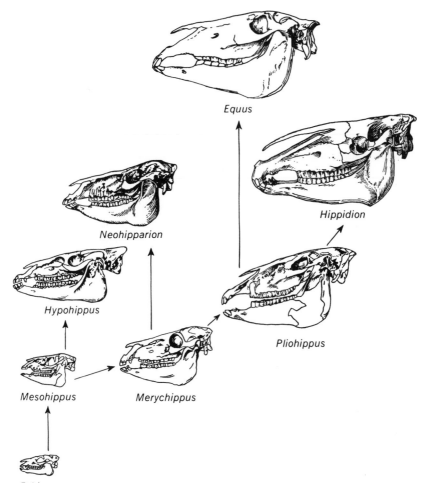

Equus

Neohipparion

Hippidion

Hypohippus

Pliohippus

Mesohippus   *Merychippus*

Eohippus

Figure 28.5. Evolutionary trends in the horse skull. Shown to same scale. (From G. G. Simpson, *Horses*, copyright 1951, © 1979, Oxford University Press, New York; reproduced by permission)

did have a rather small brain. In the subsequent history of the family the brain became larger and more complex, with a development of regions associated with intelligence in modern mammals. Progressive development of the equid brain and intelligence took place gradually throughout the Tertiary and Quaternary.

Trends in Horse Teeth

The evolutionary trends in horse teeth are very complex and cannot be fully described without recourse to a highly technical terminology. It is sufficient for our present purposes to present the main outlines of horse-tooth evolution with a minimum of terminology.

The teeth of *Equus* exhibit the following advanced features: (a) differentiation of chisel-like front teeth for biting and rear cheek teeth for grinding; (b) large-sized cheek teeth; (c) long (or high-crowned) cheek teeth; (d) an increased number of functionally grinding teeth (six on each side of each jaw); (e) highly developed ridges (crests) composed of enamel on the grinding surfaces; and (f) development of hard cement in pockets between the crests.

The corresponding primitive features in the teeth of *Hyracotherium* are: (a) less differentiation between front and rear teeth; (b) smaller teeth; (c) short (low-crowned) cheek teeth; (d) no grinding teeth (three molars on each side of each jaw, but these were not true grinders); (e) a simple pattern of low enamel crests on the grinding surfaces; and (f) absence of cement. See figures 28.5 and 28.6.

These dental differences reflect differences in diet. *Hyracotherium* was a forest-inhabiting browsing animal, whereas *Equus* is a grazing animal of grassy plains.

The various dental character differences between *Equus* and *Hyracotherium* were brought about by a series of trends, e.g., the grinding teeth became larger (trend b), higher-crowned (c), with more elaborate crest patterns (e), and so on. The increase in number of grinding teeth was achieved by the gradual transformation of three of the premolars of *Hyracotherium* into molar-like teeth in *Equus*, a process known as molarization.

The changes in the various tooth characters were not simultaneous but occurred at different times. Molarization (trend d) took place mainly during the Eocene and Oligocene and leveled off thereafter. Increase in height of crown (trend c) took place mainly in the Miocene, and, moreover, exclusively in the then new grazing groups such as *Merychippus*, while the Miocene browsing horses remained low-crowned.

The changes in tooth characters were gradual, with one type grading into the next. The intergradation can be seen in cases where a phyletic line is represented by fossil teeth in a series of geological horizons of closely spaced ages. Consider the lineage *Parahippus-Merychippus-Nannippus* of Miocene-Pliocene age in figure 28.6. Figure 28.7 portrays crown height and grinding

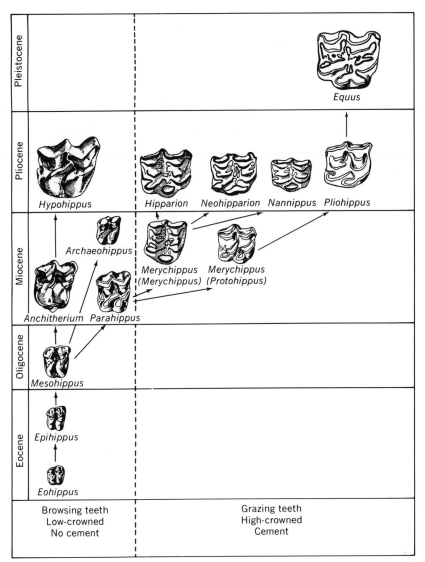

Figure 28.6. Evolutionary trends in molar teeth of horses. Crest patterns on grinding surface of upper molars are shown in drawings. Drawings to same scale. Other features are indicated in legends. (From G. G. Simpson, *Horses*, copyright 1951, © 1979, Oxford University Press, New York; reproduced by permission)

surface in a more complete series of stages in this same lineage. A gradual and nearly continuous increase in crown height is evident.

We can next amplify the Late Miocene segment of the series shown in figure 28.7. Downs (1961) has made a detailed quantitative study of the differences in teeth between two populations of *Merychippus* in the Mascall and Coalinga formations, respectively. The two populations are separated by 1–2 million years of time. The similarities and slight differences in their enamel crest patterns can be seen in figure 28.8.

Downs (1961) measured a number of tooth characters in population samples of adult individuals of *Merychippus* from the Mascall and Coalinga formations, and treated the data statistically. Some of his results are given in table 28.2. We note first that there is much individual variation within each population; and second, there are slight differences between populations in means and ranges. Most of the differences in means (including some not shown in table 28.2) are significant statistically. Nevertheless, there is a broad overlap between the two populations, as shown by the ranges and standard deviations in table 28.2. The statistical nature of the differences between the

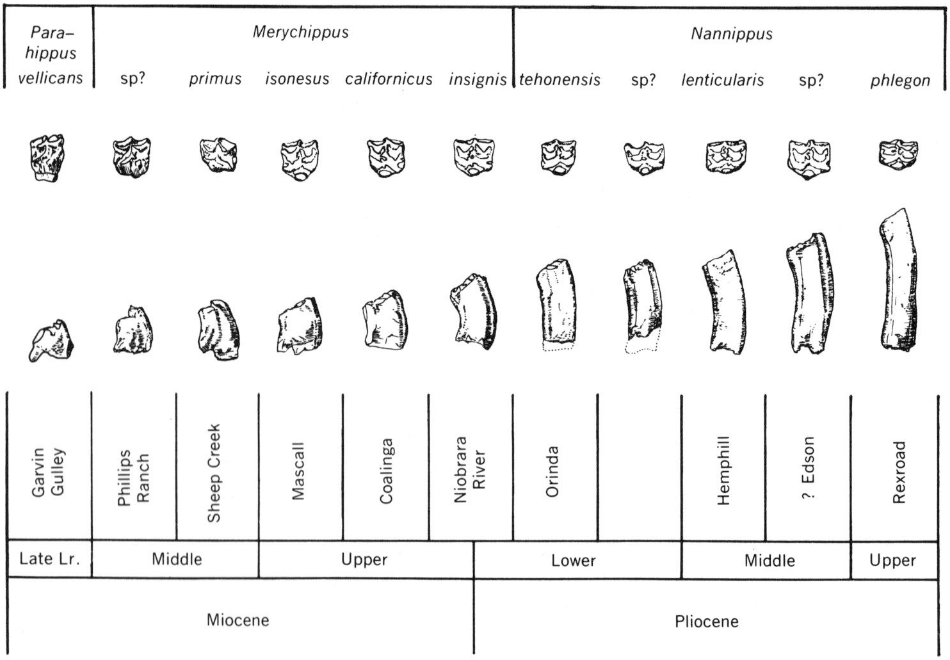

Figure 28.7. Evolutionary trend in height of crown of cheek teeth in a lineage of horses in the Miocene and Pliocene. (Redrawn from Stirton 1947)

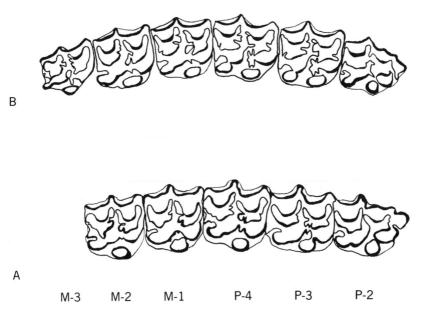

B

A

M-3   M-2   M-1   P-4   P-3   P-2

Molars   Premolars

Figure 28.8. Enamel crest pattern in upper cheek teeth of two Late Miocene species of *Merychippus*. (A) *M. seversus*, Mascall formation. (B) *M. californicus*, North Coalinga formation. (Redrawn from Downs 1961)

Mascall and Coalinga populations bridges the gap between macroevolutionary and microevolutionary changes.

Microevolutionary Aspects of Trends in Horse Teeth

Some interesting cases are known in which a tooth character appears as a polymorphic variant in an older fossil population, and becomes fixed as a constant character in a later population.

One such case concerns a particular crest, known as the crochet, on the upper cheek teeth in the *Mesohippus-Equus* line. This crest was usually lacking in *Mesohippus*. But some populations of *Mesohippus* contained a few mutant individuals with tiny crochet crests. Some derivative species of Miohippus in the Oligocene were relatively constant for the presence of a tiny crochet; other later derivative species of *Parahippus* in the early Miocene

Table 28.2. Variation in individual upper cheek teeth of adult horses (*Merychippus*) from two Late Miocene formations (Downs 1961)

Character and Formation	Range	Mean	Standard Deviation	Number of Teeth	Statistical Significance (p)
Height of crown of unworn premolars and molars, mm					
N. Coalinga (younger)	30.5–39.6	34.9	2.23	39	
Mascall (older)	24.6–32.3	29.4	2.45	18	<0.01
Height of crown of worn molars, mm					
N. Coalinga	23.4–33.9	27.9	2.53	57	
Mascall	18.1–27.5	22.9	2.37	16	<0.01
Total plications on molars, mm					
N. Coalinga	2–17	7.6	2.02	66	
Mascall	5–11	7.3	1.52	19	>0.05

possessed a small crochet as a constant feature. The crochet later increased to a prominent size in the line leading to *Equus* (Simpson 1944:59–60; 1953:105–106). Tooth cement makes its first appearance in a similar way in the *Parahippus-Merychippus* line. Cement is typically lacking in *Parahippus*, but occurs as a polymorphic variant in some populations of *Parahippus*, and is later fixed in monomorphic condition in *Merychippus* (Simpson 1953:106–107).

Here again a macroevolutionary trend is reduced to microevolutionary changes.

Selection could readily account for the observed rate of evolution in horse tooth characters. The observed rate of change is not great. It is estimated that only about 1–2 selective deaths per million individuals per generation would be necessary to bring about the known changes in tooth characters in the time period avaiable (Lande 1976).

Adaptive Nature of Evolutionary Trends in the Horse Family

The adaptive nature of the trends described above can be inferred with reasonable confidence from the circumstances of the animals. Habitat, diet, and defense were undoubtedly important factors in equid evolution. Indirect as well as direct effects of these and other factors must be taken into account.

Forests were very widespread in the early Tertiary, but climatic changes during the Miocene caused these extensive forests to break up and be partially replaced by open savannas and prairies. In the Pliocene the area of open grassland continued to spread at the expense of forest.

The Early Tertiary Equidae were forest-dwelling herbivores with a browsing mode of feeding, as noted earlier. In the Miocene, however, a branch line of the horse family tree moved into the open savanna and grassland habitats and adopted a grazing mode of feeding. This change in habitat and food habits took place in North America. It must have exposed the branch line to strong new selective pressures, which can account for many of the observed evolutionary trends.

Grass is a harsh, abrasive type of food material that wears out teeth. Most of the changes in tooth characters in Miocene and Pliocene horses are initial adaptations and improved adaptations for grazing. The front biting teeth, the long rows of cheek teeth, and the enamel crests and cement on the grinding surfaces of the cheek teeth are all functional for nipping off and chewing grass. The long crowns of the cheek teeth give these teeth, and their owners, a longer life span.

Some of the changes in the head also appear to be related to the adoption of a grazing habit. Elongation of the muzzle helps to accommodate differentiated sets of front nippers and rear grinders. Wide jaw bones are needed to house high-crowned cheek teeth. Strong wide jaws are required to grind large quantities of grass.

The teeth of *Equus* operate in concert with a specialized digestive system. This digestive system, with its cecum and rapid passage of food material, is adapted for processing large amounts of high-fiber grass, and for extracting sufficient food from sparse low-quality vegetation. This opens up steppe habitats for *Equus*. *Equus* can make a living in barren steppes which will not support most other types of ungulates (Janis 1976).

Grazing animals in open savannas and plains are more exposed visually to predators than are forest animals. Increase in body size and strength is one effective means of defense against predators in land mammals. Intelligence is another. And speed in running is still another. Trends in these characteristics in the horse family can probably be related in large measure to the requirements for defense in a plains animal.

Large body size solves some problems but creates other new ones. A large grazing horse needs bigger, harder, more durable teeth for its support than does a small grazing horse. Therefore, as Simpson (1951) points out, the trends in tooth characters are probably correlated adaptively with increase in body size as well as with change in diet.

Increase in body size also has side effects on running speed. The heavy body of a large horse is a handicap for fast running, as compared with a smaller horse, and calls for improved leg and foot mechanisms to overcome this handicap. *Hyracotherium* was built for fast running. Changes in the proportions of the legs, feet, and toes were necessary in the later, larger-bodied horses just to maintain the speed of the ancestral *Hyracotherium*, let alone to surpass it (Simpson 1951).

Adaptive Aspects of Cope's Rule

Many groups of animals exhibit a phylogenetic increase in body size. This common, but not universal, trend is known as Cope's rule, as noted earlier.[2]

2. For discussions see Cope (1896), Newell (1949), Rensch (1960b:206 ff.), and Stanley (1973).

Among mammals the size-increase trend is found in the marsupials, carnivores, perissodactyls, artiodactyls, primates, and other groups. Parallel trends occur in reptiles, arthropods, molluscs, and other major groups.

Large body size is selectively advantageous in several ways, in land animals at least. Four advantages will be listed; others could be added.

Body size is an important factor in predator-prey interactions. A predator must possess the size, weight, and strength to bring down its prey animal; the prey, conversely, must have the size and strength to ward off its predator. Some predator-prey combinations probably become locked into armament races that lead to ever-increasing body size in both parties. Horses and big carnivores may have interacted in this manner and with this effect.

Intelligence depends upon a large and complex brain, and the latter requires a large head, which in turn calls for a large body. Selection for intelligence may entail selection for increased body size as a correlated character.

Strength in combat is another correlated aspect of body size. In struggles between males for possession of females, such as occur in many species of ungulates, there is a selective premium on superior size and strength. Some size trends may be products, in part, of sexual selection.

Large animals have still another advantage over small ones in the retention of body heat. Large bodies are more efficient than small ones for heat retention because of a smaller and hence more favorable surface-to-volume ratio in the former.

The size-increase trend has never been universal. Some phyletic lines have increased in size, while other related lines decreased in size, during the same time span. And the same phyletic line may exhibit a size-increase trend during one time period and a reversed, size-decrease trend in another period. Both deviations from Cope's rule are found in the horse family, as well as in elephants, deer, and other groups.

This is a reflection of the relativity of the selective advantage of body size. Increase in body size is advantageous—under certain conditions. It is not necessarily advantageous, and may indeed be disadvantageous, under other conditions. Island animals generally become smaller than their mainland ancestors as a result, in part, of selection for ability to survive on a strictly limited food supply.[3]

3. See Carlquist (1965: ch. 7; 1974:24) for examples.

CHAPTER 29
Evolutionary Rates

Measurement of Rates

Evolutionary rate is the amount of evolutionary change per unit of time. The measurement of the rate is beset with numerous technical and practical difficulties.

One question involves the selection of time scale. Should we use chronological time, or biological time, that is, the number of generations? A consideration here is that if we want to compare rates in different groups we will have to use chronological time.

Another question concerns the type of evolutionary change to be considered and measured: whether it is phyletic, speciational, or a mixture of both. Practical necessity provides an answer to this question. The obtainable evidence pertains to phyletic evolution in some groups and to speciation and branching in others. We have to work with the evidence available. Comparisons of rates between groups will obviously be most meaningful, however, if they are made for one type of evolutionary change or the other.

The next problem is to decide what aspect of the changing lineage to measure. Three broad possibilities present themselves: genetic, biochemical, and exophenotypic. The amount of genetic change over time would provide the best measure of evolutionary rate, if it could be obtained, but this measure cannot be obtained for changes of the sort documented by paleontological evidence. Biochemical characters are close to the genetic determinants, and might be the next best choice, but they usually cannot be read in the fossil record either.

We are left then with exophenotypic character change as an aspect that can be measured in practice in fossil series. Two types of exophenotypic rates are recognized by Simpson (1953), morphological rates and taxonomic rates.

Morphological rates measure changes in a single character or character complex over time. For single-character changes, Haldane (1949) suggested a unit of measurement known as the darwin; 1 darwin = increase/decrease in size of 1/1000 per 1000 years. This unit has not been employed extensively.

One of the objectives of rate studies is to find a measure of the amount of exophenotypic change that will permit rate comparisons between different groups. Morphological rates are ordinarily group-specific. It is notoriously difficult to measure the amount of morphological similarity and difference between different groups which possess different characters. This is where taxonomic rates become useful. Taxonomic rates make use of the experienced judgment of taxonomists. The recognition of taxa *is* the recognition of differences in character complexes; and the assignment of taxonomic level is a recognition of the magnitude of those differences.

There are several types of taxonomic rates. One type which has proven very useful is the origination rate. This is the number of new taxa, usually genera or families, appearing per time period. One can express the origination rate for each period or epoch in the history of a group, or the average origination rate for its entire duration. Genus origination rates are used extensively by Simpson (1949, 1953, 1967) and will be used also in this chapter.[1]

It is possible to estimate the age of a living group without a fossil record if that group is both endemic and autochthonous in an area whose geological history is known. Thus a particular race or species may be endemic in a local area that became habitable at some recent date; for example, an isolated mountain range covered with ice until several thousand years ago. A terrestrial group of larger taxonomic rank may be confined to an isolated large island or island continent that the group could have reached only by migrating over land connections that existed at some known previous time, say, several million years ago. In such cases the age of the autochthonous and

1. More detailed reviews of evolutionary rates are given by Simpson (1953: chs. 1, 2, 10), Dobzhansky et al. (1977:327–336), and Stanley (1979: chs. 4, 5).

endemic taxon can be inferred. The age of the endemic group sets limits for estimates of evolutionary rate, but does not necessarily indicate the maximum rate actually attained, for the group could have diverged rapidly from the ancestral stock during the early period of colonization and undergone little or no change thereafter.

Variation in Rates

Group to group differences in evolutionary rates are well documented. This is illustrated by the average genus origination rates for three groups as listed below (Simpson 1953):

Ammonites	0.05 gen/10^6 yrs
Chalicotheriidae	0.13 "
Equidae	0.13 "
(*Hyracotherium-Equus*)	

It is illustrated again by the different slopes of the curves representing the genus origination rates of three orders of mammals in the Tertiary (Simpson 1967 (figure 29.1).

During the same time span some groups have evolved rapidly and others have remained static. Thus during the Tertiary there was a succession of eight genera in the equid line from *Hyracotherium* to *Pliohippus*, and other hoofed mammals show similar rates in the Tertiary. But during the same period, indeed over a much longer time, the opossum lineage from *Alphadon* to *Didelphis* underwent relatively little change, Cretaceous opossums being very similar to the modern ones (Simpson 1967).

Some classic examples of conservative groups which have undergone relatively little change from earlier geological periods to the present are the opossum *(Didelphis)*, the rhynchocephalian reptile *Sphenodon*, the crocodiles *(Crocodylus)*, the oyster *(Ostrea)*, the horseshoe crab *(Limulus)*, and the marine brachiopod *Lingula*. The time spans during which these and other groups have remained unchanged or nearly so are given in table 29.1. One does not have to be reminded of all the evolutionary waters that have flowed over the dam while these conservative groups have plodded on in their age-old ways.

Paleontological evidence also points to the conclusion that evolutionary

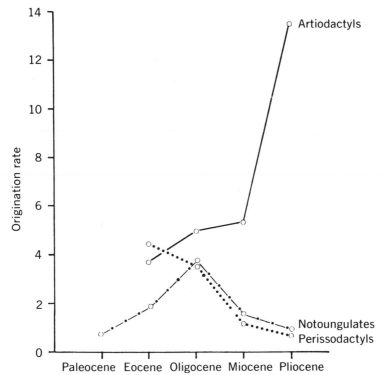

Figure 29.1. Variation in genus orgination rates in three orders of mammals through the Tertiary. Ordinate indicates number of new genera per million years. (Redrawn with modifications from G. G Simpson, *The Meaning of Evolution,* © 1967 Yale University Press, New Haven, Conn.; by permission)

rates have not remained constant in any one phyletic line throughout its whole known history. This is clearly indicated by Simpson's (1967) graph of genus origination rates in brachiopods during different periods of the Phanerozoic (figure 29.2), as well as by the graph of origination rates in mammals during the Tertiary (figure 29.1).

Simpson (1967:102) points out that living fossils such as the opossum and horseshoe crab, though conservative today, were progressive when they first appeared in the geological record. Prior to this first appearance, when the groups were evolving from their ancestral stocks in a limited time period, their evolutionary rate must have been relatively high. And the rate then leveled off to low or nil after the initial period of formation.

Table 29.1. Ages of some living fossils. Animal examples are from Simpson (1967, and pers. comm.), plant examples from Stewart (1983) and Carole Gee (pers. comm.)

Genus or Closely Related Pair of Genera	Little change between given period or epoch and Recent
Animals	
Alphadon-Didelphis	Late Cretaceous
Crocodylus	Late Cretaceous
Ostrea	Cretaceous
Homoeosaurus-Sphenodon	Jurassic
Limulitella-Limulus	Triassic
Lingula	Silurian (or Ordovician?)
Plants	
Sequoia	Cretaceous
Elatides-Cunninghamia	Mid Jurassic
Ginkgo	Jurassic
Ginkgoites-Ginkgo	Late Triassic
Araucaria	prob. Triassic
Equisitites-Equisetum	Mid Permian

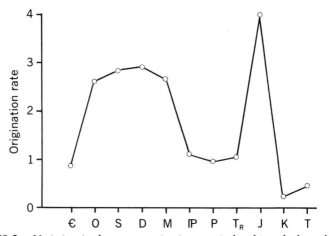

Figure 29.2. Variation in the genus orgination rate in brachiopods throughout the Phanerozoic. Ordinate indicates number of new genera per million years. (Redrawn from G. G. Simpson, *The Meaning of Evolution*, © 1967 Yale University Press, New Haven, Conn.; by permission)

Classification of Rates

A few groups of organisms, such as carnivores, pelecypods, and diatoms, are diverse enough and well enough known paleontologically to permit a meaningful plotting of the frequency distribution of their evolutionary rates. When the frequency distribution of rates is plotted in such cases, the histogram or curve turns out to deviate from a normal distribution by being leptokurtotic and strongly skewed. There is a high peak, and this peak lies near the fast end of the range of rates. The leptokurtotic and asymmetrical pattern of distribution of rates may well be general in many or most large groups (Simpson 1944, 1953:314–319).

The pattern of rate distribution provides a natural basis for classifying rates. Three categories of rates are recognized: normal rates (horotely), low rates (bradytely), and high rates (tachytely) (Simpson 1944, 1953).

The high peak in the distribution histogram or curve sets the norm for evolutionary rates in any given group. The leptokurtosis means that most members of a group have similar rates, rates that are characteristic of the group in question. These are the horotelic rates. The skewness of the distribution indicates further that the horotelic rates are relatively rapid.

A smaller number of members of the group have bradytelic rates, and a still smaller number, tachytelic rates. Living fossils would of course fall at the low end of the bradytelic range. Rapid changes at particular stages of horse evolution would exemplify tachytelic rates.

It should be noted that paleontological methods and evidence, by their nature, are best suited for dealing with bradytelic and horotelic rates. Direct paleontological evidence of tachytelic rates requires a very good fossil record, and is probably limited to the slower part of the tachytelic range. Evolutionary changes taking place not in millions but in thousands or hundreds of years usually leave no fossil record. Yet we have experimental and field evidence in plants of speciational shifts occurring in scores or hundreds of years. Tachytelic rates may sometimes be more rapid than is indicated by the paleontological evidence.

The Environmental Factor

A number of factors have been found to affect evolutionary rates, and additional factors have been suggested. A list of the known and possible factors would include: (a) stability of the environment, (b) population structure,

(c) position of the organism in the hierarchy of nutrition, (d) its position in the hierarchy of reproduction, (e) the supply of genetic variability, and (f) length of generations. The environmental and populational factors are of overwhelming importance and will be discussed in this and the next section. The other factors will be considered later.

If the environment is relatively stable, evolution slows down or stops when a population becomes well adapted to that environment. When the population reaches a high adaptive level in a stable environment, any new mutations or recombinations are likely to be selectively disadvantageous. Almost any change in the population is opposed by stabilizing selection (Simpson 1944).

A new variation may appear occasionally that has a positive selective value in some other available habitat with different ecological conditions. Then a branch phyletic line may arise. But the central group remains constant. This happened in the history of the opossum, which gave rise to four major branch lines of family rank in South America (i.e., the rat-opossums or Caenolestidae, etc.), but remained little changed itself (Simpson 1953:332).

Bradytelic or low-rate animals occur in and are broadly adapted to habitats that have had great stability through geological time. They are found in tropical and warm-temperate forests (opossum), major tropical rivers (crocodiles), the sea (*Ostrea, Lingula, Limulus*), and other similarly long-lasting habitats. Bradytelic groups are not found, by contrast, in impermanent lakes, high mountains, volcanic deposits, arctic tundra, etc.

A population in a changing environment has three alternatives: extinction in situ, migration, or evolution. Bradytely is impossible in the area of environmental change. If the population is to persist in the area undergoing environmental change, it must evolve. The rate of evolution now depends first on the rate of environmental change, and second on the ability of the population to keep pace with the external changes.

This ability of the population to sustain horotelic or tachytelic rates depends, in turn, on other factors to be considered next.

The Evolutionary Potential of Different Population Structures

Wright (1931, 1949, 1960) described the evolutionary characteristics of different types of populations. Three types that are relevant to the present discussion are: (1) small isolated populations, (2) large continuous outcrossing populations, and (3) a system of semi-isolated small populations (colonial or island-like population system).

1. Small isolated population.   Here most polymorphic genes are fixed quickly by drift or selection-drift. Consequently little genetic variation remains available in storage.

2. Large continuous outcrossing population.   Selection is relatively ineffective on recessive alleles at low frequencies, including most new mutations. New gene combinations do not persist but are swamped out by wide outcrossing. Furthermore, gene systems promoting homeostatic buffering properties tend to develop, and these homeostatic systems resist change. Selection-drift does not take place.

3. Colonial population system.   The individual colonies may be monomorphic or nearly so, but the population system as a whole has a decentralized store of genetic variation, distributed among the different colonies; and this variability can spread from colony to colony by occasional interdeme gene flow. Each colony can respond rapidly to local environmental conditions by an interplay between the existing variability and selection-drift. Favorable new gene combinations can be fixed in a colony, as well as single gene alleles, because of the protection from wide outcrossing afforded by semi-isolation.

The population factor interacts with the environmental factor. Let us recognize three modal types of environmental condition: (a) stable environment, (b) slowly changing environment, and (c) rapidly changing environment. We can now explore the important population-environment interaction by presenting the expected performance of each type of population (1–3) in each type of environment (a–c).

1. Small isolated population.   (a) In a stable environment it quickly reaches the best state of adaptedness with the variation on hand, and then becomes evolutionarily stagnant. (b, c) In a slowly or rapidly changing environment, the population is handicapped by lack of variability, and cannot make an adequate evolutionary response. It may migrate to a refuge, or it is likely to become extinct.

2. Large continuous outcrossing population.   (a) In a stable environment the large population maintains the existing adaptations but does not change (e.g., opossum). (b) In a slowly changing environment the population evolves slowly and steadily (e.g., horse). (c) But it cannot keep up with a rapidly changing environment. For the various reasons given in a preceding paragraph, the large continuous population is not able to respond suc-

cessfully to intense selection. It may break up into small relictual populations (e.g., sequoia) or suffer extinction.

3. Colonial population system.   (a) In a stable environment the colonial population system becomes differentiated into local races by selection or selection-drift. (b) In a slowly changing environment some local races are likely to be favored selectively, and will expand, while other local races contract. Some interdeme selection is postulated to occur (see chapter 15). (c) A rapidly changing environment evokes the same type of response. And other processes may also occur. Expansion of some colonies may bring them into closer proximity, promoting increased rates of interdeme gene flow, and perhaps interracial hybridization followed by the segregation of new types. Rapid evolutionary changes can take place within favored colonies as a result of the operation of selection-drift, and the possibility of evading the cost-of-selection restriction on evolutionary rate (see chapter 17).

In summary, there are limitations on the evolutionary potential of both very small and very large populations. As a result of these limitations—lack of variability in the one case and incapacity for selection-drift in the other—neither type of population possesses the necessary conditions for sustained rapid evolutionary rates. The colonial population system, however, does not have either one of the above limitations. It avoids the disadvantages found in small isolated and large continuous populations, and possesses certain advantages of its own. Conditions are favorable in the colonial population system for rapid evolutionary change (Wright 1931, 1949, 1960).

The Hierarchies of Nutrition and Reproduction

Schmalhausen (1949) observed a correlation between the position of an organism in the food pyramid and its evolutionary rate. Organisms standing at the bottom of the pyramid tend to have low rates, and those at the top, high rates.

Schmalhausen divides the food pyramid, or hierarchy of nutrition as he calls it, into four levels: (a) The lowest position is occupied by organisms whose only defense against predation is rapidity of reproduction (e.g., plankton, bacteria, green algae); (b) The next level consists of organisms with purely passive means of protection (e.g., molluscs, heavily armored animals, higher plants); (c) Above those in the hierarchy are animals that escape from aggressors by rapid locomotion; (d) The top position is occupied by animals that are themselves predators.

Schmalhausen points out that low evolutionary rates are characteristic in organisms belonging to levels a and b in the nutrition hierarchy; by contrast, rapid rates are often found in organisms occupying levels c and d.

The next problem is to identify the causal factors responsible for the observed correlation. Schmalhausen's (1949) view is that an organism's relation to other organisms, as reflected in its position in the nutrition hierarchy, determines the mode of selection that prevails, and this in turn determines the rate of evolution.

In the case of organisms that stand low in the food chain and are defenseless, elimination by predators is largely non-selective, and under these conditions evolution remains slow and does not follow progressive trends. Predatory animals at the top of the chain (level d), on the other hand, compete with one another for food. Success or failure in this competition depends to a considerable extent on the native ability of the individual animals. Elimination is highly selective, in other words, and permits rapid evolution of more efficient characteristics. Elimination is also highly selective in animals that depend on their own activity to escape from predators (level c); such animals also tend to develop adaptive specializations at rapid rates (Schmalhausen 1949).

The observed correlation between nutritional level and evolutionary rate is open to alternative interpretations. The factor of population size and structure is an uncontrolled variable in Schmalhausen's nutrition-evolution correlation. Organisms at levels a and b frequently exist in very large populations, which are capable of only slow responses to selection, while animals belonging to level d often have colonial populations, which can be changed rapidly by either selection or selection-and-drift. This raises the question as to whether the causal connection is between nutritional level and evolutionary rate, or between population structure and evolutionary rate, or is a mixture of both.

Higher plants all occupy the same low level (level b) in the hierarchy of nutrition, but differ markedly in method of reproduction, and Stebbins (1949) has arranged these differences in a hierarchy of reproduction parallel to Schmalhausen's nutritional hierarchy: (a) the lowest level contains plants that rely on sheer numbers and passive transport for dispersal of their spores, pollen grains, or seeds (e.g., wind-pollinated conifers, Amentiferae, grasses); (b) the middle level contains plants with spores, pollen grains, seeds, or fruits protected by tough and resistant coats (e.g., hard-seeded pines, oaks, legumes); (c) the highest position is occupied by plants that exploit animals for pollination, seed dispersal, or both (e.g., animal-pollinated legumes, orchids).

Evolutionary rates in organs involved in pollination or dissemination are correlated in a general way with the foregoing levels (Stebbins 1949). Thus the wind-pollinated conifers, Amentiferae, and grasses (level a) show little evolutionary diversification in their pollen-producing organs; but the animal-pollinated legumes and orchids (level c) have evolved advanced specializations, apparently at rapid rates, in their flowers. Parallel differences are found between the seeds, cones, or fruits of wind-dispersed plants (level a) and animal-disseminated plants (level c).

Variability

The possibility has been considered that the mutation rate or amount of polymorphic variation might be a limiting factor on evolutionary rates in some cases. The available evidence does not support this idea for species with normal large populations. Morphological evidence points to a normal or high degree of genetic variation in populations of the bradytelic opossum (Simpson 1944). Electrophoretic evidence likewise points to ample stores of genetic variability in such bradytelic groups as *Limulus* and the fern ally *Lycopodium* (Selander et al. 1970; Levin and Crepet 1973).

Natural interspecific hybridization is a source of recombinational variation in many groups of plants. Furthermore, the hybrid derivatives often become established in new habitats created by environmental disturbances.[2] A plant group possessing decentralized stores of potential variation—decentralized in two or more species capable of hybridization—probably has an advantage over a strictly monotypic group in responding successfully to rapid environmental change.

Length of Generation

Another characteristic that has been considered to be a limiting factor on evolutionary rate is length of generations. Since environmental change takes place on a chronological time scale, whereas natural selection operates on

2. See Grant (1981a).

generation time, it is logical to expect that organisms with short life cycles could achieve more rapid evolutionary rates than long-cycle organisms. But actually there is little paleontological evidence to support this logical deduction.

Among mammals, short-cycle opossums are bradytelic, whereas slow-breeding elephants have had rapid evolutionary rates. Short-cycle rodents and long-cycle ungulates have evolved at approximately the same rate in South America since the Pliocene and Pleistocene. Long-cycle carnivores have evolved much faster than short-cycle pelecypods or bivalve molluscs (Simpson 1944:20, 25, 63).

The evidence in plants, though mixed, is generally in agreement with that in animals. For example, in the California coastal and foothill regions, annual herbaceous plants have evolved rapidly, but so also have long-lived woody plants such as *Quercus*, *Ceanothus*, and *Arctostaphylos* in the same areas (Stebbins 1949, 1950). However, the woody plants in the California chaparral can be subdivided further into two categories on the basis of reproductive method and generation time. The woody chaparral plants that reproduce sexually by seeds have only moderately long generations, whereas those that reproduce vegetatively by crown sprouts have exceedingly long generations; the former category, as exemplified by *Ceanothus* and *Arctostaphylos*, shows a much higher rate of speciation than the latter (Wells 1969).

There probably are some rapid evolutionary rates—rates measured in historical time rather than geological time—that are possible only in short-cycle organisms. Short-cycle and long-cycle organisms may be equally able to respond evolutionarily to some rates of environmental change. But let the rate of environmental change become ever more rapid. Sooner or later a threshold is reached that separates the long-cycle organism from the short-cycle organism on ability to generate the evolutionary rate needed to keep pace with the environmental change.

Length of generation is probably a factor affecting evolutionary rates, therefore, but it is a factor that comes into play only in the fast end of the rate spectrum, where it is not detected by paleontological methods.

CHAPTER 30
Evolution of Major Groups

Origin and Development
Saltations vs. Gradual Divergence
The Role of Quantum Evolution
Adaptive Radiation
Adaptive Radiation in the Hawaiian Honeycreepers
The Class Aves

Attention is focused in this chapter on the evolutionary patterns involved in the development of groups of medium or high taxonomic rank. It is useful to recognize two phases in the evolution of such groups: an early formative phase and a subsequent period of expansion and proliferation. This breakdown of the developmental process is useful because different evolutionary patterns appear to predominate in the different phases. Accordingly, in our treatment of the problem here, we will maintain the same distinction between the origin of a major group and its development.

Origin and Development

In order to discuss the origin of major groups it is necessary to define the essential features of a major group. One could define a major group as a phylogenetically natural group of substantial taxonomic rank. This, however, focuses attention on the wrong end of the problem for our present purposes. Assignment of taxonomic rank is a retrospective decision reached after a group has evolved up to the present time, the era of human taxonomists. We are interested in the beginning stages as well as the eventual flowering of the group.

One could also define a major group in terms of its key character(s). This approach is also unfruitful. A major group may or may not possess a key character when it is originating.

A more useful criterion of a major group is a state of adaptedness to a particular broad zone on the adaptive landscape. The *origin* of the group coincides with the occupation of a new adaptive zone (Wright 1949; Simpson 1953:349). The group *becomes* a major group later by development and proliferation within this zone.

The successful initial occupation of the adaptive zone can come about in two main ways.

1. A colonizer species may move into a more or less uninhabited territory. The new territory presents no particularly novel environmental conditions other than relative emptiness at the time of colonization. Nor does the original colonizer species necessarily differ much from its ancestral stock. But in the new zone it finds an ecological opportunity for expansion and diversification.

2. Occasionally a species may acquire a character that opens up a new way of life and preadapts it for a previously unexploited habitat. Thus, in the evolution of vertebrates, the development of lungs and legs opened up the land for colonization, and the development of wings opened up the air. The preadapted colonizer species, in possession of its new pivotal character, invades a new habitat and undergoes further evolution there (Huxley 1943; Wright 1949).

In some cases the new character of pivotal importance had evolved earlier in relation to ancestral environmental conditions, but turns out to be preadapted to and useful in the new type of habitat. Thus the lungs and paired fins of crossopterygian fishes were the forerunners of the air-breathing lungs and crawling legs of primitive land vertebrates.

In both modes 1 and 2, the event of colonization of the new adaptive zone is usually followed by an adaptive radiation in that zone. The original colonizer species usually gives rise to a diverse array of descendant lines occupying different subzones. It is this adaptive radiation that makes the group a recognizable major group. But we must bear in mind that the adaptive radiation represents the successful exploitation, at a later stage, of the ecological opportunity seized by the original colonizer species.

Mode of origin 1, above, is simpler than mode 2; conversely, mode 2 probably accounts for more important evolutionary developments than does mode 1. Mode 1 is a source of endemic groups of medium taxonomic rank, as exemplified by the Hawaiian honeycreepers (Drepanididae) discussed later in this chapter. Whether higher categories of continental or cosmopolitan distribution originate by mode 1 is open to question. Mode 2 seems to be the main source of truly major and widespread groups, such as the amphibians and the birds.

Saltations vs. Gradual Divergence

A new group usually leaves no paleontological record of its period of origination. The absence of paleontological evidence leaves the field open for speculation. Some evolutionists (Goldschmidt 1940; Gould 1977) have proposed that new groups arise by saltations. Drastic and sudden mutational changes take place in a pre-existing organism to produce the ancestral form of the new group.

Saltationism in this sense is highly unlikely and inconsistent with the evidence of genetics. The genotype is composed of numerous genes that must interact in a harmonious fashion if the organism is to have normal viability and fertility. Drastic mutations tend to upset the internal genic balance in ways that are selectively disadvantageous. A genotype that has been built up by selection in stages cannot be remodeled successfully overnight; it has to be remodeled in stages.

Let us note parenthetically that we are not ruling out the possibility of a reorganization of the genotype in a relatively few generations or the establishment of macromutations. In fact there is experimental evidence for both processes in plants (see chapters 6 and 24). Such rapid changes have been referred to as "saltations" by some authors, although this application of the term may be inappropriate since the changes are stepwise rather than abrupt. In any case, rapid stepwise changes are not what Goldschmidt and his followers postulate.

Although the paleontological evidence leaves the question of the origination of new groups open in most cases, it is not always silent on this issue. In some cases where the fossil record is very complete, gradual divergence, and not saltationism, is clearly indicated.

Recent carnivores (Fissipeda) and Recent ungulates (Perissodactyla and Artiodactyla) are well differentiated major groups. These groups, in the Recent fauna, are differentiated with respect to an extensive series of characters. One thinks of the generally carnivorous diet, clawed feet, and large canine and carnassial teeth of carnivores, in contrast to the herbivorous diet, hoofed feet, and grinding molar teeth of ungulates. These and other character differences are listed in table 30.1.

Table 30.1. Character differences between Recent carnivores and ungulates (Simpson 1953)

Character	Carnivores (Fissipeda)	Ungulates (Perissodactyla and Artiodactyla)
Habitat	Terrestrial, arboreal, or semi-aquatic	Terrestrial
Mode of walking	Digitigrade or plantigrade	Usually unguligrade, sometimes digitigrade
Number of digits	4–5	1–4, rarely 5
Toenails	Well-developed claws, usually hooded	Well-developed hoofs (except hyraxes)
Diet	Generally carnivorous, sometimes omnivorous, rarely herbivorous or insectivorous	Generally herbivorous, sometimes omnivorous
Dental formula	$\dfrac{3.1.4-1.3.1}{3.1.4-1.3.1}$ (Molars exceptionally $\frac{4}{5}$)	$\dfrac{3-0.1-0.4-2.3}{3-0,1-0.4-2.3}$
Canine teeth	Large	Usually small or absent, sometimes large
Premolars	Usually simple, not molariform	Generally complex, often molariform
Molars	Usually simple, without grinding pattern	Complex, with grinding pattern
Carnassial teeth	P^4 and M_1 teeth enlarged and carnassial	No carnassials
Crown height	Brachydont	Brachydont to hypsodont

Nevertheless, they have a common ancestry in the insectivores of the Cretaceous. Their respective ancestors of Early Paleocene time had diverged to the level of closely related and poorly demarcated orders, the archaic meat-eating placentals (Creodonta) and the primitive hoofed mammals (Condylarthra). The creodonts and condylarths of the Early Tertiary were generally similar, and some genera have been classified first in one order and then in the other. The Paleocene-Eocene condylarth, *Phenacodus* (Phenacodontidae), about the size of a sheep, looked like a primitive carnivore, with its large canine teeth and full set of toes (Simpson 1953; Romer 1966; Colbert 1980).

The carnivore character combination and the carnivore order thus developed in the historical course of divergence from its Paleocene ancestors; likewise, the ungulate character combination and the two existing ungulate orders developed by divergence from their Paleocene ancestors. This divergence was gradual.

Likewise gradual was the origin of the ungulate order Artiodactyla out of its ancestral condylarth stock. A character in the tarsus of the foot, giving the animal powerful and speedy locomotion, is diagnostic of the Artiodactyla. The transition from the primitive condylarth foot to the artiodactyl foot took place gradually, with no abrupt phase, within a 15 million year time span (Schaeffer 1948).

The Role of Quantum Evolution

The period of origin of a major group usually represents a relatively small part of the total duration of that group. It was during this relatively short period, moreover, that the distinctive features of the group were being developed. Rapid or tachytelic rates are thus required by the time element for the early history of many or most major groups (Simpson 1944, 1953).

It should be noted that rapid evolutionary rates, as indicated here, are not inconsistent with gradual stepwise divergence, as concluded in the preceding section.

The later paleontological history of major groups typically consists of a long recorded sequence during which horotelic or bradytelic rates and large population size prevailed. Such rates are expected in large continuous populations, as we have seen in chapter 29.

Population systems containing small populations are the likely sites for the inferred tachytelic rates during the origination of major groups. The small population size, besides agreeing with population-genetical theory (see chapter 29), would account for the poor fossil record of major groups in their origination phase (Simpson 1944, 1953).

Simpson (1944, 1953) proposed the term quantum evolution for the mode just described, namely, for the rapid change of an organism from an ancestral to a new adaptive character state in a small population system. Quantum evolution is thought to be the normal mode of evolution in the origin of major groups. We will return to this mode in chapter 31.

Fossil evidence of quantum evolution is expected to be scanty, by the nature of the case, and it is. Two examples are the adaptive shifts from a pad-footed to a springing foot mechanism and from browsing to grazing food habits in the Equidae (Simpson 1951, 1953). Other examples include the origin of the subfamily Stylinodontinae in the Taeniodontia, an extinct edentate-like

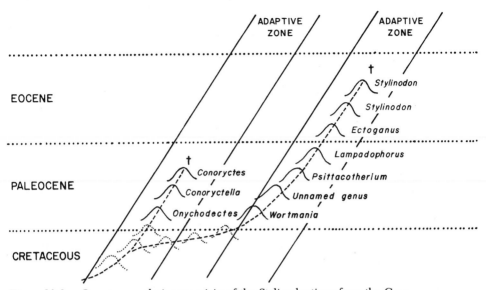

Figure 30.1. Quantum evolutionary origin of the Stylinodontinae from the Conoryctinae (Taeniodontia, Stylinodontidae). Dashed line indicates course of phylogeny. (Patterson 1949)

order of placental mammals, and the origin of lungfishes (Dipnoi) (Patterson 1949; Westoll 1949; Wagner 1980) (figure 30.1). The origin of the temperate North American tribes of the plant family Polemoniaceae from their neotropical ancestors can also be interpreted as a quantum shift (Grant 1959).

Adaptive Radiation

When a species succeeds in establishing itself in a new territory or new habitat, it gains an ecological opportunity for expansion and diversification. The original species may respond to this opportunity by giving rise to an array of daughter species adapted to different major niches within the territory or habitat. These daughter species become the ancestors of a series of branch lines when they, in turn, produce new daughter species. The group enters its second phase of development, the phase of proliferation.

Adaptive radiation is the pattern of evolution in this phase of proliferation. And speciation is the dominant mode of evolution in adaptive radiation. The first generation of speciational events gives rise to the primary branches in the now growing phylogenetic tree. These primary branches correspond to different primary ecological niches. Second, third, and later generations of speciation in each primary branch then parcel out the available ecological niches in the territory or habitat into a larger number of more highly specialized species.

The new territory or habitat fills up with species, and as it does, the group becomes a major group. And the major group comes to have a taxonomic structure organized along adaptive lines.

At one stage of development the group has the taxonomic rank of a family. Its primary subdivisions are then tribes and genera with different primary adaptive modes; each generic line consists of species groups, representing secondary subdivisions of the adaptive zone; and each species group consists of species representing tertiary subdivisions of the adaptive zone. At a more advanced stage of development the group may be an order. Now the primary divisions, still along adaptive lines, are families, and the second-, third-, and fourth-order subdivisions appear as genera, species groups, and species, respectively.

The pattern of adaptive radiation is efficient for a major group, since it permits the maximum exploitation, by that group, of the available ecological niches in its adaptive zone.

Adaptive Radiation in the Hawaiian Honeycreepers

A good example of adaptive radiation is furnished by the bird family Drepanididae in the Hawaiian Islands. This family has been studied by a series of workers over the years. The monographs of Amadon (1950) and Baldwin (1953) are the basic works; supplementary information and interpretations are given by Bock (1970) and Carlquist (1974).

The family consists of nine genera grouped into two subfamilies: the Psittirostrinae, with yellow or green plumage; and the Drepaniinae, with black and red plumage. The Psittirostrinae contains the following genera with number of species indicated parenthetically: *Loxops* (5), *Hemignathus* (4), *Pseudonestor* (1), and *Psittirostra* (6). The Drepaniinae contains: *Himatione* (1), *Palmeria* (1), *Ciridops* (1), *Vestiaria* (1), and *Drepanis* (2) (Amadon, 1950). Unfortunately some of these species have become extinct or very rare in modern times.

The Drepanididae is monophyletic and endemic to the Hawaiian Islands. It belongs to the suborder Passeres of the Passeriformes, and is related to the true honeycreepers (Coerebidae) and/or the tanagers (Thraupidae) (Amadon

1950; Baldwin 1953; Bock 1970). Since the family is endemic to a geologically young oceanic island chain, and has affinities to mainland passerine groups, its phylogenetic roots must lie in the latter, but the identity of the ultimate mainland ancestor remains uncertain. The ultimate ancestor, whatever it was, colonized the Hawaiian Islands and developed into the immediate ancestor of the family.

This immediate ancestor is inferred to have been a short-billed bird with a diet consisting of insects and nectar (Baldwin 1953). This diet, bill type, and the corresponding type of tongue are found in the more primitive living members of both subfamilies. A great diversity of feeding habits and bill types has evolved in the Hawaiian drepanids. This diversity is much greater than is normal for a passerine bird family. There are warbler-like, creeper-like, finch-like, fruit-eating, flower-feeding, and other types within the limits of a single family.

The probable phylogeny of the family down to the genus level is shown in figure 30.2 (from Amadon 1950; see also Bock 1970 for additional details). The two subfamilies, Psittirostrinae and Drepaniinae, represent the primary branches in the family tree, and the genera, the secondary branches. Third-order branchings produced the species in the non-monotypic genera such as *Loxops* and *Hemignathus*.

The branchings constitute an adaptive radiation. The adaptive radiation in the Hawaiian honeycreepers is related mainly to feeding habits and is manifested in bill and tongue characters as well as in other features. The main adaptive zones that have been occupied are insect-eating, nectar-eating, fruit-eating, and seed-eating. Some of these zones were further subdivided. Thus, among insect-eating drepanids there are short-billed foliage gleaners and long-billed bark probers (Baldwin 1953).

The subfamily Drepaniinae consists mainly of flower-feeding nectar eaters. Some are short-billed nectar eaters, e.g., *Himatione* (figure 30.3A); others are long-billed nectar eaters, e.g., *Vestiaria* and *Drepanis* (figure 30.3B). *Ciridops* in this subfamily is a fruit eater (figure 30.3C). Some of the bill types in the flower-feeding drepanids are correlated with various flower forms in the Hawaiian lobelioids (Campanulaceae, subfamily Lobelioideae).

The subfamily Psittirostrinae consists of insect, fruit, and seed eaters. The genus *Loxops* contains five species of insectivorous birds with small or medium-sized bills. The small-billed species glean insects from leaves (e.g., *L. coccinea*, figure 30.3D), while the species with medium bills probe for insects in bark crevices. *Hemignathus lucidus* (figure 30.3E) is a sickle-billed insect eater that obtains insects from wood more or less in the fashion of a

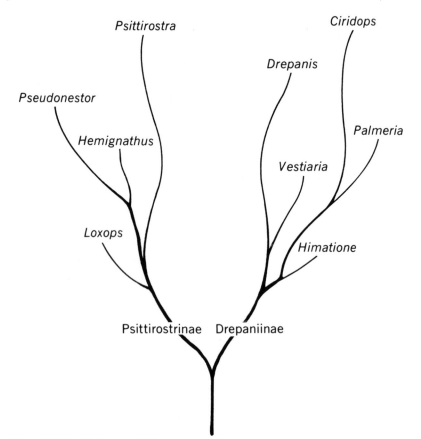

Figure 30.2.   Phylogeny of the Hawaiian honeycreepers (Drepanididae). (Amadon 1950)

creeper. Other sickle-billed species of *Hemignathus* have been reported to feed on flowers.

Pseudonestor in the same subfamily Psittirostrinae has a parrot-like bill that is used for crushing twigs and prying loose bark to obtain wood-inhabiting insects and grubs. The related genus *Psittirostra* (figure 30.3F) also possesses powerful bills, used in this case for crushing fruits and cracking seeds.

The Class Aves

The two phases in the evolution of a major group—the occupation of a new adaptive zone and subsequent radiations in that zone—are well illus-

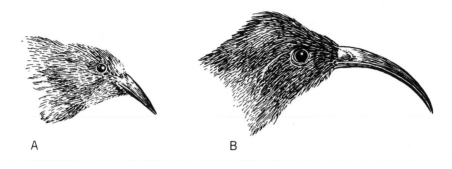

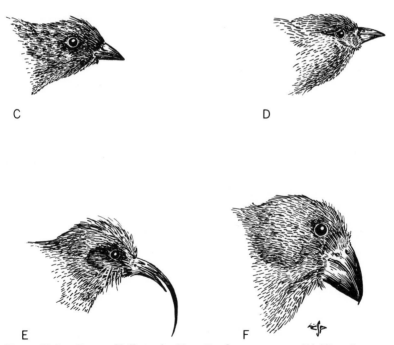

Figure 30.3. Types of bills in the Hawaiian honeycreepers. (A) *Himatione sanguinea* (nectar feeder). (B) *Drepanis pacifica* (nectar feeder). (C) Ciridops anna (fruit eater). (D) *Loxops coccinea* (insect eater). (E) *Hemignathus lucidus* (insect eater). (F) *Psittirostra palmeri* (fruit and seed eater). (Drawn from photographs in Amadon 1950; drawings by Charles Papp)

trated by the class Aves. The pivotal characteristic of birds is controlled winged flight, which gave them access to the air above land and open water. Birds are the most successful of flying vertebrates.

The ancestry of birds lies in or near the dinosaurs. There are two serious candidates for the ancestral stock. It may have been a pseudosuchian reptile (order Thecodonta) of Triassic age, ancestral to the dinosaurs; or, more likely, a coelurosaurian dinosaur (order Saurischia) of late Triassic or Jurassic age (Feduccia, 1980). In either case the ancestral form was small, carnivorous, and bipedal or partially so (see Colbert 1980).

The oldest known bird is the famous *Archaeopteryx* of Jurassic age. It is known from five specimens from a limestone formation in Bavaria. *Archaeopteryx* was bipedal, small (about the size of a crow or roadrunner), and possessed feathers, wings, and a bill. It also exhibited reptilian features, such as teeth, a long bony tail, and a dinosaur-like pelvis.[1] Its feathers were modified reptilian scales. Because of its mosaic of reptilian and avian characters, it is placed in a special subclass of Aves, the subclass Archaeornithes.

It is generally agreed that *Archaeopteryx* was only an indifferent flyer by modern standards. It was carnivorous or insectivorous, and is thought to have been homoiothermic.

Archaeopteryx is a natural focal point for theorizing about the origin of flight in birds. Two classical theories, dating back to the last century, are the arboreal and cursorial theories. According to the arboreal theory, the development of bird flight begins with an original adaptation for arboreal life, followed by subsequent adaptations for gliding downward. Gliding then changed to flapping flight. According to the cursorial theory, the sequence begins with bipedalism and running, and continues with leaping up from the ground. Flapping flight was gradually perfected during such leaps.

Among recent authors, Feduccia (1980) presents arguments in favor of the arboreal theory, and Ostrom (1974) argues for a modified version of the cursorial theory in which the forelimbs were used as insect nets.

The problem has recently been examined from the standpoint of aerodynamics by Caple et al. (1983). These authors point out that bird flight involves several components: lift, thrust, directional control, and braking action. Gliders have no thrust and only limited body control. Therefore the arboreal theory is unlikely. The use of forelimbs as insect nets is not a precursor of controlled flight, and therefore Ostrom's (1974) hypothesis also seems

---

1. For reviews of the characters of *Archaeopteryx* see Dorst (1974), Ostrom (1976), Colbert (1980), and Feduccia (1980).

unlikely. A running and jumping habit in a bipedal insectivore, in which the forelimbs are used for balance, would, however, set the stage for gradual improvements in lift, thrust, and body orientation. Aerodynamic considerations favor the cursorial theory (Caple et al. 1983).

The original function of feathers in *Archaeopteryx* is also debated. Ostrom (1974) contends that feathers originally functioned to prevent heat loss from a homoiothermic body, and later became modified for flight. But Feduccia (1980) points out that feathers are aerodynamically efficient for flight and seem to be designed primarily for this function. In flightless birds, the feathers often lose their typical form and become hair-like.

We have, then, a bipedal, cursorial, predatory vertebrate with leaping habits. Bipedalism released the forelimbs for development as flapping wings. Feathers and various skeletal modifications made possible a high level of flight capability. Homoiothermy provided a metabolic level to sustain active flying movements.

Birds became more modern in their skeletal features and increased in diversity in the Cretaceous. They apparently achieved dominance in their adaptive zone, the air, in the Cretaceous period. At any rate the other group

Table 30.2. The order of Recent birds (Dorst 1974)

Sphenisciformes; penguins	Lariformes; gulls, skuas, auks
Struthioniformes; ostriches	Columbiformes; pigeons, doves, dodo,
Rheiformes; rheas	Rodriguez solitaire
Casuariiformes; emus, cassowaries	Psittaciformes; parrots
* Aepyornithiformes; elephant-birds	Cuculiformes; cuckoos, turacos
* Dinornithiformes; moas	Strigiformes; owls
Apterygiformes; kiwis	Caprimulgiformes; nightjars, oilbird
Tinamiformes; tinamous	Apodiformes; swifts, hummingbirds
Gaviiformes; divers	Coliiformes; mousebirds
Podicipitiformes; grebes	Trogoniformes; trogons
Procellariiformes; albatrosses, petrels	Coraciadiformes; kingfishers, motmots, bee-
Pelecaniformes; pelicans, cormorants,	eaters, hornbills
gannets, frigatebirds	Piciformes; woodpeckers, toucans, barbets
Ciconiiformes; herons, storks, ibises	Passeriformes:
Phoenicopteriformes; flamingos	Eurylaimi; broadbills
Anseriformes; ducks, geese	Tyranni; tyrant flycatchers, ovenbirds, ant-
Falconiformes; hawks, falcons, vultures,	birds
condors	Menurae; lyrebirds
Galliformes; curassows, grouse, pheasants,	Passeres (Oscines); swallows, crows, tit-
quail, turkeys	mice, wrens, warblers, vireos, sunbirds,
Gruiformes; cranes, rails, bustards, kagu,	Hawaiian honeycreepers, orioles, tana-
takahe	gers, finches, etc.
Charadriiformes; plovers, woodcocks, avocets,	
phalarops, jacanas	

* Became extinct in human times in the Holocene epoch.

Table 30.3. Examples of flightless birds.
See table 30.2 for ordinal relationships.
(Based on Feduccia 1980)

Terrestrial habitats of continental areas
    Ostrich. Africa
    Rhea. South America
    Emu. Australia
    Cassowary. Australia–New Guinea

Terrestrial habitats on islands
* Dodo. Mauritius
* Solitaire. Rodriguez
* Elephant-bird. Madagascar
    Rails. Many island species
    Kagu. New Caledonia
* Flightless ibis. Hawaii
* Flightless goose. Hawaii
* Moa. New Zealand
    Kiwi. New Zealand
    Owl-parrot. New Zealand.
    Takahe. New Zealand.

Marine habitats
    Penguins. Southern hemisphere
* Great auk. North Atlantic
    Flightless cormorant. Galapagos

Freshwater aquatic habitat
    Flightless grebe. Lake Titicaca, Peru

* Became extinct during human times.

of flying vertebrates, the pterosaurs, became extinct by the end of the Cretaceous, perhaps as a result in part of competition from birds (Colbert 1980).

Birds underwent various radiations in the Tertiary, on their modern structural groundplan, to give rise to the great diversity found in that period and in the Quaternary (see figure 27.1). The best way to portray this diversity in a brief form is to list the 31 orders of Recent birds (in table 30.2).

A successful dominant group often produces lineages which leave an ancestral adaptive zone and occupy other zones to which they were not originally adapted. Such evolutionary reversals are represented in birds by flightlessness, which has evolved in many different groups (Dorst 1974; Feduccia 1980). Examples are listed in table 30.3. The flightless birds listed there belong to 14 orders (Sphenisciformes, Struthioniformes, Rheiformes, Casuariiformes, Aepyornithiformes, Dinornithiformes, Apterygiformes, Podicipitiformes, Pelecaniformes, Ciconiiformes, Anseriformes, Lariformes,

Columbiformes, Psittaciformes). It is a measure of a group's evolutionary success if it can maintain dominance in its own broad adaptive zone, and go on to invade other zones, such as that of large terrestrial herbivores (rheas) and that of seals (penguins).

CHAPTER 31
Modes and Concepts in Macroevolution

Approaches and Guidelines

Macroevolution involves changes of far greater magnitude, both phenotypically and genotypically, than those found in microevolution. Changes of macroevolutionary extent occur in the development of the characteristics that distinguish major groups such as genera, families, and orders. Such developments take place on a scale of geological time.

The methods of research in macroevolutionary studies are necessarily different from those employed in studies of macroevolution. Genetics, ecology, and minor systematics are the key approaches in the latter field, whereas our knowledge of macroevolution has been gained by paleontology and comparative morphology. Work in population genetics and related fields has yielded the direct evidence concerning the evolutionary forces. But we owe our knowledge of the broad sweep of organic evolution to work in paleontology and comparative morphology.

Different groups of research workers are necessarily engaged in the macro and non-macro areas of evolutionary biology, and they often differ in their perspectives on experimentalism, the time scale, and other matters. Many older paleontologists and morphologists failed to appreciate the significance of genetic findings for evolutionary theory; conversely, neontological population biologists are apt to have only a vague feeling for the time element involved in macroevolution. Many controversies in evolutionary biology can be traced to characteristic differences between macroevolutionists and microevolutionists in research methods, habits of thinking, and background knowledge.

A question in evolutionary biology is whether macroevolution is an extension of microevolution on a grander scale, or whether there is some essential difference between the two. Most evolutionists, including all advocates of the synthetic theory, believe that our interpretations of macroevolutionary phenomena must be based on our understanding of processes at the microevolutionary level. This position has been well stated by Dobzhansky (1937a:12):

> Experience seems to show . . . that there is no way toward an understanding of the mechanisms of macro-evolutionary changes, which require time on a geological scale, other than through a full comprehension of the micro-evolutionary processes observable within the span of a human lifetime and often controlled by man's will.

The above position has, however, been opposed by a minority faction of evolutionists in almost every generation. Goldschmidt (1940, 1955) claimed that macroevolutionary change is determined by special processes—systemic mutations and saltations—not found operating at the microevolutionary level. In recent years a basic distinction between microevolution and macroevolution has been maintained by Stanley (1975, 1979) and Gould (1980, 1982). Gould argues that the attempt to base macroevolution on microevolution is reductionism and extrapolationism, and that this is an error.

There is, contrary to Goldschmidt (1940), no gap in nature between microevolution and macroevolution. Speciation phenomena fill that alleged gap. Evolutionary divergences, in a broad sample of phyletic groups, appear at every level from race through species and species group to genus and the higher categories. Organic evolution is a unified whole.

The great difference between the time scales of micro- and macroevolution in their extreme forms does make it necessary to guard against oversimplified extrapolations from the one level to the other. Many microevolutionary changes are repeatable and predictable. Macroevolution, on the other hand, is a historical process.

When we consider the long time available for macroevolutionary changes, we also have to consider the possible occurrence of unique events that have far-reaching consequences. A primitive organism may acquire, by infection, a permanent symbiont that enables it to carry out photosynthesis. A terrestrial organism may colonize a distant vacant island only once and then embark upon a new course of evolution in its adopted territory.

The known evolutionary forces and factors may have operated in different ways at different times in the history of a group. Perhaps the mutation rate was different in the past as compared with the present. The environment might have been changing rapidly in the past, but is stable today, or vice versa. A group might have consisted of numerous sympatric species, capable of limited hybridization, at one stage in its history, but is monotypic with greatly reduced variability at present.

Finally, what aspect of microevolution should we use as our point of departure in extrapolating to macroevolution? Evolutionists of the early period of the synthetic theory extrapolated from microevolution proper, that is, from evolution at the population level, and regarded speciation as something of a side issue. However, there are reasons for believing that speciation is more centrally involved in macroevolution than was formerly thought. We will return to this question later in the chapter.

The Adaptive Landscape

The metaphor of an adaptive landscape is a useful way of representing the interactions between organism and environment. This metaphor has been employed by Wright, Dobzhansky, Simpson, Stebbins, and other evolutionists.

Consider a landscape consisting of hills and valleys. The topography symbolizes the distribution of adaptive fields. The hilltops are adaptive peaks, and the valleys between them constitute a no-man's-land for organisms, a series of adaptive valleys. Populations and species occupy the various adaptive peaks by virtue of their adaptive character combinations and the underlying gene combinations. Some hilltops or adaptive peaks are narrow, and others broad, corresponding to the relative breadth of specialization. The hilltops also vary in height; this suggests that some adaptive peaks are easier to climb and occupy than others (see chapter 1).

The metaphor can be extended so as to embrace species groups, genera, and larger phyletic assemblages. The peaks on the adaptive landscape are not randomly distributed, but are clustered in separate ranges. A range of hills would represent the adaptive zone of a genus.

So far we have considered only the statics of the situation. What about the dynamics? Both organisms and environment are subject to change.

How do species come to occupy their adaptive peaks, and genera their mountain ranges? Evolutionary trends are the pathways to the adaptive peaks and ranges. And evolutionary rates correspond to the rate of ascent. Adaptive peaks of different height may call for different combinations of trends and rates.

The adaptive landscape does not necessarily remain static, but like a real landscape, is subject to the forces of erosion and mountain-building. These forces may act slowly or rapidly. This represents the role of environmental change in evolution. An adaptive peak or range can move, at one rate or another, and the species or genus inhabiting that peak or range must then try to keep pace with the movement.

Irreversibility

It is often stated that evolution is irreversible. This generalization is known as Dollo's law. There are some important qualifications to this generalization.

In the first place, irreversibility is not a characteristic feature of microevolution, where genetic changes are relatively simple and changes in direction can occur readily. A particular allele or chromosome segment may increase in frequency at the expense of a competing allele or segment in one environmental regime, but decrease relative to the other allele or segment if the environment changes in the opposite direction. Irreversibility becomes a feature of evolution at the levels of speciation and macroevolution.

Here a further distinction must be made. Macroevolutionary changes are irreversible as regards anatomy, and the genetic determinants of anatomy, but not as regards ecology. Many groups of terrestrial reptiles, terrestrial mammals, and aerial birds have returned to the aquatic habitat of their primitive vertebrate ancestors, but their anatomical relationships to reptiles, mammals or birds are not erased by the change. Body size in various mammal groups, horses for instance, may increase during one time span and decrease during another. But the derived small types are not the same anatomically or genetically as the ancestral small types.

The crux of the matter is the historical nature of macroevolutionary change

(Simpson 1953:310–312). Genetic changes are piled up in successive layers over long periods of time, so that a return to the original state is impossible. In the evolution of a complex genotype over a long time, history does not repeat itself exactly. But this does not preclude the possibility of reversals in adaptive morphology.

Anagenesis and Cladogenesis

Two modes of evolution have long been recognized on the basis of the direction of evolutionary change. The first mode is change through time in a species population; the second is branching or divergence at the species level and higher. Rensch's (1947 1960b) terms, anagenesis and cladogenesis, are widely adopted for these two modes respectively. Darwin's terms were "descent with modifications" and "origin of species." Other evolutionists have used the expressions, phyletic evolution and evolutionary divergence.

Mayr (1982a) makes the interesting point that the early evolutionists tended to see, or at least to emphasize, only one mode or the other. Thus Lamarck was primarily an anagenicist and Darwin a cladogenecist.

Evolution no doubt takes place in both of these modes. But macroevolution may not be as simple as this terminological dichotomy suggests. Such additional modes as reticulate evolution and quantum evolution, to be discussed later in this chapter, complicate the picture. Furthermore, the distinction between anagenesis and cladogenesis breaks down in certain types of evolutionary trends, namely, speciational trends, also discussed later in this chapter.

Adaptive Radiation

Cladogenesis prevails in this mode so as to diversify a group and fit its members into the various habitats and niches in its territory (see chapter 30).

This mode was named and described by Osborn (1910) with reference to mammal evolution. Osborn recognized five main lines of specialization in the limbs and feet of mammals, fitting their possessors for diverse modes of locomotion and habitats, as follows: (a) swift running (cursorial) in terrestrial habitats; (b) digging habits (fossorial) for underground life; (c) swimming in amphibious and aquatic mammals; (d) climbing in arboreal mammals; and (e) parachuting and flying in aerial groups. Other divergent lines of specialization were recognized in the teeth and diet: insectivory, omnivory, herbivory, carnivory, myrmecophagy. Each main line was subdivided, e.g., the cursorial habit into digitigrade and unguligrade, and herbivory into browsing, grazing, frugivory, etc.

The orders of mammals, according to Osborn (1910:25), represent different combinations of these sets of characteristics, and different branch lines in a pattern of adaptive radiation.[1]

Simpson (1977) notes that Darwin had recognized this mode long before Osborn. So indeed he did (see Darwin 1859: ch. 4). Darwin's discussion emphasizes "divergence of character" in relation to interspecific competition and ecological diversity. Perhaps we can say that Osborn developed one aspect of Darwin's discussion further, made it more explicit, and gave it the name we now use.

Convergence and Parallelism

Convergence is the opposite of adaptive radiation. Convergence is the development of similar characters separately in two or more distantly related lineages as a result of adaptation to similar ecological conditions or challenges. The common ancestry of the lineages is not a contributing factor to the character similarity (paraphrasing Simpson 1961:78–79).

Examples of convergence are legion. Often cited in this connection are the various adaptive types of marsupials in Australia and their placental counterparts on other continents; and the stem-succulent euphorbs and cacti of hot dry regions in Africa and America respectively.

Parallelism is similar to convergence in the end product but differs in the phylogenetic pathway. Parallelism is defined as the development of similar characters separately in two or more related lineages where the similarity is channeled by the characteristics of the common ancestor of those lineages (Simpson 1961:78).

Parallelism is also widespread. One thinks of the increase in intelligence in the different orders of mammals. Some other examples were mentioned in chapter 28.

A new adaptive zone is often colonized concurrently by two or more un-

---

1. For a recent statement of the adaptive nature of mammalian orders, see Van Valen (1971). The modern treatment of adaptive radiation in mammals at all taxonomic levels is Eisenberg (1981).

related or distantly related groups. When the previously unoccupied adaptive zone comes within reach, that is, when preadapted organisms exist, two or more such preadapted groups frequently seize the opportunity at about the same time. This results in convergence or parallelism on a grand scale.

Thus in the Mesozoic the adaptive zone of flying vertebrates was occupied by birds and pterosaurs. During the same era the adaptive zone of insect-pollinated plants was occupied by angiosperms, cycadeoids, and probably other groups. The mammal life form was developed in this era by marsupials and placentals and by the remotely related monotremes. Similarly the angiosperm life form was developed in parallel by the dicots and monocots. In an earlier era the adaptive zone for medium or large-bodied marine plants was colonized independently by the green, brown, and red algae.

Multiple entry into a new adaptive zone leaves its stamp on the taxonomy of some high-level categories. A major taxon often contains an anomalous subgroup that does not fit comfortably in the taxon in question, but does not indisputably warrant segregation as a sister group either. The monotremes are something of a side issue in the mammals, as are the brown and red algae in the plant kingdom, and the sponges in the animal kingdom. Parallelism at high levels introduces a certain amount of asymmetry and untidiness into the taxonomic system.

Orthogenesis vs. Orthoselection

Many older evolutionists favored orthogenesis in one version or another as an explanation of long-term evolutionary trends (see chapter 2). This theory was applied inter alia to trends in the Equidae. The extensive evidence from the Equidae can be used to test this theory. Orthogenesis flunks the test.

If an evolutionary trend is the result of orthogenesis, it should be general and undeviating within a phyletic group. In the Equidae, however, there were no trends that continued throughout the whole history of any phyletic line, or that took place in all lines at any one time. Body-size increase was a general trend in the Equidae, but it was subject to reversals in some lines and at some times. Horse phylogeny was actually highly ramified with many side branches, and some of the trends, including those in the line culminating in *Equus*, occurred in side branches (Simpson 1951, 1953) (see chapter 28).

The alternative interpretation of the synthetic theory holds that the observed trends in horse evolution are determined by the organism-environment interaction. If environmental selection continues to operate in a given direction for a long time, it can be called orthoselection, and it will produce an evolutionary trend (Simpson 1944).

The evidence in the Equidae is consistent with the interpretation of orthoselection. The observed trends are clearly adaptive. Furthermore, they are not undeviating, but exhibit signs of opportunism, continuing under one set of conditions, but shifting direction when the conditions change. Finally, some trends in tooth characters can be traced back to polymorphisms in ancestral populations (Simpson 1951, 1953) (see chapter 28).

Reticulate Evolution

In many groups of plants the most inclusive unit of gene exchange at any one time level is a set of occasionally hybridizing species, a syngameon, and the evolutionary lineage through time is a shifting combination of hybridizing species. The unit of phyletic evolution is not single species, but anastomosing networks of species. Anagenesis and cladogenesis are intertwined. Macroevolution is reticulate. This pattern is widespread in vascular plants.[2]

Quantum Evolution

It will be recalled from our brief discussion in chapter 30 that quantum evolution is the relatively rapid shift of a lineage from an old adaptive zone to a new one and from the ancestral character state to a distinctly different one adapted to the new zone. Our objective in this section is to examine the concept of quantum evolution.[3]

Simpson (1944, 1953) proposed quantum evolution as a distinctive mode of evolution coordinate with phyletic evolution and branching. In some cases it grades into phyletic evolution. It may also be involved in speciation. The aspect which Simpson emphasizes, however, is the close relation with phyletic evolution; quantum evolution "is a special, more or less extreme and

2. Space does not permit a full exposition of this pattern here; for more on this subject see Grant (1981a).
3. Since this chapter was written an interesting review of quantum evolution has appeared (Laporte 1983). Laporte's approach is different from and complementary to the present discussion.

limiting case of phyletic evolution" (Simpson 1953:389). Although quantum evolution grades into these other two evolutionary modes, it is, in its well developed form, different from either, and thus worthy of recognition as a third mode according to Simpson.

The site of quantum evolution is small populations. Small isolated populations were postulated in the original version (Simpson 1944), subdivided populations in the revised version (1953:356). The role of drift in quantum evolution is problematical. Drift probably occurs in some cases but not in others (Simpson 1953).

Simpson (1944, 1953) felt that the populations in transition from one adaptive zone to the other would be in a poorly adapted or unstable state. At any rate the transition phase would be of relatively short duration.

Rapid evolutionary rates are characteristic of quantum evolution. But how rapid? The rates are viewed on a paleontological time scale. In one passage Simpson (1953:356) suggests that a transition from one adaptive zone to another may require time of the order of 1–10 million years.

Quantum evolution has an episodic quality, occurring as an exceptional event in the history of a lineage. A large ancestral population persists through time with horotelic or bradytelic rates up to the moment of quantum evolution. Then rapid rates occur concomitantly with small population size. When the quantum shift is completed, the lineage is likely to return to large-sized populations with horotelic or bradytelic rates (Simpson 1944, 1953) (figure 31.1).

This sequence is in agreement with paleontological evidence. The fossil record of many groups consists of a long recorded sequence when horotelic

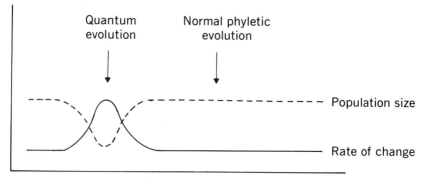

Figure 31.1. Relationship between population size and evolutionary rate in quantum evolution and normal phyletic evolution. (Redrawn from Simpson 1953)

or bradytelic rates prevail, preceded by a much shorter unrecorded time span during which the origin of the group must have taken place. The fossil record gives evidence of large population size accompanied by horotelic or bradytelic rates during periods of normal phyletic evolution. Conversely, the absence or rarity of fossils during the period of origin and main development of the group is consistent with the postulate of small population size in quantum evolution. The postulated features of quantum evolution provide an explanation for the great rarity of so-called missing links in the fossil record (Simpson 1944, 1953).

The product of quantum evolution is a new group of organisms, and often a new major group. Quantum evolution is thought to be involved in the origin of most new major groups (Simpson 1944, 1953).

Much progress has been made in our understanding of evolution in the decades since Simpson presented his concept of quantum evolution. Some reevaluation of the features originally proposed for this mode is in order now. Our comments will be focused on five aspects: the paleontological picture, population size, drift, evolutionary rate, and relation to other modes.

The first aspect, the paleontological picture, has stood the test of time very well. Many groups, once formed, tend to be very conservative thereafter. Examples of long-term species stability in mammals and beetles are given by Stanley (1982) and Coope (1979).

The question as to whether drift does or does not occur in quantum evolution would be better rephrased today in terms of the selection-drift combination. Selection-drift is the force best suited to bring about a rapid adaptive shift in small populations.

The original idea of small population size was definitely on the right track, but can be spelled out more specifically today. The evolutionary potential of small isolated populations is strongly limited, as noted in chapter 29. The most likely sites for quantum evolution would be colonial population systems and sets of peripheral founder populations (see chapter 16).

This consideration leads to a further deduction as to the type of terrain in which new major groups are likely to originate. Quantum evolution requires island-like population systems. Such population systems are not found commonly in plains country. They do occur commonly in country with broken terrain and a diversity of habitats, such as mountains for terrestrial plants and animals. Country of this sort could be expected to be a center of origin of major groups.

With regard to rates in quantum evolution, it should be noted that tachytely covers a rather wide range of rates. Morphological changes taking place in a 10-million-year period are regarded as tachytelic by paleontologists. Evolutionists working with living plant populations see evidence of substantial changes in much shorter time spans measured in scores or hundreds of years. Simpson and other early students of quantum evolution (Schaeffer 1948; Patterson 1949; Westoll 1949) were paleontologists and viewed this mode on a paleontological time scale.

One wonders if this is too conservative, if we are missing something. Perhaps the paleontological examples of quantum evolution are intermediate in the direction of phyletic evolution, as Schaeffer (1948) indeed suggested for the case of the artiodactyl foot mechanism studied by him. It is probable that some significant quantum shifts are actually taking place on a neontological time scale but are undocumented by paleontological evidence.

In Simpson's original view, speciation was not necessarily involved in quantum evolution, but might be in some cases. I would go further and say that speciation is normally, even necessarily, involved in quantum evolution where it takes place on a neontological time scale. It seems unlikely that a really rapid quantum shift from one adaptive zone to another could take place unaccompanied by a speciational shift. The mode of speciation which I termed quantum speciation is particularly well suited to bring about rapid adaptive shifts on a neontological time scale.

The quantum evolution concept in its original (1944, 1953) formulation encompassed a rather wide range of conditions as to rate and direction of change. Phyletic evolution was emphasized but branching was allowed. Tachytely on a geological time scale was envisioned but rapid rates on a neontological scale would be included too. In the light of our present understanding it may be desirable to subdivide quantum evolution into more homogeneous modes. The suggested modes are: (a) tachytelic phyletic evolution on a geological time scale; (b) formation of branch lineages at geologically tachytelic rates; and (c) quantum speciation.

Quantum Speciation

It will be recalled from our earlier discussion (chapter 24) that quantum speciation is the budding off of a daughter species from a large, polymorphic, outcrossing, ancestral population via the intermediate stage of a small local race. The local race is founded by one or a few individuals from the ancestral species population. It is spatially isolated or semi-isolated from the ancestral species; it usually occurs on the periphery of the old species area, and it may occur in a new ecological zone.

The gene pool of the local race is affected by drift and inbreeding at the beginning of its existence, and later by the selection-drift combination. The gene pool may also go through a genetic revolution (see below). These factors are capable of bringing about rapid and sometimes drastic changes in ecological preferences, morphological characters, and fertility relationships. Therefore the local race may diverge rapidly to the species level, and furthermore, the daughter species may occupy a new adaptive zone beyond the ecological range of the ancestral species.

A rapid adaptive shift during quantum speciation is further facilitated by a population structure which can evade the cost-of-selection restriction on multiple-gene substitution (see chapter 17).

The concept of quantum speciation is essentially a synthetic one. It represents a synthesis of selected parts of previous concepts of Wright (1931), Simpson (1944), and Mayr (1954), and also of the concepts emerging from Lewis' work on plant evolution (Lewis and Raven 1958; Lewis 1962). But no single previous concept fits the description of quantum speciation exactly. One gets selection-drift from Wright (but not from the other authors), quantum shifts (but not speciation) from Simpson, and rapid speciation (but not selection-drift) from Mayr and Lewis. The synthetic concept of quantum speciation combines these ingredients.

Genetic Revolutions

Mayr (1954, 1963) suggested that populations would undergo a profound genetic reorganization on passing through a bottleneck of small size. Mayr's thesis is based on known facts regarding homeostasis in experimental organisms, which will be summarized briefly first.

Outcrossing organisms in *Drosophila*, mice, chickens, *Primula*, *Zea*, and other groups are highly heterozygous and homeostatic, exhibiting phenotypic constancy in the face of changing environmental conditions. The heterozygosity and the homeostasis are connected; the latter depends on the former. Furthermore, the well-buffered normal phenotype is produced, not by any one heterozygous combination, but can be arrived at by numerous pathways involving different heterozygous combinations, and may in fact depend on heterozygosity at numerous loci. When the multilocus heterozygosity is lost or greatly reduced by inbreeding, one result is the production of offspring with aberrant characteristics, known as phenodeviants (Dubinin 1948; Lerner 1954).

Mayr (1954) applies these considerations to natural populations. He contrasts the internal genetic environment in large widespread populations of an outcrossing species with that in small isolated derivative populations of the same species.

A local subpopulation in a large continuous population is exposed to a stream of gene flow from neighboring and occasionally from distant subpopulations. The more or less frequent immigration of alien genes into the local subpopulation in question sets up selective pressures favoring genes with good combining ability. The gene pool comes to consist of arrays of alleles that give rise to normal viable products in a wide variety of heterozygous gene combinations. The pool of "good mixer" genes that develops in this genetic environment also, as a corollary, has homeostatic buffering properties (Mayr 1954).

This gives a certain conservative character to the variation pattern of the large continuous population. It exhibits clinal variation and geographical racial variation. But deviant or novel forms of the species cannot and do not persist as racial entities.

Now consider what happens when a few emigrant individuals from the large cross-fertilizing population succeed in founding a small daughter colony on the periphery of the species area. In the first place, the founder population will contain only a small non-random sample of the gene pool of the parent population. But furthermore, it will be isolated from the stream of gene flow. This changes the genetic environment. The selective values of the genes in the isolated daughter gene pool are altered; "good mixer" genes are no longer favored, whereas genes that produce viable types in homozygous condition now have a positive selective value (Mayr 1954).

The passage from a large cross-fertilizing population to a small isolated daughter colony thus in itself changes the genetic environment and alters or even reverses the selective values of the genes at many loci. The change in internal genetic environment affects many gene loci simultaneously, like the change brought about by hybridization or polyploidy, and it may be the beginning of a so-called genetic revolution (Mayr 1954).

The population may not be able to make the passage through the bottleneck from large to small size. It may not be able to stand a genetic revolution; in this case it will soon become extinct. But in a series of trials, as where numerous independent founder events take place, there is a chance that one or a few founder populations can survive. Survival in this situation depends on the founder population inheriting a gene pool that can tolerate homozygosity at many loci and that is preadapted to some available new ecological zone.

The genetic revolution in a successful daughter colony results in a more or less radical change in ecological preferences and morphological characters. Novel types, analogous to phenodeviants in experimental organisms, arise. The novel types of interest here are adapted to new ecological zones beyond the frontiers of the large conservative ancestral population. Taxonomic evidence supports the thesis to some extent. Some species and species groups in birds and plants are represented by novel types in their isolated peripheral populations (Mayr 1954) (see chapters 23 and 24).

In short, isolated peripheral populations derived by founder events from a large cross-fertilizing ancestral population may be sites of rapid evolution and of the origin of novel forms (Mayr 1954).

The divergent population does not necessarily remain small, isolated, and peripheral forever. It may build up in size, spread in area, and eventually reinvade the territory of the ancestral species. In time it too may become conservative in its variation pattern. But then it can give rise to a new generation of isolated daughter colonies, which break out of the ancestral mold once again (Mayr 1954).

Mayr did not relate his concept of genetic revolutions to Wright's views on subdivided populations or Simpson's quantum evolution. Nevertheless, good congruence is evident between the three concepts as regards the expected end results in small populations.

Differences of opinion exist with regard to the evolutionary force involved in the divergence of the small daughter populations. Mayr attributes the evolutionary change to the founder effect, and rejects genetic drift as a factor (Mayr 1954, 1963). However, the founder effect results from accidents of sampling occurring when one or a few emigrant individuals from a large ancestral population establish a new daughter colony, and can, therefore, be regarded as a special case of drift. Moreover, drift in itself is not the issue. The controlling force determining the divergence of the small daughter populations is best viewed as the selection-drift combination (see chapter 16).

Phyletic Trends and Speciational Trends

Let us next take a closer look at the nature of evolutionary trends. Treatments of macroevolution in the early modern period (1930s through 1950s)

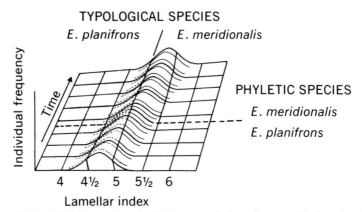

Figure 31.2.   Phyletic trend in the Pleistocene elephant lineage; *Elephas planifrons-E. meridionalis*. The lamellar index is a measure of the amount of enamel in the molar teeth. Phyletic species = successional species, and typological species = taxonomic species, in our terminology. (Redrawn from Simpson 1953)

have generally assumed or stated that evolutionary trends are products of anagenesis or phyletic evolution. Many trends probably are. However, it has been suggested that some or many trends are products of speciation. A number of speciational events may be linked together in a time series to form an evolutionary trend (Grant 1963, 1971; Stanley 1975, 1979).

Evolutionary trends, according to this viewpoint, fall into two broad classes: phyletic trends and speciational trends (Grant 1963, 1977a, 1977b).[4] We can distinguish the two types as follows.

A phyletic trend is a gradual unidirectional evolutionary change within a species lineage. It is a product of phyletic evolution exclusively. The stages in the trend are represented by successional species rather than by contemporaneous biological species.

An example is the trend in the structure of the molar teeth in a lineage of extinct European elephants of Pleistocene age. The lineage includes a pair of successional species, the older *Elephas planifrons* and the younger *E. meridionalis* (figure 31.2). Another example of a phyletic trend is *Homo erectus–H. sapiens*.

A speciational trend is a stepwise progression resulting from a succession

4. The concept of speciational trends was presented in Grant (1963:566–568, 570; also 1971:42–43). The term was not used there, however, in order to economize on new terminology. The paired terms, speciational and phyletic trends, were later deemed desirable for clarity, and were introduced (Grant 1977a, 1977b).

of speciational changes. Each new species in the series advances further in the direction of the trend. The trend is the line P–Q in figure 31.3. The stages in the trend are represented by biological, not successional, species. Here the predominating evolutionary pattern is speciation, rather than phyletic evolution, and the trend is the resultant of successive speciations.

In a phyletic trend there is only one biological species at any given time level. In a speciational trend the conservative ancestral form and the advanced descendant coexist temporarily as two contemporaneous biological species. The line P–Q in figure 31.3, although it consists of a succession of species in the literal sense, does not contain successional species; the line P–Q, therefore, is not a phyletic trend, but rather a speciational trend, by the criteria proposed here.

Evolutionary rates in a phyletic trend will be slowest where a large continuous population is exposed to mass directional selection. The rates can be faster if the population is subdivided and is worked over by the selection-drift combination (see chapter 29). The evolutionary rates that can be sustained in the latter situation, however, probably generally fall short of those to be considered next.

We noted in chapter 25 that new adaptive gene combinations can be fixed

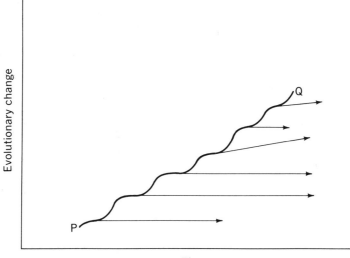

Figure 31.3. Speciational trend. The trend (line P–Q) is the resultant of successive speciational changes in a given direction. (From V. Grant, *The Origin of Adaptations*, © 1963 Columbia University Press, New York; reproduced by permission)

most quickly and expeditiously by speciation. If speciation is the best course to follow at time level 1 it may be the best course also at time levels 2, 3, and 4. Speciational trends are thus capable of bringing about evolutionary changes of substantial magnitude in a relatively short time period. Theoretically, they can proceed at more rapid rates than phyletic trends.

The orientation in a speciational trend develops in relation to some gradient or unidirectional change in the environment. The environment controls the trend through the medium of interspecific selection. Assume that a well-established ancestral plant species is adapted for a warm moist climate, and that the climate is becoming progressively colder. The only successful daughter species that this parent species can give rise to will be a cold-adapted form. Derivative forms with other ecological preferences will be nipped in the bud. The first-generation cold-adapted species can then go on to produce a daughter species of its own which is even more cold-tolerant, and so on. Speciational shifts to ever colder conditions are the only way open, under the given environmental gradient, and so the speciational trend develops in this direction (Grant 1963, 1977a, 1977b; Stanley 1979).

Speciational trends could be common. It is difficult to say how frequent they are, however, owing to the problem of identifying a speciational trend as such whenever the fossil record is incomplete, as it typically is. The fossil record of a group, being collected from different deposits, is usually scattered with respect to the real phylogenetic tree. The published phylogenetic tree based on this record—that is, the phylogenetic hypothesis—has gaps and uncertainties in it. In fact, different phylogenies can often be drawn from the same incomplete fossil record, depending on how one connects the branch lines in the unrecorded gaps. And consequently a speciational trend can easily be misread as a phyletic trend, and vice versa.

An example of a speciational trend has recently been reported in the Ordovician genus *Pydogus* (Conodontophoridae). The younger *P. anserinus* replaces the older *P. serrus* in some vertical columns, suggesting at first an anagenetic series, but when more localities are studied it becomes evident that *P. anserinus* arose from *P. serrus* by allopatric speciation and was contemporaneous with its ancestral species for awhile (Fahraeus 1982).

Two probable examples of speciational trends in plants are inferred from the comparative biology of recent forms. Such inferences have their weaknesses but also their strengths, since the species can be studied in a living condition and their interrelationships determined.

A trend in *Clarkia* (Onagraceae) in Pacific North America runs from relatively mesic to xeric forms. The ancestral mesic form was a species close to the present *Clarkia amoena* or *C. rubicunda*. The intermediate stage in the series is represented by the existing *C. biloba*, and the terminal xeric stage by *C. lingulata*. The transition from *C. biloba* to *C. lingulata* was brought about by quantum speciation (Lewis and Lewis 1955; Lewis and Roberts 1956).

The trend in *Polemonium* (Polemoniaceae) in western North America runs from mild coastal temperate forms to alpine types. The stages in this trend are represented by the following species:

Polemonium carneum	Moist coastal lowlands
P. caeruleum	Coniferous forest zone
P. californicum-delicatum	
group	Subalpine forest zone
P. pulcherrimum	Alpine zone
P. eximium group	High peaks in alpine zone

This series of species occurs in the zones indicated, on transects from coast to high mountains in western North America. The series does not manifest itself as a continuum on such transects, as is theoretically conceivable, but as a stepwise succession of distinct species, suggesting a speciational trend (Grant 1959, 1963).

It seems likely that speciational trends, because of their capacity for substantial changes and tachytelic rates, may be involved in the formative period of evolution of major groups.

Quantum Speciational Trends

The rate of evolution in speciational trends is probably correlated with the mode of speciation involved. Successions of geographical speciations in continuously large populations would be expected to be capable of moderate or normal rates, but not tachytelic rates. Successive quantum speciations, on the other hand, could theoretically generate rapid rates in an evolutionary trend.

If the special advantages of quantum speciation can be exploited once, at one time level, they can be exploited again and again. Quantum speciational shifts can occur repeatedly in cycles. And these shifts can have an overall trend, like the line P–Q in figure 31.3. A quantum speciational trend of this sort is a means for *sustained* rapid evolution in a given direction. It is perhaps the main way by which an evolutionary trend can proceed at a high tachytelic rate.

Let us return to the metaphor of the adaptive landscape introduced at the beginning of this chapter. Imagine an unoccupied high adaptive peak and a population camped at its base. There are different possible pathways from base to peak. Some pathways ascend gradually and are consequently long and slow. These routes symbolize phyletic trends and geographical speciational trends. They are adequate for the ascent if sufficient time is available. Other pathways are steep but short and potentially quick. If the population at the base of the mountain must climb the peak in a short time, because the time available for ascent is limiting, the population will have to take the short steep route, that is, follow a quantum speciational trend if it can.

Punctuated Equilibria

The punctuated equilibrium model has received much attention in recent years. The basic concepts were presented in four key publications (Eldredge and Gould 1972; Gould and Eldredge 1977; Stanley 1975, 1979).[5] Eldredge and Gould, on the one hand, and Stanley, on the other, agree on essential theoretical points, but differ notably in scientific style.

Eldredge and Gould's thesis contains the following main points: (a) Most evolutionists since Darwin, including advocates of the synthetic theory, have viewed evolution as a state of slow and even change, as a case of "phyletic gradualism." (b) Actual macroevolutionary changes come in spurts; there are short episodes of rapid change alternating with long periods of constancy or "stasis." This is "punctuationalism." It is regarded as new and noteworthy. (c) The episodes of substantial change coincide with events of speciation.

Point a is a misrepresentation of the synthetic theory, and point b has no novelty. Evolutionists of the synthetic school—Wright, Simpson, Rensch, Mayr, Stebbins, etc.—have been advocating special episodes of rapid change for fifty years. Most of the relevant literature is not cited in the two papers of Eldredge and Gould. Quantum evolution, for example, is not mentioned at all in the 1972 paper, and is mentioned only once and then incidentally in the 1977 paper.

Eldredge and Gould are probably right about point c, but they have not caught up with the previous literature on this subject either.

The same lack of knowledge of the literature is evident in their treatment

---

5. Much additional literature pro and con has accumulated since these papers. For review, critique, and bibliography see Grant (1982, 1983).

of various subsidiary topics. Interspecific selection ("species selection"), for example, is presented as a "previously unrecognized mode of operation for natural selection" (Gould and Eldredge 1977:139), although this process has been well known in evolution and evolutionary ecology for many years.

Stanley has developed essentially the same thesis, as already mentioned, in several papers and one book. Stanley's (1979) book is well documented, as to both the factual evidence and the evolutionary literature, and therefore warrants serious consideration. The case for a close connection between speciation and macroevolutionary trends is strengthened by this book.

I would have to add that Stanley (1979, 1981, 1982), like Eldredge and Gould, credits most recent evolutionists with an affinity for "gradualism" (meaning constant slow evolutionary rates) which they do not have. Gradualism has become something of a straw man and whipping boy for the punctuationalists.[6]

In general, the attempt to reduce all or nearly all paleontological series to two main conditions—stasis and punctuated equilibria—seems to be a step in the direction of oversimplification. The fossil record is biased in favor of static and slowly changing groups. Even so, a wide range of evolutionary rates can be found in the record (Simpson 1953; Gingerich 1983). Punctuated equilibria and stasis are probably best viewed as a couple of points in a continuous spectrum.

6. This and some other areas of disagreement between myself and Stanley are discussed in Grant (1981b, 1982).

PART VII
Macroevolution: Special Aspects

CHAPTER 32
Molecular Evolution

Several biochemical approaches make it possible to measure the amount of change that has taken place in homologous macromolecules in closely or distantly related species during the course of their divergence. The macromolecules investigated are usually proteins but may be DNA's. The molecular evidence sheds light on phylogenetic relationships. It has also generated some good controversies which will be discussed briefly here.[1]

Approaches

The main approaches in molecular evolutionary studies are: DNA hybridization (for DNA); serological tests (as used for blood types); amino acid sequencing (used for hemoglobin, myoglobin, cytochrome c, etc.); electrophoresis (used for a variety of enzymes); and biochemical systematics (utilizing various secondary substances in plants).

The direct results of the comparative biochemical study of the macromolecules can be expressed quantitatively in various ways. Among the mea-

1. For reviews of the large field of molecular evolution see Ayala (1976), Wright (1978), Goodman (1982), Nei and Koehn (1983), and Kimura (1983). Good older works are Ingram (1963) and Jukes (1966). A valuable compendium of information is given by Dayhoff (1968–1978).

sures of biochemical differentiation between different phylogenetic lines are the following. The relative degree of pairing affinity between DNA strands from different species (in DNA hybridization studies). The number and proportion of amino acid substitutions in homologous polypeptide chains (in protein sequencing studies). The probable number of point mutations involved in producing the observed differences between homologous proteins (an extrapolation from protein sequencing studies). And the proportion of the enzyme loci tested electrophoretically that are different (in electrophoretic assays).

The indirect results of the biochemical assay are of further interest. The quantitative measure of biochemical differentiation between the living species A, B, C, D, whatever the measure used, can be superimposed on a phylogenetic tree with branches terminating in A, B, C, D. The known or inferred geological time that has elapsed since the divergence of the respective branches gives the denominator. Putting the two measures together, we then have the amount of macromolecular change for a given period of time, which in turn gives an estimate of the molecular evolutionary rate.

Differences in Amino Acid Sequences

Similarities and differences in the amino acid sequences in polypeptide chains of homologous proteins belonging to different species provide a definite and quantitative measure of the amount of molecular differentiation. A large body of information on molecular homologies has now been assembled for hemoglobin, myoglobin, cytochrome c, immunoglobulin, and other proteins (see Dayhoff 1968, 1969, 1972, 1978). Only a few representative examples can be presented here.

The adult human hemoglobin molecule consists of two identical alpha polypeptide chains, two identical beta chains, and their associated heme groups. Each alpha chain contains 141 amino acids and each beta chain, 146 amino acids. In normal human hemoglobin each position in a chain is occupied by a specific type of amino acid. The sequences are known. One gene specifies the sequence in the alpha chains and another separate gene, the sequence in the beta chains. It is of interest in passing that the alpha and beta chains of human hemoglobin, though different, have similar amino acid sequences and probably arose by divergence from a common ancestral polypeptide chain (Ingram 1963).

We are more interested here in the amount of differentiation in the hemoglobin chains between different species. The differentiation can be expressed as the number or the percentage difference in the amino acid sequences. The percentage difference in the amino acid sequences between man and various other species of mammals is given in table 32.1. It is seen that man and chimpanzee have identical sequences in both the alpha and beta chains. There are only two amino acid differences between human and gorilla hemoglobins, one in each chain. Man and monkeys are close in their hemoglobin structure. Other orders of mammals show greater divergence from man in hemoglobin, with percentage differences ranging from 10–26% (table 32.1).

The hemoglobin of man is differentiated from that of frogs and carp to a greater degree, as would be expected. The percent difference in sequence is 46% for the beta chain in the human-frog comparison, and 50% for the alpha chain in the human-carp comparison (Dayhoff 1972).

Another respiratory protein, cytochrome c, is located in the mitochondria of eukaryotic organisms, and is well suited for comparative biochemical studies of members of different phyla and kingdoms. Table 32.2 gives some data on molecular differentiation in cytochrome c. Man is taken as the standard for one set of comparisons and *Drosophila* for another. Here again we see a gen-

Table 32.1. Differences in the amino acid sequences of hemoglobin in man and various other mammals (Dayhoff 1972)

Species Pair	Percent Difference in: Alpha Chain	Beta Chain
Human–chimpanzee	0	0
" –gorilla	1	0
" –Rhesus monkey	3	5
" –macaque	5–7	–
" –mouse	13–15	18
" –rabbit	18	10
" –dog	16–17	10
" –horse	13	17
" –llama	16	14
" –pig	13	16
" –cow	12	17
" –sheep	15	18
" –goat	14–16	18–20
" –Barbary sheep	15–16	21–23
" –gray kangaroo	19	26

Table 32.2. Differences in the amino acid sequences of cytochrome c in man and various other organisms (Dayhoff 1972)

Species Pair	Percent Difference
Human–Rhesus monkey	1
" –horse	12
" –cattle, sheep	10
" –dog	11
" –rabbit	9
" –chicken, turkey	13
" –pigeon	12
" –snapping turtle	14
" –rattlesnake	13
" –bullfrog	17
" –tuna fish	20
" –dogfish	23
" –fruit fly	27
" –screw-worm fly	25
" –silkworm moth	29
" –wheat	38
" –*Neurospora*	44
Fruit fly–screw-worm fly	2
" –silkworm moth	14
" –tobacco hornworm moth	13
" –dogfish	24
" –pigeon	23
" –wheat	42

eral correlation between amount of molecular differentiation and closeness or remoteness of phylogenetic relationship.

Dayhoff (1969) and her coworkers point out that the observed number of amino acid differences between homologous proteins is not necessarily equal to the number of actual amino acid substitutions during their evolutionary divergence. Where many amino acid differences occur between two polypeptide chains, the evolutionary distance becomes greater than the observed differences. Dayhoff and coworkers have devised a unit of evolutionary distance, called the PAM unit ("accepted point mutations per 100 links" in the chain), which is designed to give a corrected estimate of molecular evolutionary divergence. The relation between the observed number of amino acid differences per 100 links in a chain and the evolutionary distance in PAM units is as follows (Dayhoff 1969):

Observed Differences	PAM Units
1	1
5	5
10	11
25	31
50	83
75	208
85	370

Other Biochemical Evidences of Relationship

DNA hybridization techniques involve the mixing of single-stranded fragments of the DNA of two species. The proportion of the total DNA in the mixture that reassociates to form double-stranded helices, and the rate of the reassociation, are measures of the degree of genetic affinity between the species. An example of the results obtained by this technique in *Drosophila* is shown in table 32.3. *Drosophila melanogaster* and *D. simulans* belong to the same species group. We see that 80% of their DNA forms pairs in DNA hybridization tests. By contrast, the DNA of *D. funebris*, which belongs in a different subgenus, mostly does not pair with that of *D. melanogaster* or *D. simulans* (Laird and McCarthy 1968).

Hubby and Throckmorton (1965) used electrophoretic methods to determine the similarity or difference in proteins within the *Drosophila virilis* group. Ten species in this group (*D. virilis*, *D. americana*, *D. texana*, etc.) were compared with respect to numerous proteins and their underlying gene determinants. They found that 60% of the proteins sampled are common to the whole species group. Another smaller fraction is common to members of closely related subgroups. The remaining fraction of the proteins tested is

Table 32.3. Proportion of DNA of *Drosophila* species that pairs in DNA hybridization tests (Laird and McCarthy 1968)

Species Combination	Percent of DNA That Pairs
D. melanogaster × *simulans*	80
D. funebris × *melanogaster* or *simulans*	25

unique in each species. This fraction amounts to 2.6% for D. *virilis*, 5.3% for D. *americana*, and ranges up to 28.2% in D. *littoralis*. Another series of studies using gel electrophoresis compared four sibling species of the *Drosophila willistoni* group with respect to 14–28 enzyme loci (Ayala et al. 1970; Ayala and Tracey 1974). The species turn out to be different at about half of the loci sampled.

Electrophoretic data for several or many loci in two populations or species can be converted into an index known as genetic identity (*I*) (Nei, 1972). This index gives a measure of the proportion of the genes in the two populations or species that are identical. *I* ranges in value from 0 to 1; *I* = 1 indicates that the same alleles occur in the same frequencies in the two populations; while *I* = 0 indicates that the populations have no alleles in common (Nei 1972).

An example of the genetic identities of related species in the plant genus *Tragopogon* is presented in table 32.4 (Roose and Gottlieb, 1976). Species pairs in other plant genera often show similar values of *I*, and sometimes have high values of *I* such as 0.9 (see Gottlieb 1977).

A close serological relationship exists between man and apes in the ABO blood groups. The following blood types are known to occur in four species:

Man	A, B, AB, O
Gibbon	A, B, AB
Orangutan	A, B, AB
Chimpanzee	A, O

The agglutination reactions of chimpanzee blood are indistinguishable from those of human blood, and the same is true of the blood of the gibbon. The gorilla, however, differs from man and the other great apes and resembles monkeys in its ABO blood groups (Wiener and Moor-Jankowski 1971).

Table 32.4. Genetic identity (*I*) between species of *Tragopogon* (Compositae) (Roose and Gottlieb 1976)

Species Pair	*I*
T. dubius–T. porrifolius	0.50
T. dubius–T. pratensis	0.62
T. porrifolius and *T. pratensis*	0.53

Comparisons Between Electrophoretic and Morphological Differentiation

It has been widely assumed that electrophoretic methods give a reliable estimate of the overall similarity or divergence between individuals or groups. The enzyme gene differences revealed by such methods are taken to be representative of the genotype as a whole. This assumption is implicit in Nei's (1972) coefficients of genetic identity and genetic distance, which equate electrophoretic with genetic measures of differentiation.

In some cases the assumption appears to be warranted. Thus Jain and Singh (1979) found good congruence between electrophoretic evidence and morphological and cytogenetic indicators of relationship between 15 species of the oat genus *Avena*.

In a significant number of cases, however, morphological and electrophoretic evidences are not congruent. Some examples of non-congruence between morphological and electrophoretic evidence are: pupfish, *Cyprinodon* (Turner 1974); man and chimpanzee (King and Wilson 1975); snail, *Partula* (Murray and Clarke 1980); and barley, *Hordeum* (Giles 1984). In each case the morphological differentiation between populations or species is considerable, while the electrophoretically detected enzyme differences are slight.

Findings such as these suggest that the value of electrophoretic evidence for assessing evolutionary relationships and phylogenies has been greatly overrated in many recent studies. Electrophoretic evidence should be used with, not in place of, morphological evidence. Views along these lines have been expressed by Carson (1977), Grant (1977:286), Murray and Clarke (1980), and Giles (1984).

Differences Between Man and Chimpanzee

The molecular relationships of man and the chimpanzee *(Pan troglodytes)* have been extensively explored with a variety of techniques. King and Wilson (1975) pooled the extensive molecular data from all sources. They find that humans and chimpanzees are remarkably similar at the macromolecular level, with many identical proteins and many very similar ones.

Amino acid sequencing studies of several homologous proteins in man and chimpanzee reveal that some of these proteins are completely or nearly identical. The following homologous proteins have 0 or 1 amino acid difference, as indicated below:

Hemoglobin	0
Cytochrome c	0
Fibrinopeptide	0
Lysozyme	0?
Delta hemoglobin	1
Myoglobin	1

And other proteins in the two species show only a few amino acid differences; for example, serum albumin, with about 6 such differences at 580 sites (King and Wilson 1975).

Electrophoretic analysis of 44 homologous proteins points again to close similarity between man and chimpanzee, as noted in the preceding section. About half of the proteins assayed are electrophoretically identical in the two species according to King and Wilson (1975; see also Bruce and Ayala 1979).

The amount of molecular differentiation between man and the chimpanzee is on a par with that between sibling species in other groups, such as the *Drosophila willistoni* complex; it is much less than that between non-sibling species in various other genera (King and Wilson 1975).

Yet the morphological, behavioral, and ecological differences between *Homo sapiens* and *Pan troglodytes* are sufficient to warrant placing them in different genera and families. The chromosome sets of the two species also differ with respect to a number of inversions and translocations. A discordance thus exists between the organismic evidence and the molecular evidence concerning the evolutionary distance between humans and chimpanzees.

King and Wilson attempt to resolve this paradox by suggesting that the basic genetic differences between humans and chimpanzees are due not so much to point mutations in ordinary or so-called "structural" genes as they are to changes in the regulatory systems of the two organisms. The changes in the regulatory systems would come about partly by mutations in regulatory genes and partly by chromosomal rearrangements that change the gene order and affect gene expression. "A relatively small number of genetic changes in systems controlling the expression of genes may account for the major organismal differences between humans and chimpanzees" (King and Wilson 1975:115).

The suggestion that human-chimpanzee differences are determined by relatively few structural or Mendelian genes is inconsistent with a large body of genetic evidence. Experimental hybridizations in animals and plants show that race and species differences are determined by complex multigene systems.

An alternative hypothesis seems more likely. Man and chimpanzee may have similar requirements as regards the physiological processes controlled by hemoglobin, cytochrome c, various electrophoretically detectable enzymes, etc. In other words, evolution of some physiological processes reached an adaptive plateau for large-bodied hominoids at an earlier stage in the phylogeny of the group, before such genera as *Homo* and *Pan* diverged from their common stock, and there has been little need for further change in these processes since. In a more recent phase of hominoid phylogeny, however, man and chimpanzee have evolved along divergent lines with respect to habitat, locomotion, diet, communication, etc., and the morphological differences between them represent adaptations for their respective ways of life.

The discordance between the protein and the morphological evidence of relationships in hominoids, and in other groups mentioned previously, is perhaps best viewed as an aspect of mosaic evolution. Different body parts in a group of organisms often evolve in different periods and at different rates.

The Neutral Theory

The neutral theory of molecular evolution was originally stated by Kimura (1968) and King and Jukes (1969), developed in a series of publications (Kimura and Ohta 1971a, 1971b; Kimura 1979, 1981), and has recently been given a critical book-length treatment (Kimura 1983).

The theory in its current formulation (Kimura 1983) holds that "the great majority" of molecular changes in evolution are selectively neutral or nearly so. This is not to say that the genes involved as functionless. They may or may not be so. The thought is that the different allelic forms of such genes are nearly equivalent selectively (1983:50). Nor is it contended that all molecular changes are neutral; the theory holds that most such changes are neutral (p. 54).

Two broad classes of molecular changes are envisioned: protein polymorphisms and molecular substitutions. Selection is not the controlling factor, in most instances, in either class. Most protein polymorphisms are not maintained by balancing selection. Most molecular gene substitutions are effected by drift.

Kimura (1983) makes a most persuasive case for the neutral theory. It may well be that many molecular changes are neutral or nearly neutral. Whether most are or not remains to be seen. Much more empirical evidence is needed than is now available to decide this issue.

A number of enzyme polymorphisms in *Drosophila melanogaster* and other organisms have been investigated from a physiological standpoint. The enzyme genes control important functions. And the different enzyme morphs are associated with different physiological properties in the laboratory which probably affect fitness (Johnson 1974; Lavie and Nevo 1982; Koehn et al. 1983).

Furthermore, some protein polymorphisms in *Drosophila* and other organisms do not exhibit random geographical variation, as would be expected if the alleles were selectively neutral. Instead the alleles show regular trends in geographical distribution, including clines, suggesting that the allele frequencies are controlled by selection (Ayala and Anderson 1973; Ayala and Tracey 1974; Ayala and Gilpin 1974; Bryant 1974; Koehn et al. 1983).

In a few favorable instances the same enzyme polymorphism has been investigated in both the laboratory and the field. The enzyme morphs show important physiological differences in the laboratory and regular geographical distribution patterns in nature. This is the case with alcohol dehydrogenase (ADH) and alpha-glycerophosphate dehydrogenase in *Drosophila melanogaster*, lactate dehydrogenase in the fish *Fundulus heteroclitus* (Koehn et al. 1983), and phospho-glucomutase in the shrimp *Palaemon elegans* (Nevo et al. 1984).

Admittedly, a handful of examples of selectively valuable enzyme alleles will not invalidate the neutral theory. However, we should note that it is very difficult, and entails much work, to demonstrate either the selective neutrality or the selective value of the variant forms of a gene. Should this work, and the burden of proof, be put onto the selectionists, or should it be borne by the neutralists?

Two alleles of a certain gene in some herbaceous plants differ in only one nucleotide base. One allele has adenine at a certain site, the other has guanine. It would be very easy to dismiss this difference as trivial or functionally unimportant. However, plants carrying the first allele are susceptible to the herbicide atrazine, while carriers of the second allele are atrazine-resistant (NRC Board 1984). This is obviously a selectively important difference in an atrazine-containing environment.

The absence of a known function for any given variant form of a macromolecule does not in itself justify a neutralist interpretation. Moreover, the adaptive role of the macromolecule might be expressed only in a certain environment, making it exceedingly hard to detect.

A selectionist and adaptationist interpretation of the phenomena of concealing coloration and mimicry in animals was greeted with skepticism at one time; but the accumulation of field and laboratory evidence eventually showed that the selectionist position was right. Other classes of phenomena (e.g., flower mechanisms in plants, etc.) have gone through a parallel history of skepticism, fact-finding, and triumph of adaptationism. History usually seems to be on the side of selectionists, and could be expected to repeat itself in molecular evolutionary studies (see also Clarke 1973).

My hunch at this point is that some molecular changes in evolution, but not "the great majority," will turn out to be selectively neutral. Such a finding, while less than the neutral theory is claiming, would still represent a very interesting addition to our understanding.

Molecular Evolutionary Rates

The average rate of change in protein molecules can be measured and expressed in various ways. Dayhoff (1972, 1978) uses the number of accepted point mutations per 100 amino acid residues (PAM units) per 100 million years. Kimura (1983) expresses the rate of change as the number of amino acid substitutions per site per year.

Each species of protein tends to have its characteristic rate of change. Different proteins show different average evolutionary rates. The characteristic rates in a series of mammalian proteins are given in table 32.5. Histones

Table 32.5. Average rate of evolutionary change in various mammalian protein molecules (Dayhoff 1978)

Protein	No. PAM Units per 100 MY
Kappa casein	33.0
Hemoglobin alpha chain	12.0
Hemoglobin beta chain	12.0
Myoglobin	8.9
Insulin	4.4
Lactate dehydrogenase	3.4
Cytochrome c	2.2
Glutamate dehydrogenase	0.9
Histone H2A	0.5

tend to be very conservative. Cytochrome c, insulin, and other proteins are quite conservative. By contrast, fibrinopeptides show rapid rates (90 PAM units/10^8 years) (Dayhoff 1972). Hemoglobin is intermediate.

When one compares the amount of molecular differentiation between phyletic groups for a given protein with the time elapsed since their divergence, one finds a good correlation between molecular change and time. This can be illustrated by data for the hemoglobin alpha chain in vertebrates (table 32.6). When amino acid substitutions are plotted against time on a graph, a linear relationship becomes apparent, only the chicken being a bit out of line.

Findings such as this have led to the doctrine of the constancy of molecular evolutionary rates. The rate of change is said to be constant or approximately so through time for any given type of protein (Ohta and Kimura 1971; Kimura and Ohta 1971, 1972; Jukes 1972; Kimura 1983). The term, molecular clock, is sometimes employed to express the alleged clock-like regularity of molecular evolution.

The conclusion of constant molecular rates, though supported by some empirical evidence, is completely at variance with what we know about evolutionary rates in morphological characters, which is also based on empirical evidence. Evolutionary rates in morphological characters are not constant

Table 32.6. Amino acid differences in hemoglobin alpha chains between man, dog, and other vertebrates, compared with the age of their divergence (Based on Kimura 1983:67)

Species Pair	% Amino Acid Difference	Time Since Divergence, MY*
Human–dog	16.3	90
" –kangaroo	19.1	140
" –echidna	26.2	225
" –chicken	24.8	300
" –newt	44.0	360
" –carp	48.6	410
" –shark	53.2	450
Dog–kangaroo	23.4	140
" –echidna	29.8	225
" –chicken	31.2	300
" –newt	46.1	360
" –carp	47.9	410
" –shark	56.8	450

* Ages taken from graph of Kimura.

through time (chapter 29). Some reconciliation of the two lines of evidence is needed.

Kimura (1983) regards the constancy or near constancy of molecular rates as a strong argument in favor of the neutral theory. The constancy would be difficult to explain if the molecular changes are controlled by selection, but are as expected if they are selectively neutral and are fixed by drift (Kimura 1983).

We should also note that the method of estimating evolutionary rates is significantly different in molecular and paleontological studies. Molecular rates are estimated from comparisons between recent forms, adjusted for the time of their divergence. The basic setup is this:

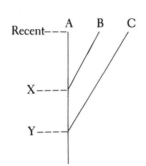

The A–B differences are compared with the A–C differences. The age of the A–B difference is Recent-X, that of the B–C difference is Recent–Y. If a regular proportion is found between the two or more horizontal measures (A–B, etc.) and the corresponding vertical ones (R–X, etc.), constancy is concluded.

Morphological and taxonomic rates are based on observations at a series of dated stages in the paleontological history of a lineage. The lineage often exhibits rapid rates in its formative stages and settles down to horotelic or bradytelic rates in its later history, as we saw in chapter 29. Many molecular changes of adaptive value could be expected to go through a similar sequence. Such variable molecular rates, however, would be difficult to detect by the standard approach in molecular evolution studies outlined above.

CHAPTER 33
Ontogeny and Phylogeny

Paedomorphosis
Recapitulation
Relative Growth Rates
Developmental Genetics
Regulation
Canalization

Embryology has a long and intimate connection with evolution. Darwin (1859) cited embryological patterns as one of the lines of evidence for evolution. Haeckel (1866) related embryology to phylogeny. Many twentieth-century evolutionists have discussed in depth the relation between ontogenetic development and phylogenetic change; e.g., Huxley (1932, 1942), Goldschmidt (1940), Rensch (1947, 1959), de Beer (1951), Stebbins (1950, 1974), Waddington (1957), Løvtrup (1974), and Gould (1977).

Two broad approaches to the subject can be discerned: that of comparative embryology (e.g., de Beer 1951; Løvtrup 1974; Gould 1977), and that of developmental genetics (e.g., Waddington 1957; Raff and Kaufman 1983). Both approaches are needed.

Paedomorphosis

Phylogenetic change in the adult organism can be viewed as the product of successively modified ontogenies (Garstang 1922; de Beer 1951; Løvtrup 1978).

Various modes of ontogenetic change have long been recognized: paedomorphosis, fetalization, neoteny, acceleration, etc. Classifications of these modes are given inter alia by Rensch (1959), Løvtrup (1974), and Gould (1977). Gould's (1977) classification of the modes of ontogenetic change and their phylogenetic results clarifies the processes and the sometimes confusing terminology.

In paedomorphosis the adult descendant resembles the juvenile stage of an ancestor. This result occurs when reproductive maturation is speeded up relative to somatic growth during development. Two processes can bring about this general result, and, accordingly, two types of paedomorphosis need to be distinguished, namely, progenesis and neoteny (sensu stricto). An absolute acceleration of maturation without a comparable acceleration of somatic growth gives progenesis. A retardation of somatic growth without a comparable retardation of maturation gives neoteny sensu stricto (Gould 1977).

Gould suggests that progenesis, bringing about early reproduction, is a life-history strategy favored in environments with r-type selection, that is, in pioneering habitats. Neoteny, on the other hand, which requires a prolonged period of growth, may be able to develop more readily in a stable biotic community with K-type selection (Gould 1977).

Neoteny has been extensively studied and discussed in animals, familiar examples occurring in the phylogenies of salamanders (*Ambystoma*, etc.) and primates (including humans). The human adult, for example, bears a much closer resemblance to juvenile chimpanzees than to adult chimpanzees in facial characteristics. Hairlessness is another neotenous trait in humans. An example of progenesis is provided by wingless aphids.

In plants, it has been pointed out that some of the main distinctive features of the angiosperms are neotenous, namely the flowers, leaves, and male and female gametophytes (Takhtajan 1959a, 1959b, 1976, 1983). Within the angiosperms, the wood of such derived life-forms as annuals, short-lived perennials, and stem-succulents is paedomorphic as compared with the corresponding tissue system in woody dicotyledons (Carlquist 1962, 1975).

Recapitulation

A parallelism exists between changes in ontogeny and those in phylogeny. Haeckel (1866) made this the basis for his famous law that "Ontogenesis is the recapitulation of Phylogenesis." Haeckel's recapitulation theory has been widely discussed (see Garstang 1922; Rensch 1959; Løvtrup 1978), and often rejected, but as Løvtrup (1978) notes, it has a measure of truth in it.

Phylogenetic changes sometimes occur by the addition of new terminal stages to the ancestral ontogeny (Rensch 1959; Gould 1977; Løvtrup 1978). Thus the fish *Belone* develops very long forceps-like jaws by proceeding further in ontogeny than more conservative related forms such as *Atherina*. (Rensch 1959). Acceleration of development may be combined with the addition of new terminal stages to keep the length of ontogeny more or less constant (Gould 1977).

Where this type of ontogenetic alteration occurs, the ontogeny of the advanced form does pass through stages corresponding to the adults of its phylogenetic ancestors. Ontogeny can then be said to recapitulate phylogeny.

As Løvtrup (1978) notes, Haeckel's law of recapitulation holds true where the evolutionary change involves a terminal addition in development. This is a special case, and apparently also a common one, but it is not universal. Therefore Haeckel's conclusion is not a universal law, nor is it discredited, but it stands as a useful generalization.

The parallelism between ontogenetic and phylogenetic changes had also been noted at an early date by von Baer (1828), who stated and interpreted it in a different way. Von Baer's concept has become widely adopted.

Von Baer (1828) noted that embryos of related groups resemble each other more closely than do adults of the same groups. A developing embryo passes through a series of stages representing the body plan of the various groups to which it belongs, and furthermore, passes through them in a sequence from the more inclusive groups to the smaller ones. Ontogeny proceeds from the more general to the more specific characters; from characters of class rank through those of family and generic status to the final species-specific characters (Garstang 1922; Løvtrup 1978).

An element of recapitulation is present in von Baer's concept. Løvtrup (1978) proposes, therefore, to recognize and distinguish a von Baerian recapitulation and a Haeckelian recapitulation. Of the two the von Baerian recapitulation is the more general.

Løvtrup states the relationships between the ontogenies of divergent phyletic lineages as follows: "In the course of their ontogeny the members of a set of twin taxa follow the same course of recapitulation up to the stage of their divergence into separate taxa" (1978:352).

Relative Growth Rates

Morphological divergence between a derived group and its ancestor, or between sister lineages, can be resolved into ontogenetic alterations, as we have seen, and these in turn can be resolved into differences in the growth rates of different body parts.

D'Arcy Thompson (1917, 1942, 1961) used Cartesian coordinates to compare the divergent body forms in related genera belonging to various major groups of animals: crustaceans, hydroids, fishes, reptiles, birds, mammals. Examples from fishes are shown in figure 33.1. The standard type in each related pair is on the left in the figure, and the divergent or "deformed" type is on the right.

In figure 33.1A, it is seen that the standard type (on the left) can be transformed into the divergent type (on right) by inclining the vertical axes of the coordinates at a certain angle. In figure 33.1B, the divergent type can be related to the standard type by deforming the coordinates from rectilinear to circular. Other modes of transformation are seen in figures 33.1C and 33.1D.

These comparisons suggest that the actual phylogenetic change entailed a change in relative growth rates of different body parts during ontogeny. Thus in figure 33.1C the adult form of the divergent type is a result of an increased growth rate in the anterior end and a decreased growth rate in the tail, as compared with the standard type.

A further development along this line of thought is embodied in the concept of allometric growth (Huxley 1932; see also Rensch 1959). The parts of the body have their characteristic and often constant growth rates during ontogeny. Different body parts often have different growth rates. Consequently, the proportions of the body are determined by the length of the growth period and the size attained in the adult stage.

The principle of allometry can be extended to comparisons between related species or genera. Related groups with different adult sizes can be expected to show different body proportions.

The antlers of male deer, for example, have positive allometry, so that as body size increases, antler size increases, not only absolutely, but also relative to body size. Small-bodied deer species have very small antlers, medium-sized deer have large antlers, and the largest of recent deer, the extinct Irish elk, had gigantic antlers (Huxley 1932; Simpson 1949, 1967).

A progressive development of the horn is seen in titanotheres from Early Eocene to Early Oligocene, as shown in figure 33.2 (Osborn 1929). The Early Eocene titanothere, *Eotitanops*, had no horn (figure 33.2A). Small bony protuberances were present on the faces of Middle and Late Eocene titanotheres (figures 33.2B and C). Moderately large horn-like protuberances occurred in one Early Oligocene genus, *Megacerops*, and very large horns in another Oligocene genus, *Brontotherium* (figure 33.2D).

The trend can be explained in terms of allometric growth rates. Titanothere horns show positive allometry, increasing in relative size as the body in-

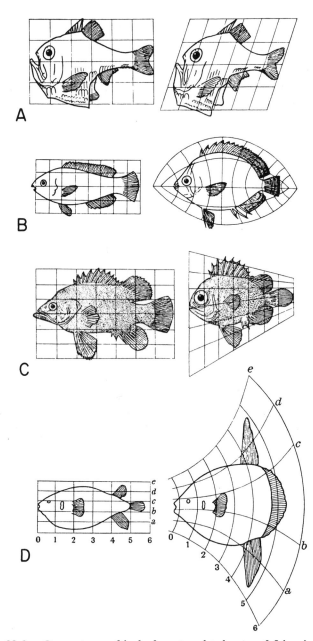

Figure 33.1. Comparisons of body form in related pairs of fishes by the use of Cartesian coordinates. The standard type in each pair is on the left, the "deformed" type on the right. (A) *Argyropelecus olfersi* (left) and *Sternoptyx diaphana* (right). (B) *Scarus* (left) and *Pomacanthus* (right). (C) Polyprion (left) and *Pseudopriacanthus altus* (right). (D) *Diodon* (left) and *Orthagoriscus mola* (right). (Rearranged from D'Arcy Thompson, *On Growth and Form*, copyright 1942, © 1961 Cambridge University Press, London; reproduced by permission)

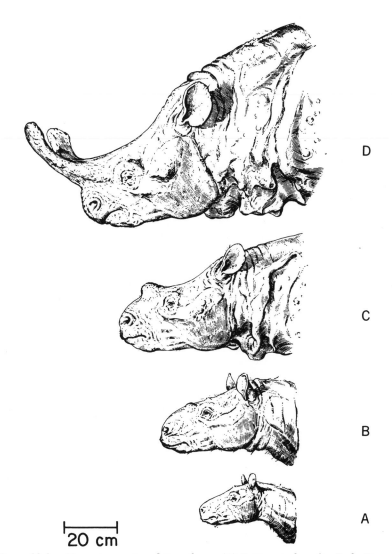

20 cm

Figure 33.2. Horns in a series of titanotheres. (A) *Eotitanops borealis*, Early Eocene. (B) *Manteoceras manteoceras*, Middle Eocene. (C) *Protitanotherium emarginatum*, Late Eocene. (D) *Brontotherium platyceras*, Early Oligocene. (Osborn 1929)

creases in absolute size. An increase in body size is apparent in Eocene and Oligocene titanotheres (figure 33.2). Below a certain body-size threshold no horns developed. In the larger animals the horns were very large. Selection for body size would bring about a correlated, relative size increase in the horns (Huxley 1932, 1942; Stanley 1974).

Developmental Genetics

Developmental genetics carries the analysis of ontogenetic alterations and their phylogenetic effects to a deeper level of causation, that of the genes. As Huxley (1932) and Goldschmidt (1938) emphasized at an early date, many genes produce their phenotypic effects by governing the rates of developmental processes. These processes may be growth rates, rates of production of growth substances, timing of differentiation, and so on. Rate genes form an intermediate link between selection and phyletic change.

A classical example of gene-controlled development involves fruit shape in cultivated varieties of squash, *Cucurbita pepo* (Sinnott and Hammond 1930; Sinnott 1936). The fruit is disk-shaped, spherical, or elongated in different varieties of squash. The various shapes can be resolved into different balances between growth rate in width and growth rate in length.

The relative rates of growth along the different axes of growth are controlled by interactions between the gene *I* and the gene system *A-D*. Certain allelic forms of the genes, *A*, *B*, *C*, and *D* act to produce a short longitudinal axis and hence a disk-shaped fruit. The dominant allele *I* of the gene *I* acts in the opposite direction to bring about growth in length. The two series of genes work in balance. Several allele combinations containing a preponderance of axis-shortening genes (e.g., *AABBii*) yield disk-shaped fruits. Other allele combinations in which *I* predominates yield elongated fruits. And genotypes containing a mixture of shortening genes and elongating genes (e.g., *AAccii*) produce spherical fruits (Sinnott and Hammond 1930).

Many genes are active at only one stage or another in the life cycle. This is illustrated by the genes determining storage proteins in the seeds of soybean *(Glycine max)*. These genes are active only in the embryonic stage. They are present but inactive in adult plants (NRC Board 1984).

Regulation

It is a short step from the concept of regulation of development to that of regulatory genes. Both concepts are receiving much attention in current discussions of ontogeny and phylogeny. There can be no denying the evolutionary importance of genetic control of development.

The term, regulatory gene, as used in the current literature means more than this, however. Regulatory genes sensu stricto stand in contrast to structural genes. Structural genes determine the synthesis of proteins, being more or less equivalent to classical genes, while regulatory genes control the action of structural genes. This distinction between structural and regulatory genes has emerged from work on bacteria where it fits. It has been extensively applied in its original form to higher organisms where it does not fit the situation.

Consider the situation in higher plants. Here there is no single class of regulatory genes. Regulation is brought about by a heterogeneous assemblage of controlling genes. This assemblage includes inhibitor genes, minus modifiers, oppositional gene systems, heterochromatin blocks, and controlling elements (which may be heterochromatin blocks). Furthermore, in higher plants the distinction between regulatory and structural genes breaks down. Some types of controlling genes *are* structural genes, inhibitors being a good example (see Grant 1975: ch. 8).

When, therefore, an author who accepts the structural-regulatory dichotomy concludes that evolution in higher organisms depends primarily on regulatory mutations and only secondarily on structural gene changes, the conclusion does not rest on a rigorous analysis of the genetic elements in eukaryotes. Yet just this conclusion is stated by a number of authors (e.g., Wilson 1975; King and Wilson 1975; Bush 1975; Gould 1977).

Regulation of development is also attributed to side-effects of chromosomal rearrangements in recent literature discussions (e.g., Wilson 1975; King and Wilson 1975). Changes in segmental arrangement are said to bring about evolutionary changes by altering the pattern of ontogenetic development. This is a modern version of Goldschmidt's (1940) concept of pattern effect which in turn is an extension of the known phenomenon of position effect.

The known position effects are usually deleterious, however, and do not look like suitable raw material for evolution. Also, chromosomal rearrangements have other genetic effects besides position effect. Furthermore, some groups have managed to evolve without detectable chromosomal rearrangements, e.g., the family Fagaceae (see Grant 1981: ch. 14). The thesis that pattern effect plays an important role in evolution by regulating development is possible, even plausible, but requires much more supporting evidence than is available at present.

Canalization

The course of development is canalized; it resists external pressures to divert it from the normal pathway (Waddington 1957; Whyte 1965). If the phenotypic product of development is adaptive, selection has probably favored canalized genotypes, genotypes which give rise to the same character in different environments.

Canalization of development is thus a conservative force in evolution. The canalized ontogenetic development is resistant to radical change. Gene mutations or recombinations which radically alter the normal development will be eliminated. The only genetically determined changes of ontogeny that can persist are those that alter the developmental pathway in relatively small ways. The preexisting morphological structure restricts the range of gene mutations and recombinations that can be adaptively useful (Waddington 1957; Whyte 1965; Stebbins 1974).

Confronted with an environmental change calling for a new and different character, the conservative force of developmental canalization may well prove fatal. Extinction is indeed common. It is not universal, however, and there are some ways out.

When a group faces a new environmental challenge, it may be more feasible developmentally to change the function and shape of a preexisting organ or organ system than to produce a new one. This is the principle of preadaptation. An organ or organ system originally adapted for one set of conditions can sometimes be modified successfully for a new use.

The history of life affords numerous examples. Thus the paired fins of crossopterygian fishes, which were adaptive for swimming, happened to be suitable for crawling on land and developed into the limbs of the early semiterrestrial and terrestrial tetrapods. Normally green calyces and inflorescence bracts are capable of being modified as the colored attractive organs of flowers. The calyx performs the attractive function in *Mirabilis*, and the bracts do so in *Bouganvillea* (both Nyctaginaceae).

Juvenile stages of ontogeny usually lack the specializations found in adults. Gene-controlled changes in rate of development leading to paedomorphosis, therefore, can release a group from its present specialization, and enable it to evolve along new lines (Huxley 1942; de Beer, 1951; Hardy 1954; Takhtagan 1959a, 1959b). The wide occurrence of paedomorphic characters among animals and plants suggests that this course has been followed fairly frequently.

CHAPTER 34

Specialization and Progress

Specialization

Specialization in organisms is a relative matter and is therefore difficult to define precisely. All organisms are specialized. It is the degree of their specialization that varies, and this varies in a spectrum from broad to narrow.

Specialization is an aspect of adaptation, which also ranges from relatively narrow to relatively broad. Specialization may be viewed as the set of adaptations and tolerances of a medium or high-level group for its adaptive zone, or of a species-level group for its habitat or niche. If the organism is adapted for a narrow zone, habitat, or niche, it is highly specialized; if it is adapted for a wider range of environmental conditions, its specializations are relatively broad.

Thus, the koala, restricted to a diet of *Eucalyptus* leaves, has a narrow food niche, while the omnivorous raccoon has a relatively broad one. The two types of mammals show narrow and broad dietary specializations, respectively. Most orchids are specialists in their pollinator relationships, while many Compositae are generalists, and the two groups exhibit narrow and broad specializations, respectively, for pollination.

It is perhaps well to avoid using the terms "unspecialized" and "overspecialized" in this connection, although these terms are often seen in the evolutionary literature, because both terms have misleading connotations. Simpson and others have pointed out that an unspecialized organism, if such can be imagined, could scarcely make a living. What constitutes overspecialization, on the other hand, depends on the pragmatic test of survival during unpredictable future changes. No group of organisms is overspecialized as long as it is able to survive; any group can be called overspecialized after it has become extinct. It is preferable to think in terms of relative degrees of specialization.

A narrowly specialized organism is more vulnerable to extinction during a change in the environment than an organism with broad specializations, other factors being equal. If one type of food should disappear, koalas would become extinct, whereas raccoons would not. Vulnerability to extinction is a weakness inherent in narrow specialization.

Narrow specialization is promoted by interspecific competition. A biotic community containing several or many species with similar ecological requirements is likely to develop into an array of narrow specialists, as pointed out in chapter 22.

Specialization Trends

Evolutionary trends often involve an increase in specialization. Although trends are not directed by internal evolutionary forces, there is, nevertheless, a sense in which a specialization trend, once started, may become self-perpetuating as a result of internal factors.

An evolutionary trend involving a complex character or character combination, once well under way, often tends to continue in the same direction. The self-perpetuating tendency can be explained as a joint effect of orthoselection and specialization. Orthoselection is the primary force guiding the phyletic line in a certain direction. But as the group advances in the pathway of its trend it accumulates specialized characters that restrict the range of functionally useful new mutations. New mutations are selectively valuable only if they produce phenotypic changes that fit in harmoniously with the existing character combination. And the range of selectively valuable mutant types becomes restricted with an increase in specialization. Specialization reinforces orthoselection to keep the trend "on the track."

If specialization trends have a tendency to become self-perpetuating, and also increase the chances of eventual extinction, the long-term survival of a group already well advanced in such a trend might depend upon its ability to escape from specialization. The relevant question then becomes one of how a lineage can get out of the potential blind alley of a specialization trend. Paedomorphosis is one way out, and the use of existing organs for a new and different function, or preadaptation, is another (see chapter 33).

Alternatively, the highly specialized organism may *not* be able to change sufficiently to meet the challenge of changing environmental conditions, whereas other members of the same group do meet this challenge successfully. In a diversified group of medium or large taxonomic rank, the generalist members have a better chance than the narrow specialists of coping with changing conditions. The former may give rise to a new derivative group adapted to the new conditions, while the latter become restricted, relictual, or extinct.

This is probably the reason for a common pattern in phylogeny. New major groups often arise, not from advanced members of an ancestral group, but from some sector in the base of the older group.

The Concept of Progress in Evolution

The concept of progress is basically a subjective one, and in common usage it relates primarily to human affairs. A trend in human history is said to exhibit progress when a later stage is better than an earlier stage in some characteristic singled out for emphasis by an observer of the trend. The viewpoint and value system of the observer play an essential role in the identification of progress. We can readily think of trends in modern times that a land developer calls progress and that a conservationist considers to be a lamentable step in the wrong direction.

The question before us here is twofold. First, can the concept of progress be transplanted satisfactorily from human affairs to evolutionary biology? And second, if so transplanted, can it be converted from a subjective to an objective concept; or in other words, can objective criteria be found for progress in organic evolution?

In a series of essays Huxley (1942: ch. 10; 1954, 1958) has argued in the affirmative on both questions. He argues for the objective reality of a special type of evolutionary trend to be known as progressive evolution. In fact, Huxley transplants progress from human affairs to organic evolution and then brings it back to human affairs again, making mankind the culminating development of progressive evolution.

Huxley's thesis has been discussed by a number of evolutionists, usually with something less than full agreement (Simpson: 1949, 1967, 1974; Thoday: 1958; Rensch: 1960; Stebbins: 1969; Ayala: 1974; Dobzhansky et al. 1977; Grant: 1977). Nevertheless, portions of Huxley's thesis survive criticism, as will be seen later.

The evolutionary series from primitive marine animals through the vertebrates to man entails changes in a number of characteristics. If the series constitutes a progressive evolutionary trend, the changing characteristics give us a set of criteria of progressive evolution, arrived at more or less independently of subjective considerations.

Among the changes in the series from lower animals to higher vertebrates are the following (Huxley: 1942; Simpson: 1949, 1967; Rensch: 1960): (a) increase in morphological complexity; (b) increase in energy level of life processes, that is, increase in metabolic rate; (c) increase in efficiency of reproduction, including increase in care of eggs and/or young; (d) improvements in perception of signals from environment and in ability to react to environmental stimuli; and (e) increase in control over and independence of the environment.

Huxley (1942) emphasized the last characteristic—control and hence relative independence of the environment—as the most significant distinguishing criterion of progressive evolution. On this criterion, mammals rank high; certain mammals, like the beaver, very high; and man, highest of all. Huxley's criterion of independence and control of the environment comes close to the mark. But in Huxley's treatment, the term "environment" is construed broadly, and "control" narrowly, in such a way as to slant the application of the criterion toward mankind. One would prefer a criterion that is less anthropocentric and more biological.

An alternative criterion has been suggested (Grant 1977: ch. 33) and is advocated here. It is suggested that progressive evolution consists of adaptations to environments successively further removed from the original and ancestral environment of life. This criterion will be discussed in the next section.

Progressive Evolution As a Form of Specialization

Huxley (1942) draws a distinction between progress and specialization in evolution. "Specialization . . . is an improvement in efficiency of adaptation for a particular mode of life." (p. 562). But, "Specialization . . . always involves the sacrificing of certain organs or functions for the greater efficiency of others" (p. 567). Consequently the improvements incorporated into a specialization for a particular adaptive zone tend to limit the possibilities of future change to a continuation in the same adaptive zone. Specializations are self-limiting. Most of organic evolution is the development of specializations. And evolution is largely a series of blind alleys (Huxley 1942).

Progress in evolution, on the other hand, is improvement in the general efficiency of the machinery of life; it is "all-round biological improvement" (Huxley 1942:562, 567). Progressive evolution is consequently a trend in which past changes and present characteristics do not limit future possibilities. It is open-ended (Huxley 1942).

As noted earlier, Huxley (1942) equates progressive evolution with human evolution. In so doing, he has construed progressive evolution in an all too narrow sense. Organic evolution includes a very diverse array of major evolutionary trends. Certainly the biological part of human evolution is an important development in organic evolution, but important developments have occurred in other phyletic lines too. There are developments that could be called progressive in other mammals, birds, social insects, and plants.

Furthermore, the age of man will not last forever. Organic evolution will probably be making new advances in future ages, when the human species is gone from the scene. The nature of future dominant types of organisms is unpredictable at present.

We can restate the concepts of progress and specialization in terms of types of evolutionary trends. Let us recognize three modal types of evolutionary trends as classified in these terms: (a) short-term specialization trends, illustrated by a branch line in the Hawaiian honeycreepers (chapter 30); (b) long-term specialization trends, as in the evolution of the horse's hoof (chapter 28) (these two types of trends are channeled and self-limiting); (c) progressive evolutionary trends, which entail general improvements in biological organization. The series—land reptile, warm-blooded mammal, intelligent mammal, intelligent hominid—is only one example of a progressive trend; another is the series—pteridophyte, seed plant, woody angiosperm, annual herbaceous angiosperm.

We can think of progressive trends as ultra-long-term specialization trends, thus eliminating the dichotomy between progress and specialization. Progressive trends are specialization trends in which the organisms involved become specialized for adaptive zones that are common on earth but far removed from the ancestral nutrient soup.

Life began in nutrient-rich waters. The exhaustion of nutrients in the seas stimulated the development of photosynthetic aquatic forms, as well as grazers on the primitive autotrophs, and predators on the grazers. When the seas became filled, several phyletic groups colonized the land. Terrestrial organisms progressed from warm moist land areas to dry and cold lands. Several terrestrial groups also colonized the aerial adaptive zone. Man is the culmination of one progressive trend; but rodents, birds, winged insects, desert shrubs, and desert annual plants are the culminations of other such trends.

The stepwise conquest of environments that are widespread on earth but basically inhospitable to life, such as nutrient-poor water, dry land, and air, is the common theme running through those long-term adaptive trends that we call progressive.

Succession of Grades

A characteristic feature of progressive evolution is a succession of grades. A grade is a structural-functional level of organization (Huxley 1958). A grade may be a natural taxonomic group (bony fishes), but is not necessarily so (worms, pteridophytes).

A given grade is often the dominant type in its kingdom during a given era. And a succession of such dominant types occurs through the course of time. Progressive evolution takes place through such a succession of grades.

A progression of organizational grades has occurred throughout the history of life. We will pick up the story at the beginning of the colonization of land habitats by originally aquatic organisms. The conquest of the land took place as a succession of grades in both the vertebrates and the vascular plants.

The main grades of terrestrial vertebrates are outlined below.[1] The characteristics listed are subject to exceptions.

1. Amphibians. Two pairs of crawling legs; air-breathing lungs; eggs laid in water and fertilized there. Wetlands. Derived from crossopterygian fishes, a group of lobe-finned, partially air-breathing, bony fishes, in the Devonian. Expanded in the Carboniferous. Dominant in the Permian.

2. Reptiles. Body covered with drought-resistant scales; internal fertilization; amniote egg with nutritive yolk, internal membranes, and protective shell. The earliest reptiles were amphibious, but went on land to lay eggs, so the terrestrial habit developed in the reptilian egg before it did in the adult animal (Romer 1967). Appeared in late Carboniferous and expanded in Permian. Dominant through the Mesozoic.

1. For details see Romer (1966, 1967) and Colbert (1980).

3. Birds. Homoiothermy; feathers; bipedal; controlled winged flight; highly active; parental care of eggs and nestlings. Appeared in Jurassic and expanded in Cretaceous. Large terrestrial flightless birds were widespread in the early Cenozoic but were reduced in numbers and restricted in area later, probably by mammal competition (Colbert, 1980). Dominant flying vertebrates through the Cenozoic.

4. Mammals. Homoiothermy; protective coat of hair; quadrupedal with good running ability; intelligence; internal gestation; provision of milk for young. Basically terrestrial but have radiated into other zones. Appeared in late Triassic time, developed in the Jurassic, and expanded in the Cretaceous. The Tertiary was the Age of Mammals.

The colonization of land habitats by vascular plants, which paved the way for land animals, also took place as a succession of grades.[2] The characteristics listed below are subject to exceptions.

1. Rhyniophytes. Plant body simple, consisting of erect, photosynthetic, dichotomously branching axes, arising from a horizontal rhizome with root hairs; aerial axes containing vascular tissue; sporangia borne on tips of some axes. Life cycle probably included a free-living gametophyte and motile sperm as in similar living pteridophytes. May have been derived from green algae. Silurian-Devonian.

2. Ferns. Plant body consisting of leafy shoot, rhizome (usually), and adventitious roots; megaphyllous leaves; sporangia usually of one type; life cycle typically including a mesophytic free-living gametophyte; motile sperms requiring a film of water in which to swim to reach the eggs in the gametophyte; embryo and young plant developing on the gametophyte. Mostly in mesic habitats. Derived from Trimerophytopsida in late Devonian. Expanded in the Carboniferous and were abundant in the Mesozoic and Cenozoic.

3. Gymnosperms. Plant body with shoot and well-developed root; sporangia transformed into ovules and pollen sacs; ovules and stamens mostly borne in strobili or cones; gametophytes reduced and included within these organs; ovules pollinated by pollen grains; wind pollination mostly; internal fertilization; ovule with embryo and stored food developing into a naked seed. Mesic to cold and xeric habitats. Arose in late Devonian or Carboniferous, depending on the group in question, and expanded in the Carboniferous. Dominant in Mesozoic.

4. Flowering plants. Vascular tissue including vessels in most members;

2. For details see Bold et al. (1980) and Stewart (1983).

ovules borne in closed carpels; carpels and stamens borne in flowers, which are primitively showy; animal or wind pollination; gametophytes greatly reduced; seed containing a new type of nutritive tissue, the angiospermous endosperm; seeds maturing in and often dispersed in a fruit. Wide range of habitats including tundras and deserts. Appeared in Lower Cretaceous and expanded in Upper Cretaceous. Dominant through Cenozoic.

CHAPTER 35

Physical and Biotic Factors

Types of Factors

The environment of an organism consists of several sets of factors. There are physical factors of the environment; biotic factors stemming from other species in the same community; and influences produced by other individuals of the same species (see Schmalhausen 1949).

The biotic aspect of the environment is itself a complex of diverse elements. The biotic environment of a given species may consist of: (a) host or prey species; (b) other members of a mutualistic association to which the species belongs; (c) commensals with neutral effects; (d) competitor species; and (e) pathogens, parasites, predators, and grazers.

The species in question derives benefits from the first two classes of biotic elements (a and b). The last two classes (d and e), on the other hand, are detrimental to it, and tend to evoke counteradaptations. This suggests that the evolutionary effects of the two sets of biotic factors, the beneficial and detrimental ones, should be discussed separately.

It is clear at the outset that an organism must be and normally is adapted to the various different facets of its environment. Thus a cactus plant in a dry habitat has one set of adaptations for storing water and reducing transpiration, a response to an adverse physical factor; it is also protected by spines against browsing animals, a response to a detrimental biotic factor; and its flowers enter into mutualistic associations with animal pollinators, a response to a beneficial biotic factor.

Beyond this obvious point lie other questions of deeper interest. What are the characteristic evolutionary effects of physical factors, of the different types of biotic factors, of combinations of physical and biotic factors, of combinations of different biotic factors?[1]

Differences Between Physical and Biotic Factors

Physical factors tend to be density-independent in their mortality effects and second-order selective effects, whereas biotic factors are often density-dependent in their impact. The population density of a species is irrelevant when a wave of cold or drought hits it. A wave of infectious disease organisms, on the other hand, may have only slight mortality and selective effects at low population densities, but severe effects at high densities. Hard selection prevails in the one case, with density-independent physical factors, and soft or flexible selection in the other case, with density-dependent biotic factors (see chapter 13).

An important difference between physical and biotic factors is that the biotic environment *evolves*. A given species affects other species around it in positive or negative ways, and these make an adaptive response, within the limits of their stored variability and population structure. A species and certain elements in its biotic environment may follow a pathway of coevolution.

Physical factors do not behave in this way. Certainly a species produces changes in its physical environment, and vice versa, but the changes are not mutually adjusting or coevolutionary. Adaptation here is a one-way street of species to physical environment.

Adaptations to physical environments that diverge from the ancestral habitat of life are on a pathway of progressive evolution, according to the criteria proposed in chapter 34, whereas adaptations to biotic factors are apt to fall in the category of specializations.

Combinations of Biotic Factors

A combination of competitor species promotes a narrowing of niches and ecological specialization in each species, as noted in chapter 22.

1. Attention is called to the excellent earlier discussion of these questions by Schmalhausen (1949).

Combinations of other biotic factors lead to long-term coevolutionary sequences. It is useful to distinguish two types of coevolution on the basis of the relationship between the species involved. Mutualistic coevolution is one type. The other type includes a detrimental biotic factor as one member and is oriented toward exploitation and defense. Many examples of both types have been discussed in the literature.[2]

Mutualistic coevolution is illustrated by the development of specialized flowers and pollinating animals, animal seed dispersal, lichens, root mycorrhizae, and root nodules. Well-known modes of exploitation-defense coevolution are predator-prey series, herbivores and plant defense mechanisms, host-parasite series, and host-pathogen series.

Good examples of exploitation-defense coevolution are provided by the changing relationships between plant herbage and herbivorous insects, mainly butterflies. The following sequence of stages is reconstructed from comparative evidence: (a) phytophagous insects expand in response to a valuable food resource, herbage, and have destructive effects on the plants; (b) some plant species develop poisons that deter insect herbivores, and, being shielded from destructive herbivory, undergo expansion and radiation; (c) some insect species develop resistance to the plant poisons, giving them access to a plant food resource from which other herbivorous insects are excluded; it is now their turn to expand and radiate; (d) some of these resistant insects use the ingested toxic chemical substances as deterrents against their own bird predators (Ehrlich and Raven 1964; Brower and Brower 1964; Rothschild et al. 1970; Berenbaum 1983).

Detrimental biotic factors probably have an important pervasive influence in mature (climax and subclimax) communities. The environment of a species in such a community is constantly worsening due to the appearance of new forms of pathogens, parasites, predators, and aggressive competitors. This effect is described in Van Valen's (1973) Red Queen hypothesis; a species has to keep running just to stay in the same place.

Much evolutionary change must go on under the surface, so to speak, in mature communities as a result of the continual interplay between detrimental biotic factors and their victims. Such change goes on even though the physical environment remains constant.

---

2. For reviews of coevolutionary phenomena see the symposium volumes of Henry (1966–1967), Gilbert and Raven (1975), and Futuyma and Slatkin (1983).

Rare Invasion of Isolated Communities

In a mature biotic community, the available niches are more or less filled, interspecific competition is strong, and the array of pathogens and predators keep the host species in check. The pressures of stabilizing selection keep each constituent species in its niche. The same force eliminates new variant forms that appear; potential new branch lines are nipped in the bud.

Savile (1959) considers the possibilities that exist when two such biotic communities develop in geographical isolation for a long time. Each community will have its own set of checks and balances. Now let the isolation break down partially. One or a few migrants cross the geographical barrier and enter the foreign community. These will usually be eliminated by the indigenous set of checks, by competition, disease, and other detrimental factors.

Occasionally, however, an immigrant may happen to be resistant to these negative biotic factors, and may succeed in establishing itself in the foreign community. It is more or less free from the constraints that hold the native species in line, and it has escaped from its old biotic constraints. The conditions are then favorable for it to expand and radiate in its adopted community (Savile 1959).

In a later stage, after the radiation has occurred, the process of community penetration may happen again, but in the reverse direction. A descendant of the immigrant in the adopted community successfully invades the original ancestral home of the group and produces new derivative types there (Savile 1959).

An example which can be interpreted along the above lines is provided by the saxifrages (Saxifragaceae) of eastern Asia and northwestern North America. The phylogeny is inferred partly from the plants themselves and partly from their rust parasites. The original center of the group lies in eastern Asia. An ancestral form close to *Mitella* crossed via Beringia to the northern cordilleran region of North America. There it radiated, producing *Tiarella*, *Saxifraga*, and various minor genera endemic in this region. *Mitella*, *Tiarella*, and *Saxifraga* then migrated back to eastern Asia (Savile 1959, 1975).

Another example is the invasion of the long isolated island continent of South American by rodents and monkeys from other land areas in the early Tertiary. The immigrations led to the extensive radiations of the caviomorph rodents and New World monkeys (Simpson 1980).

The rare crossing of a geographical barrier separating two biota can also be achieved by a disease organism. The contact between an exotic pathogen and non-resistant hosts may lead to the decimation or extinction of whole species. A familiar example is the great reduction in range of the eastern American chestnut (*Castanea dentata*) following the accidental introduction of an alien fungus disease of chestnuts from China (Savile 1959).

Different Proportions of Physical and Biotic Factors

The relative weight of physical and biotic factors varies between different climatic regions and at different stages in an ecological succession. This can be illustrated by an inspection of contrasting types of terrestrial plant communities.

Consider first a tropical rainforest in a constantly moist tropical region. Here the physical factors of temperature and moisture are most favorable for plant growth. A lush and complex forest community develops, a community containing numerous tree species. The biotic factor of competition is strong. Furthermore, the benign physical environment is favorable, not only for plants, but also for a wide array of plant pathogens, parasites, insect herbivores, and the like. Detrimental biotic factors exert a continual strong pressure.

The diversity of tree species in lowland tropical rainforest is remarkable. It is not unusual to find 80, 90, or more species of trees in a single hectare (Dobzhansky 1950; Richards 1952). Ecological differentiation may exist between some of these species; for many others, however, there is no apparent interspecific ecological differentiation.

The latter fraction of woody species diversity could be a byproduct of the pressure of plant diseases and pests. A given type of pathogen, parasite or herbivore is often specialized for attacking certain susceptible plant species, while other plant species are resistant to it. A dense crowding of one or a few species of trees in relative pure stands would favor the development of epidemics. Attempts a few decades ago to grow the rubber tree, *Hevea brasiliensis*, in plantations in Brazil where it is native failed because of plant diseases.

A dispersion of the individuals of each tree species in the forest reduces the chances of epidemics. Any tree species which managed to achieve dominance would have its ranks thinned sooner or later by disease, and the vacant places would be occupied by other resistant species. This process, long continued, would result in a mixture of tree species which are sparsely dispersed among one another, such as actually occurs (Ridley 1930:xvi; Savile 1960; Grant 1963:420–421; Janzen 1970).

The tree species are involved in numerous other biotic relationships with parasitic lianes and large epiphytes, commensal small epiphytes, beneficial animal pollinators and seed dispersers, and so on. Biotic factors predominate here.

The situation is the opposite in a warm desert where a physical factor, drought, is limiting for plant growth. The plant community is simple. The role of the dominants is played by one or a few species of desert shrubs which have been able to adapt to the particular desert climate in question. Detrimental biotic factors are by no means unimportant. Desert shrubs typically possess defense mechanisms of one sort or another against herbivores. However, the struggle for existence of the desert shrub is, to a very large degree, directed against adverse physical factors.

A cold temperate forest presents a similar case. Here again the limiting factor for plant growth is a physical one, winter cold, and the dominant trees belong to the one or a few species which have successfully mastered the climatic conditions. To be sure, detrimental biotic factors can be serious occasionally or locally. But let us keep the priorities in order. The tree species must first achieve its adaptation to the cold temperate climate; then it can confront the plant diseases and plant pests in its area.

Analogous changes in the balance between physical and biotic factors occur during an ecological succession in temperate terrestrial plant communities. In the pioneering stage of succession, physical factors predominate. In the subclimax and climax stages the biotic factors become more important.

The varying physical-biotic conditions at different stages of an ecological succession are associated with marked differences in the reproductive system of the plants involved. In the pioneering stage, the plants tend to have small size, short generation time, high fecundity, and closed or restricted recombination systems. Subclimax and climax plants, by contrast, typically have large size, long generations, reduced fecundity, and open recombination systems (Salisbury 1942; Schmalhausen 1949; Grant 1958; 1975: chs. 23, 24; Odum 1969; Gadgil and Solbrig 1972).

Environments in which selection is predominantly by physical factors should theoretically be more favorable for rapid evolution than environments dominated by detrimental biotic selective forces. Rapid evolution requires a reproductive potential sufficient to support a high cost of selection. A species living in a mature community with strong interspecific competition will probably be unable to afford the high selective mortality rate that goes along with rapid evolution. Conversely, if high selective mortality occurs, the species is likely to lose its place in the community.

A species living in an open or pioneering habitat, and not hemmed in by strong interspecific competition, on the other hand, may occasionally be able to tolerate a high selective mortality rate and thus undergo rapid evolutionary change (see chapter 17).

The Roles of Physical and Biotic Factors in Progressive Evolution

Progressive evolution has been characterized as evolutionary trends oriented on an axis from benign ancestral environments of life to inhospitable and stressful environments (chapter 34). Selection by physical factors plays a central role in progressive evolution, thus defined. However, physical factors rarely exist in isolation. Interactions between physical and biotic factors are the norm in modern communities, as brought out by the preceding discussion in this chapter, and such interactions have probably spurred progressive evolution in the past.

The filling up of the seas, and the resulting strong competition between marine forms, probably provided a stimulus for the colonization of semiterrestrial and terrestrial habitats by several phyletic groups of plants and animals in the Silurian and Devonian. Crowded conditions in warm moist terrestrial communities probably stimulated later waves of colonization of dry and cold land areas, and also the conquest of the air.

Biotic factors may stimulate the exploration of new and unfavorable physical environments. Selection by physical factors enables the organisms to actually conquer the new environment. As the conquest of the physical environment is completed, by different groups of organisms evolving in parallel, biotic factors become significant again, and initiate a new cycle of colonizations of another new environment (Grant 1963:422–423).

CHAPTER 36
Extinction

Selected Pleistocene and Post-Pleistocene Ungulates
Sabertooth Cats
Mass Extinction of North American Mammals in the Late Cenozoic
Dinosaurs
Mass Extinction in the Late Cretaceous
The Asteroid Impact Theory
Periods of Mass Extinction

Extinction occurs in groups of varying size and inclusiveness. It is perhaps useful to recognize five modal levels of extinction: extinction of a species in a large part of its area; of a whole species; of phyletic groups of relatively low taxonomic rank, e.g., genera and families; of major groups such as orders and classes; and mass extinction, affecting many different groups in a given epoch. Examples of the various levels of extinction will be given in this chapter.

In general, groups of high taxonomic rank have a longer duration than low-ranking phyletic groups, and the latter in turn have a longer average duration than species. Extinction is the fate of the vast majority of species, but kingdoms and phyla tend to be immortal.

A further distinction is useful in considering extinction of supraspecific phyletic groups. Some such groups become extinct themselves and leave no phylogenetic descendants, for example, the pterosaurs. Other groups become extinct, but give rise to descendants that form a separate group; thus the therapsid reptiles died out in mid Mesozoic time, but gave rise to the mammals.

We can identify the cause of extinction when that cause is the direct or indirect effect of human activity in historical times. Man is the known cause of the extinction, for example, of the passenger pigeon, great auk, and numerous other bird species (see Dorst 1974:32); and of the American bison through almost all of its former wide range (see Hornaday 1889).

The causes of extinction are necessarily uncertain when the extinction took place in the geological past. Uncertainties surround the disappearance of particular well-fossilized species in the Quaternary, as we shall see, and it is not surprising that the problem becomes more elusive as we go back in geological time and proceed from the species level to major groups and mass extinctions.[1]

Selected Pleistocene and Post-Pleistocene Ungulates

The course of events leading up to the near extinction of a great herbivore by a new predator, the white man, is illustrated by the historically documented and sad story of the American bison *(Bison bison)*. An account is given by Hornaday (1889).

The original range of the bison extended from Nevada and Oregon to Georgia and Pennsylvania, and from northern Mexico to the Northwest Territories of Canada (figure 36.1A). An estimated 50 million bisons roamed through this vast area at the beginning of the European settlement. Wild bison existed in the eastern United States until 1760–1800, in the middle west until 1810–1825, in central Texas until 1837, and in Oregon until 1838.

Hunting and habitat destruction by the white man gradually eliminated the bison from the central and eastern United States, and from the southern and far western parts of its original range, and by 1870 its range was reduced to an area in the Great Plains (figure 36.1B). About this same time (1870) the killing changed from small-scale hunting to systematic slaughter. Ten years later, in 1880, the range of the bison consisted of one large and five small disjunct areas (figure 36.1C). By the end of 1888 there were only about 830 wild bison left in six small scattered localities (figure 36.1D) (Hornaday 1889). In 1889 the number had dropped further to 541 individuals (Walker 1975). Conservationists then moved to save the bison from complete extinction, and protected herds now exist in several national and state parks.

The Irish elk *(Megaloceros giganteus)*, a giant deer of Pleistocene and early Holocene time, ranged from Ireland across Europe to Siberia and China and south to northern Africa. It flourished in Ireland during a phase following the retreat of ice at the end of the last glaciation. It became extinct in Ire-

1. Attention is called to several treatments of extinction that are quite different from the one presented here, and also different inter se (Simpson 1953: ch. 9; Van Valen 1973: Stanley 1979; Ehrlich and Ehrlich 1981; Martin and Klein 1984).

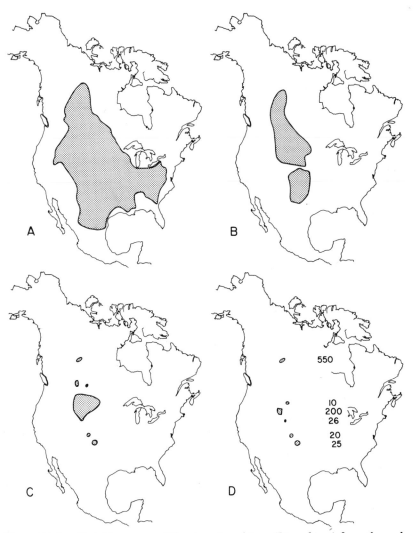

Figure 36.1. Shrinking range of the American bison *(Bison bison)* from the early 18th to the late 19th century. (A) Original range. (B) Range in 1870. (C) Range in 1880. (D) Range and number of individuals at end of 1888. (Rearranged and redrawn from Hornaday 1889)

land about 11,000 years ago. It may have survived into early Holocene time in parts of continental Eurasia (Gould 1974).

Most of the specimens of the Irish elk have been obtained from lake deposits and peat beds in Ireland. The specimens are those of a gigantic deer, standing about 2 meters tall at the shoulders. The stags bore huge antlers weighing up to ca. 25 kg and spanning a width of about 3 meters. These are maximum dimensions. Detailed measurements are given by Gould (1974). The antlers are the biggest known in the deer family.

Various authors have noted that the large and heavy antlers of the Irish elk, which were shed and regrown each year, must have been both a physical burden and a physiological drain on the animals. The older explanations of the extinction of the Irish elk focused attention on the probable disadvantages of the huge antlers, which were simply assumed to be non-adaptive. The antlers of the irish elk may well have been a contributing factor in the animal's extinction, but, as Simpson (1949, 1967) pointed out, there is no reason to believe that the antlers were inadaptive when they developed and while the species was flourishing.

Simpson (1949, 1967) pointed out further that antler size in the Irish elk should not be considered as an isolated character in itself, but rather as a product of relative growth rates in different parts of the body during phylogeny. In the members of the deer family, the allometric growth rate of the antlers is much greater than that of the body, so that the largest species of deer have the proportionately largest antlers (see chapter 33). The large-bodied Irish elk has antlers of about the expected size for a very large deer.

It may be, therefore, that selection promoted large body size in the irish elk, and large antler size was a side effect (Simpson 1949, 1967). Increase in body size is a common trend in mammals, and possesses various selective advantages, as discussed in chapter 28. An alternative possibility is that sexual selection favored increased antler size in the Irish elk because of the advantage of large-antlered males in mating, and large body size was the side effect (Gould 1974).

The causes of extinction of the Irish elk, in Ireland at least, and perhaps elsewhere, probably should be sought in other aspects of its life. The habitat of the Irish elk was apparently open grassy country with scattered woods. This type of vegetation flourished during one phase in the warming trend at the end of the last ice age, and the Irish elk flourished with it. In a later phase of the same climatic trend, however, the grassland was replaced by forest, and the Irish elk, which was not a forest deer, probably could not survive in dense forest. Climatic and vegetational changes in later Glacial time, unfavorable to the Irish elk, are the probable cause of its extinction (Gould 1973, 1974).

The horse, *Equus*, became extinct in North America at the beginning of the Holocene epoch, about 8,000 to 10,000 years ago. North America, as will be recalled from chapter 28, had been the main theater of evolution of *Equus*, and the source of migrant stocks of *Equus* on three other continents. Vast herds of horses, belonging probably to several species of *Equus*, had existed in North America during the Pleistocene. Then *Equus* became completely extinct in North America (Simpson 1951, and personal communication).

Millennia later, in 1519, the European domestic horse, *Equus caballus*, was introduced in North America by the Spaniards. The introduced horse was very successful and established wild herds in the ancient homeland of its forebears.

The reasons for the extinction of *Equus* in North America are unknown. And the mystery is compounded by the fact that *Equus* was able to reestablish itself successfully in the wild in North America several millennia after its extinction here.

Various explanations have been suggested for this extinction, but none of them seem to be completely satisfactory. Perhaps the native North American horse was a victim of the ice age. But *Equus* had survived through the Pleistocene, and, moreover, it occurred in areas beyond the limits of the glaciation. Perhaps the horse was hunted to extermination by the early American Indians, with whom it overlapped in time. This is possible, but the Indians hunted the bison and did not exterminate it. Perhaps the horse fell prey to a disease epidemic. This is possible too. However, there is no evidence of a disease epidemic in the fossil record, although this does not rule out the possibility of such an epidemic (Simpson 1951).

A parallel series of events took place in South America. There also, *Equus* flourished during the Pleistocene, became extinct in the early Holocene epoch, but was introduced from Europe in historic time and subsequently ran wild. The reasons for the extinction in South America are also unknown (Simpson 1951).

Sabertooth Cats

The sabertooth cats were one of the two main branches of the Felidae, the subfamily Machairodontinae, the other branch being the feline cats (subfamily Felinae). The sabertooths differed from the feline cats in the great size of the upper canine teeth and in other related features of the jaws and teeth (figure 36.2).

Sabertooth cats occurred in North American and Eurasia throughout most

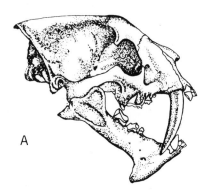

A

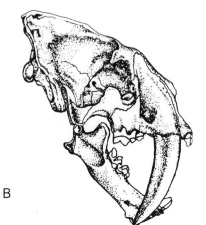

B

C

Figure 36.2. Skulls of two sabertooth cats (A, B) and one feline cat (C). Not shown to same scale. (A) *Hoplophoneus*, Oligocene, skull 15.9 cm long. (B) *Smilodon*, Pleistocene, skull 30.5 cm long. (C) *Pseudaelurus*, Pliocene, skull 14.6 cm long, showing its canine teeth for comparison. (Redrawn with modifications from Romer 1966)

of the Tertiary from the early Oligocene on, and reached South America in the Pleistocene. A general increase in canine size and body size occurred through the Tertiary. *Hoplophoneus*, a characteristic Oligocene member of the group in North America, and *Smilodon*, a larger Pleistocene member in North and South America, are shown in figure 36.2A, B). Then in the Pleistocene the whole group of sabertooths became extinct (Romer 1966).

It was commonplace at one time to attribute the extinction of the saber-tooths to the enlarged canines themselves, which, like the antlers of the Irish elk, were supposed to be inadaptive, and indeed were thought to interfere with biting. As a matter of fact, the large canine teeth did not interfere with biting. The jaw was capable of opening wide and closing tight.

The sabertooth cats were specialized for preying on large, slow-moving, thick-hided herbivores, such as mastodons and giant sloths. Their large canine teeth, with the jaw in an open position, would have served as effective stabbing and tearing organs, backed up by a strong neck and shoulders.

During the Pleistocene, mass extinction thinned the ranks of the category of thick-hided herbivores. The reduction in numbers and eventual extinction of this type of prey animal shrank or closed the adaptive zone of the sabertooth cats. Meanwhile the alternative method of ambush-and-pursuit hunting of fast-moving prey animals had become occupied by the feline branch of the cat family. With their own adaptive zone closing up in the Pleistocene, and alternative adaptive zones preempted, the sabertooth cats had reached the end. This is a good example of a group which was successfully specialized at one stage, but which found itself "overspecialized" at a later stage when conditions were changed (Simpson 1953: 221–222).

Mass Extinction of North American Mammals in the Late Cenozoic

The late Pliocene and Pleistocene were a time of tectonic activity, climatic deterioration, changes in geographical distribution, and biotic turnover. Extensive extinction occurred in the mammals, which had been the dominant land animals through the Tertiary. The northern continents were more severely affected than tropical and southern land areas. Here we will focus particularly on the situation in North America.[2] The chronology of the relevant time period in North America is summarized in table 36.1.

2. For a recent symposium on Pleistocene extinctions see Martin and Klein (1984).

Table 36.1. Chronology of the late Cenozoic in North America
(From Kurten and Anderson 1980: ch. 1)

Epoch	Age	Dating (Approx.)
Holocene		10,000 B.P.–present
Pleistocene	Rancholabrean	0.5 MY B.P.–Holocene
Pleistocene	Irvingtonian	1.8–0.5 MY B.P.
Late Pliocene	Blancan	3.5–1.8 MY B.P.

Some 338 species of mammals with fossil records are known to have become extinct in North America during the late Cenozoic. Some groups became extinct in North America but survived on other continents; e.g., horses, camels, llamas, saiga antelope, yak, capybara, and speckled bear. Other groups became completely extinct; e.g., ground sloths, giant armadillos, glyptodons, giant beaver, mastodons, mammoths, woolly rhinoceros, and saber-tooth cats (Axelrod 1967; Kurten and Anderson 1980).

The extinctions occurred in 38 families belonging to 9 orders, out of 45 families and 12 orders on the continent then (Kurten and Anderson 1890:359–360).

The late Pleistocene mammal extinctions have received the most attention in both the scientific and popular literature. It is necessary to recognize that extinctions occurred throughout the whole time span of 3.5 million years (Anderson and Kurten 1980:357). The distribution of the species extinctions through time is shown in table 36.2.

The extinctions affected mammals of all size classes. it is interesting that small mammals had the highest species extinction rates in the Blancan, while large mammals had their highest extinction rate in the Wisconsinan phase of the Rancholabrean (Kurten and Anderson 1980:358).

In view of the great diversity of the extinction events, with respect to timing, systematic affinities, body size, life history, and ecogeographic province, it is probably futile to look for single specific causes (Kurten and Anderson 1980: ch. 19). An array of factors should be considered such as climatic change, vegetational change, competition, predation, disease, etc. These factors would affect each species in a different way. Some combinations of factors would have very adverse or fatal effects on certain species. If there is any single cause, it is probably not a specific one, but the general orogenic and climatic change of the period, which triggered off complex reverberations in the various ecosystems.

Any substantial climatic or other physical change is going to alter the bal-

Table 36.2. Number of North American mammal species becoming extinct in different ages of the late Cenozoic (Kurten and Anderson 1980:357)

Age and Subdivision	Number of Species	
Blancan		
Very early	12	
Early	49	
Middle	27	
Late	49	
Exact age unknown	1	
Subtotal		138
Irvingtonian		
Early	39	
Middle	14	
Late	33	
Exact age unknown	3	
Subtotal		89
Rancholabrean		
Late Illinoian	7	
Sangamonian	21	
Wisconsinan	77	
Exact age unknown	3	
Subtotal		108
Historic times		3
Total		388
Species with Pleistocene/Holocene records still surviving		229

ance between species, favoring some at the expense of others, and this will lead to species replacement.

Three special hypotheses will be passed in review next.

The overkill hypothesis of Martin (1973) attributes the extinction of large mammals in the late Pleistocene to big-game hunting by early man. The early Indians, on entering North America from Siberia about 11,500 years ago, with a hunting culture, found a rich new food resource in the large mammals then living on this continent. These mammals, as exemplified by the mastodon, mammoth, horse, camel, and bison in North America, had not previously been exposed to predation by hominid hunters. The early human population expanded rapidly and migrated rapidly through North America and to South America, on the basis of the big-game food resource, driving most of their game animals to extinction through excessive hunting. Some of these species did apparently become extinct at about the same time as the postulated early human population explosion, i.e., about 10,000–11,000 years ago (Martin and Wright 1967; Martin 1973).

The early Indians undoubtedly had an impact on the native mammal fauna, and may well have killed off some species, but it seems doubtful that, with a primitive hunting technology, they could have had the extensive effect postulated by the overkill hypothesis. It has been pointed out that a Pleistocene fauna in Australia which included a number of large animals, such as the giant kangaroo, coexisted with humans for at least 7,000 years (Gillespie et al. 1978). In North America the extinction rate in the bird fauna in the late Pleistocene was similar to that in the mammal fauna (Grayson 1977).

Axelrod (1967) emphasizes new climatic conditions, especially cold, as a cause of mammalian extinctions in the Pleistocene. Tertiary climates had been generally equable, and mammals were adapted to such equable climates. The climate of the Pleistocene was highly inequable with severe cold seasons. Severe cold brings about the death of individual mammals, by physiological effects or by starvation, and these individual deaths could accumulate to the level of species extinction. New cold climates would also have adverse indirect effects on many mammals by causing changes in the vegetation and in their habitats.

The body-size distribution of the extinctions in the late Pleistocene is in agreement with the cold hypothesis. Small mammals had relatively low extinction rates then. Most small mammals can escape a cold season by retreating into burrows, dens, or nests. Furthermore, small mammals tend to have short reproductive cycles that can be geared to a seasonal climate. In many small mammals the young can grow to maturity during a short favorable season (Axelrod 1967). Large mammals, which mostly do not have these ways of fitting into a highly seasonal cold climate, had the highest extinction rates in the late Pleistocene.

Several groups of facts are not explained by the cold hypothesis. Why did many North American mammal species become extinct only in the late Pleistocene rather than in the late Pliocene and early Pleistocene when the change to a glacial climate actually occurred? The cold hypothesis does not account for the higher extinction rate of small mammals, as compared with large ones, in the Blancan age. Another problem with the cold hypothesis is that many large mammals became extinct, not only in northern continents, but also in Africa which escaped the full force of the glacial periods (Simpson, pers. comm.).

Bryson et al. (1970) reconstruct the climatic and vegetational changes in the Great Plains during late Pleistocene time. During most of the late Pleistocene the ice sheet blocked the Great Plains off from arctic air. Tall grassland and large herbivores could exist under those conditions. But the retreat of the ice sheet in response to a warming trend at the end of the Pleistocene removed this block and exposed the Great Plains to arctic air. Cold drying winds brought about vegetational change from tall grass to short xeric grass. The combination of severe cold and poor grazing could account for the loss of many large herbivores in the Great Plains at the end of the Pleistocene (Bryson et al. 1970; M. Bolick, pers comm.).

Dinosaurs

The dinosaurs, comprising two orders, Saurischia and Ornithischia, were the dominant land animals through the Mesozoic. A great diversity of forms had developed during this era, as indicated by the classification of the group into 25 families and 218 genera (Colbert 1961:265–274). They ranged in size from various small-bodied types to giants like *Brontosaurus* and *Tyrannosaurus*, and in diet from herbivores to carnivores. The group was worldwide in distribution, and had been dominant on the land for over 100 million years.

Then in the late Cretaceous the entire group of dinosaurs became extinct. The extinctions came in a relatively short period, on the geological time scale, but appear to have been spread out over hundreds-of-thousands and millions of years (Colbert 1961; Simpson 1968; Simpson and Beck 1965: 790; Van Valen and Sloan 1977).

There are various schools of thought regarding the causes of dinosaur extinction. One of these is climatic change; it will be discussed below. A second school emphasizes biotic factors, such as mammal predation of dinosaur eggs, mammal competition, overgrazing by dinosaurs, and disease. A third postulates a physical catastrophe of global proportions, this will be discussed later.

Van Valen and Sloan (1977) studied a fossil sequence in the Hell Creek Field, Montana, ranging from late Cretaceous to Paleocene time. There is a continuous sequence of terrestrial plants and animals here with no gaps. The older Cretaceous flora was subtropical; the Paleocene flora was temperate and included deciduous dicots and ferns. The temperate vegetation came from farther north, moved in, and replaced the older subtropical vegetation. The vegetational shift indicates a climatic change from a subtropical climate to a temperate one with cold winters.

Associated with the subtropical flora was a terrestrial animal community of late Cretaceous age containing numerous dinosaurs, other reptiles, and primitive mammals including marsupials. This fauna was replaced in the early Paleocene by a different animal community containing primitive placental mammals and some reptiles but no dinosaurs. Seven dinosaur species belonging to the older community persisted during the transition period, but gradually become more rare and finally disappeared. The disappearance of the dinosaurs in Hell Creek Field took place gradually and slowly over about 100,000 to 400,000 years (Van Valen and Sloan 1977).

The evidence here suggests that the ultimate cause of the extinction of the dinosaurs in Montana was the advent of a new climate with cold winters and a temperate vegetation type. In addition, the new breed of placental mammals may have been superior competitively to the dinosaurs in the new climate and vegetation (Van Valen and Sloan, 1977).

Various authors have emphasized climate, and particularly the change from an equable to a relatively inequable climate, as the main cause of dinosaur extinction on the worldwide scale (Axelrod and Bailey 1968; Spotilla et al. 1973; Benton 1979). Equable warm climates had prevailed during most of the Mesozoic, and the dinosaurs had been evolving for about 130 million years under such climatic conditions. The latter part of the Cretaceous saw a change to more inequable continental climates with greater extremes of temperature.

The dinosaurs were probably ectothermic (Benton 1979), and not homoiothermic as has been suggested (Bakker 1975) on insufficient evidence, for ectothermy would be a successful strategy in a warm climate. In a continental climate with temperature extremes, however, an ectothermic reptile would suffer thermal stress (Spotilla et al. 1973; Benton 1979).

Some groups of reptiles in the Cretaceous could escape very cold or hot periods by burrowing or hibernating (lizards, snakes), others by submerging (turtles), and still others by inhabiting tropical and subtropical rivers (crocodilians), and these survived into the Recent. But the terrestrial dinosaurs did not have these means of escape from temperature extremes (Axelrod and Bailey 1968).

Mass Extinction in the Late Cretaceous

The late Cretaceous witnessed one of the great mass extinctions of all time. The age of reptiles came to an end. The ichthyosaurs and plesiosaurs became extinct in the seas, and the pterosaurs in the air, as well as the dinosaurs on land.

In addition, the previously successful mollusk group of ammonites died out in the seas, and so did many foraminiferans.

The controversy over the causes of dinosaur extinctions is repeated on a grander scale with regard to the mass extinctions across the board in the late Cretaceous.

The Asteroid Impact Theory

Much attention is being directed currently to the asteroid impact hypothesis of Alvarez and his colleagues (Alvarez 1983; Alvarez et al. 1980, 1984). Evidence indicates that an asteroid about 10 km in diameter hit the earth about 65 million years ago. The impact produced a cloud of dust in the upper atmosphere that lasted for two or three years. A layer of iridium, a platinum-family metal that is rare in the earth's crust but abundant in asteroids, became deposited in marine sediments at the Cretaceous-Tertiary boundary all over the world.

An extra-terrestrial origin of the layer of sediments at the Cretaceous-Tertiary boundary is not accepted by all students. Bohor et al. (1984) cite crystallographic evidence in favor of an asteroid origin, but others consider a volcanic origin to be likely (Officer and Drake 1983). For the purpose of this discussion, let us accept an asteroid origin.

A connection, or at least a correlation, between the asteroid impact and extinction is indicated by changes in fossil biota at the iridium layer. In a given vertical sequence a group may disappear at the iridium layer. This is the case for ammonites and other marine invertebrates in Denmark, foraminiferans in Italy and off Africa, radiolarians in the Caribbean, and angiosperm pollen in New Mexico (Alvarez 1983; Alvarez et al. 1984; Hsu et al. 1982).

Alvarez and his colleagues (1980, 1983) propose the following scenario to account for the mass extinction. The dust layer produced by the asteroid impact cut off light and blocked photosynthesis in both land plants and marine algae for one or several years. This led to death by starvation of marine life and land animals, first herbivores and then carnivores.

This scenario is the weakest part of the whole theory in my opinion. The following difficulties are noted.

1. The disappearance of a species at one fossil locality does not mean that the species became extinct. Other populations of the same species in other localities may have survived.

2. It remains to be seen whether the dust layer could have cut off photosynthesis sufficiently to bring about widespread starvation. Dust layers thrown up by gigantic volcanic eruptions in historical times have reduced the sunlight and temperatures in areas far from the volcano, but have not led to the complete cessation of photosynthesis or extinction of herbivore species in such remote areas (see Stommel and Stommel 1979; Stothers 1984).

3. The asteroid impact scenario does not explain why some groups died out while others survived. This is the crux of the problem, as pointed out also by Simpson (1968), Mayr (1982:620), and Colbert (pers. comm.).

4. The asteroid impact theory requires that the extinctions all came within a short period of years or decades immediately following the catastrophe. Simultaneity of the extinctions has not been demonstrated. For some groups, including the dinosaurs and foraminiferans, the available evidence indicates gradual disappearance on a geological time scale (Simpson and Beck 1965; Van Valen and Sloan 1977).

In short, a convincing case has been made for an asteroid impact, but not for its having brought about mass extinction.

The asteroid impact theory is the latest in a series of catastrophic theories of Cretaceous mass extinction. Previous suggestions have invoked lethal bursts of radiation from an exploding nova, and lethal doses of cosmic rays reaching the earth's surface during a reversal of the earth's magnetic field.

All of these catastrophic theories have been put forward by physical scientists. They are simplistic and inadequate attempts to deal with very complex biological phenomena. Most evolutionary biologists prefer some version of a climatic or combined climatic-biotic theory of Cretaceous mass extinction.

The late Cretaceous was a time of continental uplift, mountain building, volcanic activity, and climatic change. Climates became inequable by Cretaceous (though not by Quaternary) standards. A change in the physical environment will cause multiple biotic effects. The competitive balance between species will be altered, giving some species an advantage and others a disadvantage, so that species replacement takes place on a wide scale. The species extinctions will be selective, gradual, and spread out in time, as is actually found to be the case.

And what about the fossil evidence associated with the iridium layer at the Cretaceous-Tertiary boundary? It is suggested here that the asteroid impact caused local extinctions within a number of species, but not worldwide mass extinction. Big volcanic eruptions would do the same. My interpretation of the evidence is that, as far as mass extinction is concerned, the arrival of the asteroid in the late Cretaceous was not a cause but a coincidence.

Periods of Mass Extinction

It has long been recognized that the geological history of life is punctuated by episodes of mass extinction separated by long periods of gradual evolutionary change. The fossil record points to a condition somewhat intermediate between the old catastrophism of Cuvier and the strict uniformitarianism of Lyell, a condition of episodic evolution. There are periodical episodes of mass extinction, when many old groups die out, followed by the development and adaptive radiation of new groups.

Indeed, the boundaries between the geological eras and periods are based partly on the episodes of mass extinction, particularly in animals. The defining characteristic of animal episodic evolution is embedded in the terminology of the eras (Paleozoic, etc.).

A significant feature of episodes of extinction is the occurrence of high rates of extinction in independent groups at about the same time. Thus both the ammonoids and the reptiles suffered high rates of extinction in the late Permian, late Triassic, and late Cretaceous (Newell 1967).

Newell (1967) collated and analyzed the data on the first and last appearances in the fossil record of 2250 animal families belonging to all major groups. A summary of his findings is presented in figure 36.3. Figure 36.3A shows the rate of extinction of animal families, as indicated by their last occurrence, through geological time. We see that the curve for family extinction rate is very uneven. Major peaks rising above the background level of extinction occur in the late Cambrian, late Devonian, late Permian, late Triassic, and late Cretaceous; minor peaks occur at the ends of some of the intervening periods.

A recent analysis of the fossil records for about 3300 families of marine animals, both invertebrate and vertebrate, has been made by Raup and Sep-

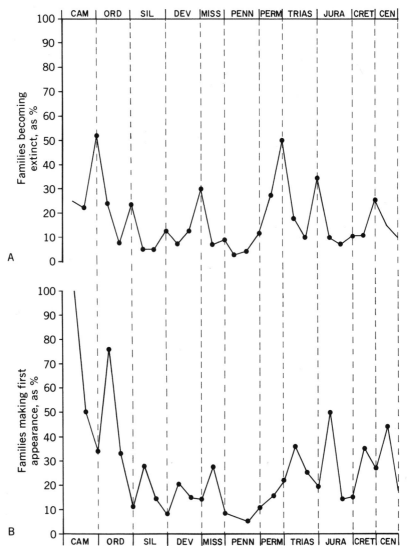

Figure 36.3. Episodes of extinction and replacement in animals. (A) Rate of extinction of families. (B) Rate of first appearances of new families in the record. (Redrawn from Newell 1967)

koski (1982). Their analysis shows five periods of mass extinction of families since the Cambrian. Four big peaks occur in the late Ordovician, late Permian, late Triassic, and late Cretaceous. A fifth smaller peak occurs in the Devonian. This survey and the earlier one of Newell yield very similar results.

Most of the periods of mass extinction listed above occur in times of continental emergence and continental climates. The conclusion drwn from this correlation by a number of students (e.g., Newell 1967; Axelrod 1974) is that the changes in the physical environment were the ultimate cause of the extinctions, particularly those in land animals.

More recently Raup and Sepkoski (1984) have assembled data on the time ranges of about 2,900 now-extinct families of marine vertebrates, invertebrates, and protozoans for the time interval from the late Permian through the Pliocene. They plotted the family extinction rate in each age against geological time. The results are shown in figure 36.4. The curve for family extinctions has 12 peaks in the interval from late Permian to Pliocene time.

A further finding of utmost interest is that the extinction peaks have a periodicity with a mean interval of 26 million years. Seven of the twelve peaks are on and five peaks are close to periods 26 million years apart, as shown by the vertical lines at the top of figure 36.4. Several statistical tests confirm the reality of this periodicity (Raup and Sepkoski 1984).

Such a regular periodicity points to some astronomical cause of the mass extinction episodes, rather than to earthbound causes, as Raup and Sepkoski

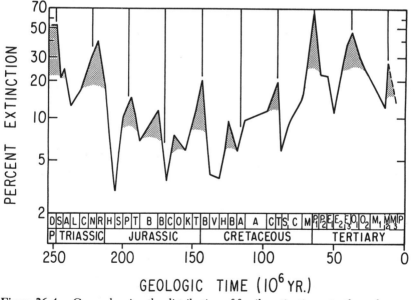

Figure 36.4. Curve showing the distribution of family extinction rates through geological time for marine animals and protistans. The time span is late Permian to Pliocene. The block letters on the abscissa are stages of the various periods. The vertical lines at the top of the graph mark 26 MY intervals. (Raup and Sepkoski 1984)

(1984) note. The astronomical factor could be solar, solar-system, or galactic. A search for an astronomical factor with a 26 MY periodicity is underway currently (see Kerr 1984).

Earlier in this chapter I rejected a proposed extra-terrestrial cause of mass extinction in the late Cretaceous, but am here presenting the case for such a cause approvingly. I am aware of the inconsistency. The reason for it is that the asteroid impact hypothesis, in its present formulation at least, does not fit all of the biological evidence, whereas the convincing evidence for 26 MY extinction cycles does suggest an extra-terrestrial factor.

That factor might only push the ultimate cause of mass extinction one step back. Some astronomical event could, for example, periodically reduce the heat received at the earth's surface, and from there on it is the familiar reaction chain. A change in the physical environment disturbs the balances within and between biotic communities, resulting eventually in numerous species replacements.

Let us return to Newell's (1967) picture of ebbs and flows in the fossil record. Figure 36.3B shows that the first appearance of new animal families in the record is uneven through time. There are peaks in the origination curve as well as the extinction curve. Furthermore, the peaks in the origination curve tend to lag behind those in the extinction curve. Episodes of mass extinction are followed by the formation and development of new groups (Newell 1967). Extinction of old groups provides ecological opportunities for the information of new groups (see chapter 30).

PART VIII

From Organic Evolution to Human Evolution

CHAPTER 37
Biological Aspects

Classification of the Primates

The order Primates is remarkable for its diversity and more especially for the range of this diversity. The range between such extreme types as a tree shrew or lemur on the one hand, and an ape or man on the other, is enormous. The diversity in other large mammalian orders, such as the Artiodactyla, is pretty much on one grade, but that in the primates presents itself in a series of grades.

The amount of diversity in the primates is indicated in table 37.1. There are 4 suborders, 11 families, 59 genera, and about 193 species. The types of animals listed in table 37.1 portray the range of the diversity.

The diversity in the primates falls into about five grades. The grades would be: (a) tree shrews, (b) prosimians (lemurs, lorises, tarsiers), (c) monkeys,(d) apes, and (e) humans.

At the lower end of the spectrum of grades, the primates approach the somewhat more primitive order Insectivora. The tree shrews are intermediate between the two orders, and are segregated out of the primates in some classifications. The primates are probably derived from the insectivores. The recorded fossil history of the insectivores goes back to the Cretaceous, and that of the primates to the Paleocene.

The fossil record indicates that the first primates were prosimians, that they were widespread in distribution, and that they underwent an adaptive radia-

Table 37.1. System of classification of Recent primates (Compiled from Colbert 1980; and Walker 1975)

Major group	Genera and Species	Geographical Distribution
Suborder LEMUROIDEA		
Superfamily Tupaioidea		
Family Tupaiidae; tree shrews	5 genera (*Tupaia*, etc.); ca. 15 species	Tropical Asia and Malaysia
Superfamily Lemuroidea		
Family Lemuroidea; lemurs	6 genera (*Lemur*, etc.); ca. 16 species	Madagascar
Family Indridae; woolly lemurs	3 genera; 4 species	Madagascar
Superfamily Daubentonioidea		
Family Daubentoniidae; aye-aye	Monotypic; *Daubentonia madagascariensis*	Madagascar
Superfamily Lorisoidea		
Family Lorisidae; lorises, galagos	6 genera (*Loris, Galago,* etc.); ca. 12 species	Old World tropics
Suborder TARSIOIDEA		
Superfamily Tarsioidea		
Family Tarsiidae; tarsiers	Single genus (*Tarsius*); 3 species	Malaysia
Suborder PLATYRRHINI		
Superfamily Ceboidea		
Family Cebidae; New World monkeys	12 genera (*Cebus, Alouatta,* etc.); ca. 30 species	New World tropics
Family Callithricidae; marmosets	5 genera (*Callithrix*, etc.); ca. 35 species	Tropical South America and Panama
Suborder CATARRHINI		
Superfamily Cercopithecoidea		
Family Cercopithecidae; Old World monkeys, baboons	14 genera (*Macaca, Papio,* etc.); ca. 66 species	Old World tropics and subtropics
Superfamily Hominoidea		
Family Pongidae; apes	5 genera; ca. 10 species *Hylobates* (gibbon), *Symphalangus* (Siamang gibbon), *Pongo* (orangutan), *Pan* (chimpanzee), *Gorilla* (gorilla)	Old World tropics
Family Hominidae; man	Monotypic; *Homo sapiens*	Cosmopolitan

tion in the Paleocene and Eocene. Some of the branch lines died out in the Eocene and Oligocene. Other branch lines persisted to the Recent. The surviving major branch lines correspond to the various superfamilies listed in table 37.1 (Simpson 1967:87–93).

The Hominoids

One of the main branches of the primates, having a separate existence since the Eocene, is the hominoid line. The Hominoidea consists of three or four families. The Recent families are the Hominidae and Pongidae. The latter is sometimes subdivided into the Hylobatidae (gibbons) and Pongidae sensu stricto (orangutan, chimpanzee, gorilla). An extinct family of apes, Oreopithecidae, includes *Oreopithecus*.

The Oligocene and Miocene fossil records of these families are fragmentary. As a result many critical points of phylogeny are uncertain. The *Oreopithecus* lineage appeared in the Miocene and died out in the early Pliocene. The true apes (Pongidae sensu lato) are represented by *Aegyptopithecus* in the Oligocene, *Proconsul* in the Miocene, and *Dryopithecus* in the Miocene and Pliocene. The ramapithecines, consisting of *Ramapithecus, Sivapithecus* and related forms, occurred in Africa, Asia, and Europe in middle Miocene and early Pliocene time. *Ramapithecus* has usually been classified in the Hominidae but now seems to be more closely allied to the pongids (Pilbeam 1984).

The Hominidae, with two genera, *Australopithecus* and *Homo*, has been separate from the apes since early Pliocene or late Miocene time (Wolpoff 1980: Pilbeam 1984). The ages of the two hominid genera are given in table 37.2 following several recent authorities. The dating is in a continuous state of flux with the finding of new fossils and use of new dating methods.[1]

The Plio-Pleistocene australopithecines are known mainly from South Africa and East Africa. They were ape-like in skull, face, and brain size (figure 37.1), but human-like in dentition, erect posture, and bipedal locomotion. Four species are recognized: the gracile A. *afarensis* and A. *africanus*, and the robust A. *boisei* and A. *robustus*. The latter two could be geographical races of one species. *Australopithecus afarensis* is the oldest and ancestral form; it was followed by A. *africanus*; and the two robust australopithecines

1. Objective reviews of fossil hominids are given by Campbell (1974) and Wolpoff (1980). A new edition of Campbell's treatment is in preparation. New findings and reinterpretations of old findings appear frequently in the general science journals.

Table 37.2. Ages of genera and certain species of Hominidae
(Data from Campbell, 1974; and Wolpoff, 1980)

Group	Epoch	Age (approximate)
Australopithecus	Pliocene to Pleistocene	3.8 to 1.3 MY B.P.
Homo	Pleistocene to Recent	
H. *habilis*	Early Pleistocene	1.8 MY B.P.
H. *erectus*	Pleistocene	1.6 MY to 300,000 B.P.
H. *sapiens*	Middle Pleistocene to Recent	300,000 B.P. to present

apparently diverged from the gracile stock. All the australopithecines died out before the end of the Pleistocene.

The genus *Homo* is a large-brained lineage. Three species are recognized: H. *habilis*, H. *erectus*, and H. *sapiens* (figure 37.1). All three existed in Africa in the Pleistocene. *Homo habilis* is the oldest member of the genus, dated as 1.8 MY old, and represents a transitional form between A. *africanus* and *Homo*. *Homo habilis*, H. *erectus*, and H. *sapiens* seem to form a series of successional species which grade into one another through time. Current estimates of their ages are given in table 37.2.

Only one species of Homo has ever existed at any one time level. However, *Homo* overlapped in time with *Australopithecus*, and H. *habilis* coexisted with the robust australopithecines (Klein 1977).

Homo is close to and probably derived from *Australopithecus*. *Homo habilis* could well be a connecting link between A. *africanus* and H. *erectus* (Wolpoff 1980). The exact structure of the phylogenetic tree is much debated, however, and there are numerous phylogenetic hypotheses to choose from.

The Arboreal Heritage

The primates are predominantly animals of warm regions (see table 37.1); they are predominantly arboreal, the ground-living forms like baboons and man being exceptional; and they are predominantly omnivorous. Life in an arboreal habitat resulted in the development of some adaptations that were later to be exploited by man.

An arboreal mammal must be able to judge distances accurately, focus on objects, hold on to branches, and hold on to food objects. The correspond-

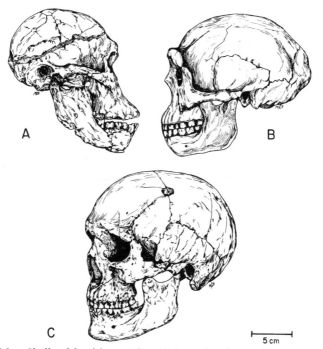

Figure 37.1.   Skulls of fossil hominids. (A) *Australopithecus africanus,* early Pleistocene, South Africa. (B) *Homo erectus,* middle Pleistocene, Peking. (C) *Homo sapiens,* late Pleistocene, Peking. All drawings to same scale. (Rearranged from Brace,
Nelson, and Korn 1971)

ing adaptations in the primates are a well-developed visual sense, grasping
hands with an opposable thumb, and a balancing tail.

The visual sense in primates is a composite of several features. The eyes
are well developed. They are placed in the front of the face, which permits
focusing on objects and judgment of distance by parallax. A spot in the ret-
ina of the eye permits fine focusing. Finally, a color sense, which is lacking
in many groups of mammals, is developed in the primates, where it probably serves to give a better sense of perspective.

The primate visual sense is an important component, in turn, of primate
intelligence. Primate intelligence is another composite trait. There is, to begin with, the brain, which is well developed in all the higher primates. But
what that brain is able to learn about the external world depends largely upon
the *combination* of primate eyes and primate hands. Other sense organs,
particularly hearing, play supplementary roles.

The primate brand of intelligence is based mainly on the combination of focusing eyes and grasping or feeling hands. This is not the only possible brand. Canine intelligence, for example, depends mainly on the combination of focusing eyes, sense of smell in the nose, and sensitive ears; bee intelligence depends on the combination of compound eyes, the insect type of color vision, and flexible, odor-detecting antennae.

The eye-hand combination was and is the basis of concentration on concrete objects in the pre-hominid primates. The same combination was to become the basis of tool making, tool use, experimentation, and problem solving in the hominids.

The Terrestrial Heritage

The apes are forest animals and are arboreal, or at least partially so. The gibbons spend most of their lives in the trees and move about by brachiating and leaping. Chimpanzees and gorillas are semi-terrestrial, and move on the ground by knuckle-walking.

Australopithecus and *Homo* in their well-developed forms were and are animals of savannas and plains. Their mode of locomotion is walking and trotting on two legs. Bipedalism and its corollary, erect posture, are distinctive morphological characteristics of the Hominidae.

This distinction between hominids and pongids is less clearcut in the case of *Australopithecus afarensis* and *Homo habilis*, which, though bipedal, probably climbed trees to a considerable extent (Susman and Stern 1982; Lewin 1983).

Forest habitats were diminishing in extent and savanna country was expanding during the middle and late Tertiary. Climatic and vegetational changes opened up new habitats. Invasion of this new habitat probably marked the divergence of the primitive hominids from the ancestral forest-dwelling apes. Bipedal locomotion and erect posture were part of the adaptive character combination fitting the hominids for life on savannas and plains. Much of the later recorded evolution of the hominids in the Pleistocene consists of a trend toward a more fully erect posture, with neck and head thrown back, and with the body modified for an improved walking gait.

Bipedal locomotion had the important consequence of freeing the hands for the making and use of tools and weapons. The earliest stone tools are found in the Pliocene, and somewhat more complex ones in the early Pleistocene (Pilbeam 1984). The handling of objects is an important component of intelligence in arboreal primates, as noted previously; and now with the change from an arboreal to a terrestrial habitat, and the release of the hands from the function of locomotion, the manipulation of things could be developed to its full potential. Primitive technology was an early result.

Intelligence is an adaptively valuable trait in the primates, and is undoubtedly favored by natural selection. An evolutionary trend toward increased intelligence can be discerned in the non-hominid primates. This trend reaches a very high level in the chimpanzee and orangutan. It continues to still higher levels in the hominids.

In the Hominidae, judging by cranial capacities, *Homo erectus* was considerably more intelligent than *Australopithecus* (see figure 37.1). Cranial capacities for the two groups show the following ranges and means; modern man is included for comparison (from Campbell 1974:272):

	Range, cm^3	Mean, cm^3
A. *africanus*	435–815	588
Homo erectus	775–1225	950
H. *sapiens*	1,000–2,000	1,330

It is quite likely that the superior intelligence and superior tools and weapons of *Homo erectus* or its immediate ancestor gave it an advantage over *Australopithecus* in direct and indirect competition between the two groups during the Pleistocene. At any rate, *Australopithecus* died out, while *Homo erectus* survived and later developed into H. *sapiens*.

Racial Variation

The modern historical age opens with *Homo sapiens* worldwide in distribution and differentiated into a series of geographical races as well as innumerable local races. The geographical races recognized by Garn (1961) are the following:

Asian mongoloids	African blacks
American Indians	Australian blacks
Indians (of India,	Melanesians
Pakistan, etc.)	Micronesians
European whites	Polynesians

These are typical geographical races, as in other species of mammals. Some of the racial characters in body size, body proportions, pigmentation, etc., are correlated with climate (see Coon 1955; Garn 1961). It is well known, for instance, that racial variation in skin color is correlated with the sunniness of the climate.

Skin color regulates vitamin D synthesis in man. Vitamin D in the right amount is necessary for proper growth, too little vitamin D being a cause of rickets, and too much, a cause of bone calcification. Vitamin D is synthesized in the skin. And its synthesis in the skin is controlled by sunlight: the more sunlight, the more vitamin D.

Dark-skinned races live in sunny areas. Here the dark skin pigments filter out sunlight and prevent the synthesis of too much vitamin D. Conversely, the white race occurs naturally in northern Europe, where it could not synthesize enough vitamin D without a pale skin, which lets in maximum sunlight. But white skin absorbs too much sunlight in sunny climates, and white persons transplanted to such climates become suntanned as a phenotypic modification to prevent this excessive light absorption. On the other hand, black races would probably die of rickets under natural conditions in northern Europe. Racial differences in skin color are thus adaptive in relation to the sunniness of the climate (Loomis 1967).

Facts such as these bespeak the role of natural selection in the development of at least some of the racial characteristics of *Homo sapiens*. Local racial variation has a haphazard pattern, in some cases, suggesting that drift or the selection-drift combination also plays a role (see chapter 16). In still other cases, the features of human racial variation point to a controlling influence of migration and gene flow (see chapter 20).

Conclusions

The aspects of hominid evolution discussed in this chapter are extensions of non-hominid primate evolution. We see evolutionary trends in hand use, intelligence, posture, etc.; we see signs of a quantum shift to a new mode of locomotion in a new type of habitat; there is interspecific selection between hominid lines; and there is geographical race formation in the surviving widespread hominid species.

These aspects of hominid evolution are a part of the larger picture of organic evolution. They can be explained as results of the operation of the evolutionary factors and forces set forth in earlier chapters of this book.

In this connection it is worthwhile to note that 600 generations or more can occur in man in a period of 10,000 years. Substantial evolutionary changes could be brought about by selection or selection-drift in this short time period. And longer periods, which are available according to the fossil record, would of course allow for more far-reaching evolutionary changes.

CHAPTER 38
Social Aspects

Social Groups in Non-Human Primates

Primates are social animals. Individuals in primates are usually members of family groups and of larger troops. Social grouping in the higher primates above the prosimian grade is promoted by two conditions. The infant and juvenile stages are relatively long in monkeys and apes, as compared with most mammals, and this leads to a prolonged association between mother and offspring. And the absence of a clearly demarcated breeding season in most higher primates and many prosimians leads to a continuous or at least prolonged association of the sexes. The combination of female-offspring bonds and female-male bonds results in family groups and troops.

There is much species-to-species variation in social structure in the primates. A detailed account of primate social organization is beyond the scope of this chapter.[1] Eisenberg and coworkers (1972) give a useful classification of the modes of social organization, which is presented below.

1. Solitary animals. Adult females and males have separate centers of activity. Maternal family of female and offspring. Found in some lemurs and loris and in the aye-aye. (See table 37.1 for primate classification system.)

1. See DeVore (1965), Eisenberg et al. (1972), Wilson (1975), and Lancaster (1975).

2. Biparental family. The family group consists of a female, a male, and young. Found in a species of woolly lemur, and in some marmosets, New World monkeys, and gibbons.

3. One-male troop. The troop consists of several maternal families and one adult male in contact with them all. The male has a strong intolerance of other mature or maturing males. Found in some New World monkeys (including the howler monkey) and Old World monkeys (including hamadryas baboons and gelada baboons).

4. Age-graded multiple-male troop. A cohesive group consisting of several females, several males, and young. There is an intermediate degree of male tolerance. This degree of tolerance permits the coexistence of several males of varying ages, mostly young males, with the dominance order corresponding to the age of the males. Found in some New World monkeys (including, again, the howler monkey) and Old World monkeys (including macaques), and the gorilla.

5. True multiple-male troop. As in mode 4, but with a high degree of male tolerance, permitting the coexistence of several adult males. These are codominant and cooperative and maintain a flexible oligarchy in the troop. Found in a species of lemur, a species of woolly lemur, in some Old World monkeys (including baboons and macaques), and in the chimpanzee.

This series of types of social organization probably corresponds approximately to an evolutionary trend. It is probable that mode 1 is primitive, mode 3 intermediate, and modes 4 and 5 derived. Going along with the changes in social structure is a trend toward larger-sized groups (Eisenberg et al. 1972).

The types of social organization also correlate roughly with habitat. One-male troops (mode 3) are found mainly though not exclusively in species with an arboreal foraging habitat. Conversely, true multiple-male troops (mode 5) tend to occur preponderantly among species with semi-terrestrial foraging habits (Eisenberg et al. 1972).

The social group serves several useful functions in the higher non-human primates. The members of a group band together in holding a territory and in defense against predators. Furthermore, and most importantly, the troop is the medium for the sharing of experiences and teaching of the young about a variety of vital matters—matters such as the recognition of edible fruits and leaves and of poisonous plants, the whereabouts of enemies, and the like.

Table 38.1. Partial vocabulary of Vervet Monkey *(Cercopithecus aethiops)* (Struhsaker 1967)

Vocalization	Peer Group Uttering Sound	Meaning
Woof, woof	all	accepts subordination
Waa	males	" "
Rraugh	juveniles, females	expresses nonaggression
Teeth chattering	males	" "
Chutter (low staccato)	juveniles, females	wants aid against intragroup threat
Squeal	juveniles, females	" "
Squeal-scream	females	does not want copulation
Aarr-rraugh	juveniles	foreign group approaching
Aarr	all	" " "
Snake chutter	juveniles, females	warning of snake predator
Uh	all	minor mammal predator near
Nyow	all	" " " "
Chirp	juveniles, females	major mammal predator near
Rraup	juveniles, females	major bird predator near or approaching
Threat-alarm bark	males	major mammal or bird predator near

Primate Language

The requirements for communication and coordination of individual activities in troops of monkeys and apes are met by body languages and by vocal animal languages. Both modes of expression are well developed in monkeys and apes.[2] Monkeys are noted for their incessant and noisy chatter.

Vocabularies are known for species belonging to the following genera of monkeys: *Cercopithecus, Erythrocebus, Macaca, Papio, Presbytis, Alouatta,*

2. See DeVore (1965) and Altmann (1967).

and *Aotus*; and for the following species of apes: *Hylobates lar, Pan troglodytes,* and *Gorilla gorilla* (Struhsaker 1967).

Struhsaker (1967) studied vocal communication in the vervet monkey *(Cercopithecus aethiops)*. At least 36 vocal sounds were recorded, which, with some synonymy, convey 23 distinct meanings. Some of these are listed in table 38.1. Walker (1975) recorded the vocal repertoire of a douroucoulis monkey *(Aotus trivirgatus)*. About 50 vocal sounds are known here. This monkey has vocalizations to distinguish between: suspicion of danger ("book") vs. warning of danger ("wook"); and curiosity without suspicion ("uuhh") vs. curiosity with suspicion ("huh" or "wheu").

Washburn and Hamburg (1965) relate the following anecdote concerning baboons in Nairobi Park, Africa. These baboons are protected and had become accustomed to people in cars. A parasitologist once shot two of the baboons from a car. Some baboons were presumably eye-witnesses of the event, but others undoubtedly were not. News of the event quickly spread throughout the baboon community, with the result that it became impossible to get close to baboons while in a car and remained so for at least eight months.

Intelligence and Learning in Non-Human Primates

A language with any substantial repertoire requires a social group above a critical size for its maintenance and perpetuation. It also requires a certain degree of intelligence and learning ability on the part of the members of the group. Complex interactions and feedback exist between the social group and the various facets of intelligence.

The non-human primates, especially monkeys and apes, are noted for their intelligence. This is shown by a large brain size in proportion to body size.

The encephalization quotient (EQ) gives a quantitative measure of brain weight relative to body weight. The average EQ for mammals is 1.0. The whitetail deer *(Odocoileus virginianus)* is close to the mammalian average with an EQ of 1.06. An EQ below 1.0 represents a low brain size relative to body size, as in the Mexican opossum *(Didelphis marsupialis,* EQ = 0.35–0.57). And an EQ above 1.0 indicates a larger than average brain for a given body size. It is interesting to see the EQ values for various primates in table 38.2. These values are all over 1.4 and, for monkeys and apes, are mostly greater than 2.0 (Eisenberg 1981).

Many examples have been recorded of the problem-solving ability of

Table 38.2. Encephalization quotients of various primates (Eisenberg 1981:500.)

Group and Species	EQ
Prosimians	
6 species	1.5–1.8
Nycticebus coucang	2.3
New World Monkeys	
Saguinus midas	2.4
S. oedipus	2.0
S. tamarin	1.9
Saimiri sciureus	3.8
Cebus capucinus	3.4
Ateles paniscus	2.6
A. belzebuth	2.6
Alouatta villosa	1.6
Old World Monkeys	
Macaca mulatta	2.3
Presbytis obscurus	1.7
Apes	
Hylobates lar	3.2
Symphalangus syndactylus	2.2
Pongo pygmaeus	1.6–1.9
Gorilla gorilla	1.4–1.7
Pan troglodytes	2.2–2.5
Hominids	
Homo sapiens	7.3–7.7

monkeys and apes. The chimpanzee is a favorite subject in this regard. Chimpanzees not only use tools for obtaining food, as do a few other mammals and birds, but actually make crude tools out of sticks for accomplishing a definite objective (Goodall 1965).

Table 38.3. Duration of stages of life in various primates (Wilson 1975)

Species	Gestation period, days	Infantile phase, years	Juvenile phase, years	Total post-natal immature stage, years	Total life span, years
Lemur	126	0.75	1.75	2.5	14
Rhesus macaque	168	1.5	6.0	7.5	27–28
Gibbon	210	2.0(?)	6.5	8.5	30+
Chimpanzee	225	3.0	7.0	10.0	40
Gorilla	265	3.0+	7.0+	10.0+	35(?)
Orangutan	275	3.5	7.0	10.5	30+
Man	266	6.0	14.0	20.0	70–75

The learning of languages and of complex skills requires time as well as native intelligence. Prolongation of the immature stages of individual development is another well-known trend in the primates, as the figures in table 38.3 show.

The Hominids

The various adaptive features of social grouping mentioned earlier for the higher non-human primates apply with equal force in the hominids. In addition, certain other adaptive advantages of social grouping have even greater force in the hominids.

The hominids diverged from the ancestral forest-dwelling stock of primates as terrestrial inhabitants of savanna and plains country, as noted in chapter 37. Primitive terrestrial hominids living in open country would have been exposed and vulnerable to large predators. A group could have put up an effective defense against such predators where one or a few individuals could not. A social organization must have had a high selective value, in relation to defense against large predators, in the newly occupied open habitat. Furthermore, groups of early hominids were probably more successful in hunting and other modes of food-getting than solitary individuals.

There is every reason to expect, therefore, that the australopithecines were social, and there is some indirect evidence to confirm this expectation (Bartholomew and Birdsell 1953; Wolpoff 1980).

The most primitive type of social group known in *Homo sapiens* is the hunting-and-gathering band. Generalized food-gathering peoples have existed in various parts of the world up to the recent past and in some cases into the present. Examples include the Shoshone and Algonquin Indians of North America, the Bushmen and Negritos of Africa, the aborigines of Australia and Tasmania, and the jungle tribes and Andaman Islanders of India (Birdsell 1958; Farb 1968).

A band typically consists of about 20 to 100 individuals and is made up of a series of biparental families. The band claims and defends a territory. Food-gathering within the territory is opportunistic, depending on the food resources available, and varies with the seasons. Both women and men participate in foraging. There is no accumulation of food reserves or of other forms of wealth. The society is basically egalitarian; there is no ruling class and no inferior sex (Birdsell 1958).

The !Kung people of the Kalahari Desert of South Africa furnish a modern example, still extant and recently described, of a primitive foraging society. The nomadic !Kung people have lived in their area as hunters and gatherers for at least 11,000 years. The situation is changing now, however, as the !Kung are being absorbed by the surrounding civilization (Kolata 1974; Lee and Devore 1976).

Hunting-and-gathering bands in primitive peoples appear to have a social organization very similar to that of multiple-male troops in higher non-human primates.

Human societies at a precivilized stage of culture are usually more complex in economic basis and social structure and larger in population size than the foraging band. One thinks of big-game-hunting tribes, pastoral tribes, and primitive agricultural village communities. Size and complexity continue to increase as we enter the stage of civilized societies. And here we can think of the wide range of sociopolitical units, from ancient agricultural civilizations to modern industrial nation-states and postindustrial super-states.

The trends in primate social evolution thus continue in *Homo sapiens*.[3] Through the various grades of the primates, from prosimian to modern man, we see an overall increase in the size of the social group, the complexity of social organization, the power of technology, the sophistication of language, intelligence, and the length of the immature stage of individual development. These are not separate, but interrelated and interacting trends; and they are not linear, but rather overall trends.

Cultural Evolution

The cultural or traditional heritage is the accumulated store of knowledge, understanding, arts, customs, and technological ability available to a human social group at any given time in its history. This body of knowledge and traditions is a product of the discoveries and inventions of preceding generations. It has been and will be transmitted from generation to generation by education in the broad sense of the word. New additions to the cultural heritage can be made in any generation, and these too can be passed on to succeeding generations by the same process of education.

The progressive development and accumulation of the cultural heritage

---

3. For a stimulating essay on the biological basis of many human social institutions and behavioral patterns see Wilson (1978).

constitute cultural evolution. The inheritance of acquired characters is a real process in cultural evolution.

The differences between twentieth-century man and Stone Age man in morphological characters, including cranial capacity, are relatively slight. But the cultural differences between them are enormous. The changes in *Homo sapiens* from the Paleolithic to the modern stage of development have been brought about mainly by cultural evolution.

The forces of organic evolution can account for the origin and early evolution of the genus *Homo*. But organic evolution alone will not account for the change from early man to modern man. The process of cultural evolution enters the picture here and provides an explanation for the changes that cannot be explained by organic evolutionary forces.

Cultural evolution has a momentum of its own that is different from that of organic evolution. And cultural evolution can be considered as a process

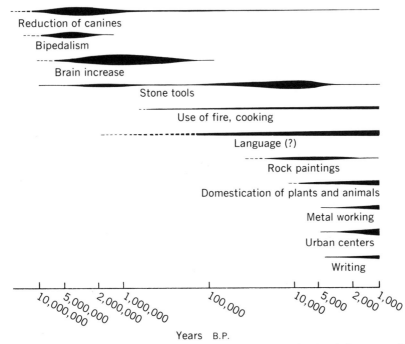

Reduction of canines

Bipedalism

Brain increase

Stone tools

Use of fire, cooking

Language (?)

Rock paintings

Domestication of plants and animals

Metal working

Urban centers

Writing

10,000,000   5,000,000   2,000,000   1,000,000   100,000   10,000   5,000   2,000   1,000

Years B.P.

Figure 38.1. Approximate timing of various organismic and cultural changes in the genus *Homo*. (Redrawn from W. F. Bodmer and L. L. Cavalli-Sforza, *Genetics, Evolution, and Man*, © 1976 Freeman, San Francisco and New York; reproduced by permission)

in itself. In practice, however, cultural evolution interacts with organic evolution. Although it is useful to isolate cultural evolution for purposes of special study, it is more correct, in any study of mankind, to view modern man as a product of the joint action of organic and cultural evolution.

A summary of the timing of the main evolutionary changes, both organic and cultural, in the genus *Homo* is presented in figure 38.1.

An accumulation and transmission of cultural elements occurs on a limited scale in various birds and mammals. Thus in some bird species the young learn a particular dialect from their parents and neighbors, and transmit it to their offspring. Specific migration routes in some migratory birds likewise represent a store of knowledge that is passed on from generation to generation (Bonner 1980). We have already mentioned the dissemination of information about enemies in monkeys. Only in man, however, among earthly beings at least, is there an accumulation of culture on a large scale, and a capacity for extensive cultural evolution.

Socialization

Man, considered as one of the mammals and one of the higher primates, is a product of organic evolution; but man as a human is a product of both organic and cultural evolution, as noted above. Moreover, the truly distinctive features of the human state, those that distinguish man from animal, are the features that have been produced by cultural evolution.

Individual organisms of *Homo sapiens* have, at birth, an animal and a primate constitution, but not a cultural endowment. The latter is acquired during individual development by socialization. During the early years of ontogeny each individual develops from a human animal condition (at birth) through a barbarian stage (as a juvenile) to the human condition, in any real or diagnostic sense of the word human. And this individual mental development takes place under the molding influence of socialization and education.

Furthermore, the individual acquires the culture that he or she is born and raised in, except in those rare instances where an individual develops the ability to think for himself and goes beyond his native cultural horizons—and to a certain extent even then.

The essential difference between man and animal lies, then, in the realm of the human mind, not in the human body, and the distinctively human aspects of the human mind are a product of socialization, as demonstrated so well by Briffault (1927).

Symbolic Language and Conceptual Thought

One of the distinctive features of human mental life is the capacity for conceptual thought, the ability to think in terms of concepts and to formulate ideas. Now conceptual thought depends on language. It depends, moreover, on a language structure more sophisticated than that of animal languages. This brings us to the question of the distinction between human language and animal language.

Animal language, including non-human primate language discussed earlier, consists of body language and vocalizations to express desires, feelings, and warnings, and to identify specific things. It is concrete. Humans have this too. Humans communicate by smiles, scowls, and other body signals, and by simple vocalizations.

But human language goes well beyond this level. The basic vocabulary is extensive. And the words can be combined into numerous phrases and sentences, making possible the precise communication of complex thoughts. ("The leaves of this plant species are good for a toothache, but the leaves of that plant species, which looks like this one, are not.") Furthermore, the vocabulary includes, not only words for concrete things, but also words or phrases for general and abstract entities and relationships between entities. ("Biological species," "interspecific competition.") A language with these properties provides a foundation for conceptual thought.

Animal psychologists have tried to teach the elements of human language to chimpanzees, with varying results. Some workers have claimed success, which would mean a narrowing of the gap between human and animal language in the case of chimpanzees, but this claim is denied by other students.

A fair summary of this controversial subject seems to be that a chimpanzee can learn a vocabulary of about 125 symbols, as represented by photographs and hand signals, and can learn to string them together in small combinations of 2–5 words. ("Baby mine," "comb black.") The language ability of the chimpanzee is on the level of a very young child. Furthermore, chimpanzees do not pick up a simple symbolic language by themselves. They learn such a language only after being tutored by humans in intensive tutoring sessions (Terrace et al. 1979; Passingham 1982).

The human type of symbolic language requires a social group above a critical size for its maintenance and transmittal, and it requires a prolonged learning period in each new generation for its acquisition. A social group above a critical size, composed of educable individuals, is essential for the development and maintenance of conceptual thought. And conceptual thought is, in turn, an essential ingredient in the higher forms of human culture.

The Evolutionary Nature of Man

The age-old question—what is man?—has been answered in a variety of ways. Man is a political animal; man is a religious animal; an ethical animal; a tool-making animal, and so on. Evolutionary biology provides an approach to this question that makes use of historical depth rather than aphoristic phrases. As seen in the light of evolutionary biology, man is a mammal, more particularly a primate, more particularly a hominoid, and a hominid of an advanced type. These successive layers or grades of organic evolutionary development are built into the human organism.

But man is more than an animal and more than a product of organic evolution alone. He is also a product of cultural evolution. Cultural evolution adds another layer, or series of layers, to man's nature. A dualistic constitution, partly biological and partly cultural, is built into mankind by the course of his evolutionary development. The top layer in the stratified human make-up, the layer implanted by cultural evolution, is the definitive one for distinguishing the human condition from the animal state. We can add to our list of aphorisms: Man is the cultural animal (see Grant 1977a).

The next question concerns the placement of the boundary line between the animal state and the human state in evolutionary development. When, in the course of phylogeny, did man enter into the human state? When did *Homo* or some earlier hominid become human in the strict sense of the word as defined above?

We must recall that the elements that make up the human condition are also found in the non-human primates; elements such as social grouping, prolonged immaturity, intelligence, language, and tool-making ability. These elements developed during the primate animal stage of evolution; they were available to the early hominids; and they were developed further to new heights in human evolution.

The transition from the animal to the truly human condition was a gradual one during a moderately long phylogenetic history. A similar gradual transition occurs in the ontogeny of every individual person. The boundary line can be drawn only in an arbitrary way in individual ontogeny, as is done in legal definitions of adulthood, and it is doubtful whether any non-arbitrary boundary can be found in phylogeny. The evolutionary change of state was historical and gradual.

CHAPTER 39
Determinants of Human Evolution

The Relation Between Organic and Cultural Evolution
Selection Modes
Social-Group Selection
Intelligence
Interspecific Selection
Mechanisms of Cultural Evolution
Interactions

Modern man is the culmination of a series of evolutionary trends that can be traced back into the non-human primates. The more important trends in human evolution were discussed in a general way in the preceding chapters. Here we are concerned with the driving forces in these trends.

It is generally agreed that human evolution involves a mixture of two sets of processes: organic evolution and cultural evolution. We know that selection plays a key role in organic evolutionary trends (see chapters 28 and 31). Cultural evolution has a momentum of its own that can explain the gradual accumulation of the cultural heritage. There is a consensus on these points.

The problems begin when we try to go beyond these very general conclusions. One problem is to identify the modes of selection involved. Another is to investigate the mechanism of cultural evolution. A third is to determine the relationship between natural selection and organic evolutionary trends, on the one hand, and cultural evolution, on the other.

The Relation Between Organic and Cultural Evolution

As regards the third question mentioned above, the simplest and historically oldest formula is that organic evolution brought the human phyletic lineage up to a certain threshold, to a level where cultural evolution was possible. At this point in human phylogeny, cultural evolution took over and brought about all subsequent developments. Organic and cultural evolution are held to be separate in time and independent in action. This view is implicit in the writings of older cultural anthropologists (e.g., Briffault 1927) and is still being expressed in some modern discussions (Kraus 1973).

Progress in both evolutionary biology and anthropology in recent decades, and the increased communication between the two formerly isolated fields, have led to a more modern view. This view maintains that natural selection has continued to operate and organic evolution has continued to occur during the era of cultural evolution (e.g., Darlington 1969; Bajema 1971). The results of cultural evolution are simply more conspicuous than those of continuing organic evolution.

A further refinement is the concept of continual interaction between organic and cultural evolution. An important effect of such interaction would be selection for educability and for the ability to assimilate culture. Organic and cultural evolution would thus often proceed hand in hand. This idea was stressed by Dobzhansky (1962).

Recently a partly mathematical treatment of "gene-culture coevolution" has been presented by Lumsden and Wilson (1981) and a primarily mathematical one by Cavalli-Sforza and Feldman (1981). Their approach is quite different from the one adopted here and in my previous treatment (Grant 1977a).

Selection Modes

All writers on the subject apparently agree that human organic evolution is guided by selection. The problem with this conclusion is that selection is not a unitary process, but rather a set of processes (see Part 3), and the conclusion is consequently too broad.

Anthropologists dealing with human evolution, by and large, tend to be interested mainly in hominid phylogeny, secondarily in evolutionary factors such as habitat, hand use, etc., and only incidentally in evolutionary forces. With regard to the latter they are usually content to invoke "natural selection" in a general sense and leave it at that.

Population geneticists dealing with human evolution are, as expected, more interested in the forces involved, and elaborate on the various modes of individual selection and on drift as these processes apply in human populations. But certain modes are usually left out of consideration in these treatments too.

It is therefore useful to note that evidence exists for an array of selection modes in human populations and population systems, namely: directional selection; stabilizing selection; balancing selection; selection-drift; selection for parental care; sexual selection; selection along divergent lines in different areas for geographical races; social-group selection; and interspecific selection. A sample of the evidence for most of these selection modes has been presented in earlier chapters.

It is important to recognize that different selection modes often act in combination. Thus normal human behavior is a mixture of individualism, parental altruism, and solidarity of the individual with his/her group. These behavioral tendencies bespeak the action, respectively, of ordinary individual selection, selection for parental care, and social-group selection.

Drift evidently occurs occasionally in some human populations (see chapter 16). Some authors go on to suggest that the fixation of selectively neutral mutant genes by drift occurs on an extensive scale (see chapter 32). This is a possibility. In any event, however the verdict on this question turns out, it is probably safe to predict that no important human characteristics have resulted from the action of drift alone.

Social-Group Selection

The troop or band is a more or less cohesive unit of organization in the primates, including man. It is a unit of maintenance and of reproduction. The conditions thus exist in the social primates, including man, for the effective operation of social-group selection. It will be recalled from our earlier discussion (chapter 15) that we are using the term social-group selection for a special form of interdeme selection in which the competing populations are social groups.

Different troops or bands in both non-human primates and primitive hominids must have often come into competition for a foraging territory or some other necessary resource. The success or failure of a troop or band in competitive encounters and territorial disputes with rival groups would be determined, among other factors, by the intelligence of the members and/or leaders of the group, its technology and communications, and its strength in numbers. Social-group selection can come into play when the competing troops or bands differ genetically with respect to these characteristics.

It is biologically reasonable to assume that group-to-group differences in intelligence and in its products, technological ability and communications ability, are determined in part by genetic differences in both non-human and human primates.

The genetic part of this assumption in its application to humans (but not to other primates) is ideologically unpopular in some quarters. A brief digression is thus called for. We know that inter-group or interracial variation in physical characters in man has a genetic component. We know that individual-to-individual variation in behavioral traits in man likewise has a genetic component. We find some interracial differences in the behavioral phenotype. It is biologically reasonable to assume that these differences also have a genetic as well as environmental basis. There is no evidence against this assumption (see Loehlin et al. 1975); Ehrman and Parsons 1976). Indeed there is some evidence for it (Wilsom 1978; 48–50 passim). Finally, the biological situation can and should be decoupled from social policies in modern societies. For the purpose of going on with our present argument, therefore, let us accept the biologically realistic premise of genetic differences in mental and behavioral traits between some groups in human and non-human primates alike.

In a succession of group-selection events in primates, including hominids, the social groups with the best intelligence, coordination, and technology could be expected to prevail, on the average, and their genetically determined capacities would tend to spread within the species. The process of social-group selection, long continued, would bring about a gradual increase in intelligence, language, and technological ability in each surviving lineage. Social-group selection can explain the origin of a number of features of social life in primates.

One such feature is language, using this term in the broad sense to include animal language. Communication is an adaptively valuable characteristic of the troop or band. It is also a group characteristic. It seems very probable that language ability in primates, including man, has been developed by social-group selection.

Intelligence is adaptively valuable for both the individual animal and the social group. Individual selection is probably an important determinant here, but this does not rule out supplementary effects of social-group selection. A troop consisting mainly of smart monkeys, apes, or hominids would be able to cope more successfully with the problems in the world around it than a neighboring competitor troop composed of dim-witted individuals.

Even group size, which is a factor in group success, may come under the control of social-group selection to some extent. The formation of larger-sized troops in semi-terrestrial and terrestrial primates is correlated with the reduction of male intolerance, as noted in chapter 37. Large group size is adaptively advantageous in semi-terrestrial and terrestrial primates for defense, competition, and cooperative hunting. If male intolerance stood in the way of expansion of pre-existing small troops, and if some troops had a higher degree of genetically determined male tolerance than others, these latter troops and their genes for male tolerance would tend to prevail.

Group or social-group selection is not discussed at all in a number of recent books on human evolution. Some authors mention this mode briefly in connection with human evolution but do not seem to attach much importance to it. Still other authors bring kin selection more or less prominently into the picture. This is getting close to the mark. Kin selection, in one usage, is synonymous with social-group selection, but has other usages as well, making the term a constant source of confusion (see chapter 15). The process envisioned and discussed here is more aptly described as social-group selection.

It is of historical interest that Darwin in *The Descent of Man* (1883: chs. 4–7) attached importance to what we are now calling social-group selection.

In a study begun in the post-Darwin era but written and published much later, Keith (1948) developed the concept that the social group, especially the small group in primitive man, is the basic evolutionary unit in human evolution. Human evolution was supposed to result from competitive interactions between small inbreeding social groups. Keith set this idea in a framework of evolutionary theory that was out of date in 1948 and is even more so today. Partly for this reason, no doubt, Keith's 1948 book is not cited in most modern works on the subject. Notwithstanding flaws in his treatment, Keith still deserves credit for an early statement concerning the role of social-group selection in human evolution.

Intelligence

Intelligence is a complex of several different kinds of mental ability. There is problem-solving ability, mechanical ability, learning ability, communication skill, and so on. Will one single mode of selection explain the development of human intelligence? We should consider the possibility that some aspects of this complex trait have been developed by individual selection and other aspects by social-group selection.

Some components of intelligence confer an advantage primarily on the individual primate and his or her offspring. This would be the case for ability to learn from experience, memory, ability to predict results of actions, and facility in tool use. The evolutionary development of such components of intelligence can be attributed largely to individual selection.

Other components of intelligence confer their advantage in a social context. This is the case with language skill in group coordination and in education of the young. Such types of mental ability could well be produced largely by social-group selection. A logical and reasonable hypothesis, therefore, is that primate intelligence has been built up by a combination of individual and social-group selection.

Learning ability in primates is a function of the length of the immature stages of individual development. Comparative data on length of immature stages, in living primates representing different grades from prosimian to human, show clearly that the period available for learning has increased from the lower to the higher primates (see table 38.3). Rate of individual development is a genetically determined trait. Natural selection is thus implicated as a driving force in this evolutionary trend. Furthermore, since primate learning is a social process, social-group selection must have played a large role in the adjustment of development rates for greater learning ability.

Social-group selection would tend to increase the level of intelligence in all lineages of social primates. Yet the trend toward greater intelligence has obviously proceeded further in human than in non-human primates. It is probable that the special biological and social factors described in the preceding chapters—bipedal locomotion, hand use, etc.—set the stage for interactions and feedbacks between biological, social, and cultural factors that carried the development of intelligence to much higher levels in humans.

Interspecific Selection

Another mode of supra-individual selection that has evidently played a role in hominid evolution is interspecific selection leading to species replacement (see chapter 22). Early *Homo* (*H. habilis-erectus*) and *Australopithecus robustus* coexisted in Southern Africa in Pleistocene time (Klein 1977). It is probable that the two species were in direct or at least indirect competition for food or foraging territories. *Australopithecus* died out, while *Homo* survived and rose to dominance in the hominid adaptive zone. *Homo* had a larger brain and better stone tools than *Australopithecus*. These are probably two of the factors determining the course of the interspecific selection process.

Interspecific selection in the hominids differs from that in most animals in being linked to social-group selection. The contest was between social groups of early *Homo* and competing groups of robust australopithecines. Since the effective units of reproduction in the competing species were social groups, the interspecific selection takes on a special form of this process.

Mechanisms of Cultural Evolution

The standard view of cultural evolution holds that this process proceeds by the gradual build-up of the cultural heritage. Useful knowledge is first discovered, then incorporated into the general body of knowledge of the social group, and then transmitted to succeeding generations by education. Each succeeding generation adds to this store of knowledge, by its own discoveries, and these are also transmitted in the same way. Thus the cultural heritage is cumulative.

Undoubtedly cultural evolution does occur in this fashion to some extent. But does it occur only this way—does it always proceed so straightforwardly? A reading of history suggests that it does not. Current views of cultural evolution are as general, and as vague, as current views concerning natural selection in human evolution, and the former, like the latter, stand in need of critical reevaluation.

Let us again consider group size. Group selection acting on genetic group differences in male tolerance may have brought the group size up to the level of the multiple-male troop or band. This level is attained in some advanced non-human primates and in primitive hominids (see chapter 38). But the trend toward increasing size and complexity of the social group obviously continues into human prehistory and history. Bands become merged into tribes, and tribes into chiefdoms and states. States increase in size from city-states to small nation-states to large nation-states.

Degree of male tolerance has nothing to do with these latter phases of the group-size trend. Male tolerance is not noticeably different between the Netherlands and the United States. Intertribal and international wars, however, have had much to do with the size trend. Up to a point, bigger tribes or nations are superior to smaller tribes or nations in the test of war, on the average, where other factors are more or less equal, as they often are. In a succession of wars the smaller social-political units tend to disappear as independent groups, surviving in some dependent form—as a colony, a subordinate state in a federation, or a satellite country—while the larger groups emerge as the effective independent political powers. Group-size trends are determined, up to a point, by the survival of the biggest.

I say "up to a point" because very big political units, as exemplified by empires, do not have long-term durability, but always break up eventually into smaller fragments. There are reversals in the group-size trend, especially at the upper level. The size-increase trend is displayed more clearly below this level.

Here we have a social trend determined by non-genetic differences between competing groups. The non-genetic difference in this model happens to concern group size. But one can readily think of differences between groups in other cultural features—differences that, in a succession of inter-group competitive encounters, could produce a cultural evolutionary trend. Some technological trends, for example, may be a resultant of successive competitive struggles between groups with different technological capabilities. Ideological trends could be produced in the same way. One can predict the average result of numerous encounters between groups that act on erroneous preconceived notions vs. opposing groups that take a realistic and pragmatic approach to problems.

The orthodox view of general cultural evolution on a broad front is held to be too simplistic. An alternative mode that warrants consideration is that cultural evolution is often or usually a product of competition between opposing social groups with different cultures.

Interactions

Interactions between selection for anatomical and morphological characters on the one hand and selection for social traits on the other are an important aspect of human evolution.

Differences between chimpanzees and humans in language ability were discussed in chapter 38. Part of the difference lies in the voice box, which is capable of producing a far greater number of fine-tuned vocalizations in humans than in chimpanzees or any other non-human primates. Precise communication within the social group was adaptively valuable in the early hominids. That advantage was secured, inter alia, by selection for a hominid mouth and throat configuration capable of speech.

The evolution of learning ability entailed some interesting complexities.

Increase in brain size beyond a certain point was limited by the size of the birth canal through which the infant head had to pass at birth. Selection for intelligence came into opposition with selection for female viability in childbirth. The impasse was resolved by two compromises.

The first compromise was to widen the female pelvis. This entailed loss of running speed. The human female is a slower runner than the human male. This sex dimorphism does not exist in apes. A certain amount of running ability was sacrificed in the human female to bear larger-brained children.

The second compromise was to slow down the rate of development of the brain. The human brain is incompletely formed at birth, and the newborn human infant is quite helpless; the brain must undergo most of its development during infancy and childhood. Humans differ markedly in these conditions from apes and monkeys, which are more self-sufficient at birth. This second compromise had the secondary effect of prolonging the immature stages of life (Briffault 1927; Washburn 1960).

Both compromise solutions reduced the independence of the human female. She was saddled with her young for a longer period of time. And the loss of running ability made her less able to hunt for herself and more dependent on the men for game.

These third-order effects produced fourth-order effects within the primitive human social group. The band had to pick up some of the responsibility for the women and the young. The social group became a more cooperative and communal organization. The cause-effect chain went full circle. Selection for learning ability, a socially useful trait in a social milieu, brought about various morphological and developmental changes that, as a corollary, called forth a still more close-knit social organization.

Bibliography

Abel, O. 1929. Paläobiologie und Stammesgeschichte. Gustav Fischer Verlag, Jena.

Ainsworth, G. C. and G. R. Bisby. 1954. A Dictionary of the Fungi. 4th ed. Kew, England.

Alexander, R. D. 1974. The evolution of social behavior. Ann. Rev. Ecol. Syst. 5:325–383.

Altmann, S. A., ed. 1967. Social Communication Among Primates. University of Chicago Press, Chicago.

Alvarez, L. W. 1983. Experimental evidence that an asteroid impact led to the extinction of many species 65 million years ago. Proc. Nat. Acad. Sci. USA 80:627–642.

Alvarez, L. W., W. Alvarez, F. Asaro, and H. V. Michel. 1980. Extraterrestrial cause for the Cretaceous-Tertiary extinction. Science 208:1095–1108.

Alvarez, W., E. G. Kauffman, F. Surlyk, L. W. Alvarez, F. Asaro, and H. V. Michel. 1984. Impact theory of mass extinctions and the invertebrate fossil record. Science 223:1135–1141.

Amadon, D. 1950. The Hawaiian honeycreepers (Aves, Drepaniidae). Bull. Amer. Mus. Nat. Hist. 95(4):151–262.

Anderson, E. and W. L. Brown. 1952. Origin of corn belt maize and its genetic significance. In: Heterosis, ed. by J. W. Gowen. Iowa State College Press, Ames.

Anderson, W., Th. Dobzhansky, O. Pavlovsky, J. Powell, and D. Yardley. 1975. Genetics of natural populations. XLII. Three decades of genetic change in Drosophila pseudoobscura. Evolution 29:24–36.

Andrewartha, H. G., and L. C. Birch. 1954. The Distribution and Abundance of Animals. University of Chicago Press, Chicago.

Antonovics, J. 1971. The effects of a heterogeneous environment on the genetics of natural populations. Amer. Sci. 59:593–599.

Avery, O. T., C. M. Macleod, and M. McCarty. 1944. Studies on the chemical nature of the substance inducing transformation of pneumococcal types. J. Exp. Med. 79:137–158.

Axelrod, D. I. 1959. Late Cenozoic evolution of the Sierran bigtree forest. Evolution 13:9–23.

—— 1967. Quaternary extinctions of large mammals. Univ. Calif. Publ. Geol. Sci. 74:1–42.

Axelrod, D. I. and H. P. Bailey. 1968. Cretaceous dinosaur extinction. Evolution 22:595–611.

Ayala, F. J. 1969. Experimental invalidation of the principle of competitive exclusion. Nature 224:1076–1079.

—— 1974. The concept of biological progress. In: Studies in the Philosophy of Biology. ed. F. J. Ayala and Th. Dobzhansky. Macmillan, New York and London.

Ayala, F. J., ed. 1976. Molecular Evolution. Sinauer Associates, Sunderland, Mass.

Ayala, F. J. and W. W. Anderson. 1973. Evidence of natural selection in molecular evolution. Nature 241:274–276.

Ayala, F. J. and M. E. Gilpin. 1974. Gene frequency comparisons between taxa: support for the natural selection of protein polymorphisms. Proc. Nat. Acad. Sci. USA 71:4847–4849.

Ayala, F. J., C. A. Mourao, S. Perez-Salas, R. Richmond, and Th. Dobzhansky. 1970. Enzyme variability in the Drosophila willistoni group. I. Genetic differentiation among sibling species. Proc. Nat. Acad. Sci. USA 67:225–232.

Ayala, F. J. and M. L. Tracey. 1974. Genetic differentiation within and between species of the Drosophila willistoni group. Proc. Nat. Acad. Sci. USA 71:999–1003.

Ayala, F. J., M. L. Tracey, D. Hedgecock, and R. C. Richmond. 1975. Genetic differentiation during the speciation process in Drosophila. Evolution 28:576–592.

Babcock, E. B. and G. L. Stebbins. 1938. The American species of Crepis: their interrelationships and distribution as affected by polyploidy and apomixis. Carnegie Institution of Washington, Washington, D.C., Publ. 504.

Baer, K. E. von. 1828. Über Entwicklungsgeschichte der Thiere. Bornträger, Königsberg.

Bajema, C. J. 1971. Natural Selection in Human Populations. Wiley, New York.

Baker, A. E. M. 1981. Gene flow in house mice: introduction of a new allele into free-living populations. Evolution 35:243–258.

Bakker, R. T. 1975. Dinosaur renaissance. Sci. Amer. 232(4):58–78.

Baldwin, P. H. 1953. Annual cycle, environment and evolution in the Hawaiian honeycreepers (Aves:Drepaniidae). Univ. Calif. Publ. Zool. 52:285–398.

Barghoorn, E. S. 1971. The oldest fossils. Sci. Amer. 224(5):30–42.

Barrowclough, G. F. 1980. Gene flow, effective population sizes, and genetic variance components in birds. Evolution 34:789–798.

Bartholomew, G. A. and J. B. Birdsell. 1953. Ecology and the protohominids. Amer. Anthropol. 55:481–498.

Barton, K. A. and W. J. Brill. 1983. Prospects in plant genetic engineering. Science 219:671–676.

Bateman, A. J. 1950. Is gene dispersion normal? Heredity 4:353–363.

Bateman, K. G. 1959. The genetic assimilation of four venation phenocopies. J. Genet. 56:443–474.

Beadle, G. W. 1963. Genetics and Modern Biology. American Philosophical Society, Philadelphia.

—— 1972. The mystery of maize. Field Mus. Nat. Hist. Bull. 43:2–11.

—— 1980. The ancestry of corn. Sci. Amer. 242(1):112–119.

Beardmore, J. A., Th. Dobzhansky, and O. Pavlovsky. 1960. An attempt to compare the fitness of polymorphic and monomorphic experimental populations of Drosophila pseudoobscura. Heredity 14:19–33.

Benton, M. J. 1979. Ectothermy and the success of dinosaurs. Evolution 33:983–997.

Benveniste, R. E. and G. J. Todaro. 1974. Evolution of C-type viral genes: inheritance of exogenously acquired viral genes. Nature 252:456–459.

Berenbaum, M. 1983. Coumarins and caterpillars: a case for coevolution. Evolution 37:163–179.

Bergson, H. 1911. Creative Evolution. Translation. Holt, New York. Various later reprints.

Birdsell, J. B. 1950. Some implications of the genetical concept of race in terms of spatial analysis. Cold Spring Harbor Symp. Quant. Biol. 15:259–314.

—— 1958. On population structure in generalized hunting and collecting populations. Evolution 12:189–205.

Blair, W. F. 1955. Mating call and stage of speciation in the Microhyla olivacea-M. carolinensis complex. Evolution 9:469–480.

—— 1960. The Rusty Lizard: A Population Study. University of Texas Press, Austin.

Blum, H. F. 1955. Time's Arrow and Evolution. 2d ed. Princeton University Press, Princeton, N.J.

Bock, W. J. 1970. Microevolutionary sequences as a fundamental concept in macro-evolutionary models. Evolution 24:704–722.

Bodmer, W. F. and L. L. Cavalli-Sforza. 1976. Genetics, Evolution, and Man. Freeman, San Francisco.

Bohor, B. F., E. E. Foord, P. J. Modreski, and D. M. Triplehorn. 1984.

Mineralogic evidence for an impact event at the Cretaceous-Tertiary boundary. Science 224:867–869.

Bold, H. C., C. J. Alexopoulos, and T. Delevoryas. 1980. Morphology of Plants and Fungi. 4th ed. Harper and Row, New York.

Bonner, J. T. 1980. The Evolution of Culture in Animals. Princeton University Press, Princeton, N.J.

Bonnett, O. T. 1954. The inflorescences of maize. Science 120:77–87.

Boucher, D. H., S. James, and K. H. Keeler. 1982. The ecology of mutualism. Ann. Rev. Ecol. Syst. 13:315–347.

Brace, C. L., H. Nelson, and N. Korn. 1971. Atlas of Fossil Man. Holt, Rinehart & Winston, New York.

Brantjes, N. B. M. 1982. Pollen placement and reproductive isolation between two Brazilian *Polygala* species (Polygalaceae). Plant Syst. Evol. 141:41–52.

Braver, N. B. 1956. The mutants of *Drosophila melanogaster*, classified according to body parts affected. Carnegie Institution of Washington, Publ. 552A, Washington, D.C.

Briffault, R. 1927. The Mothers. 3 vols. Macmillan, London.

Brower, L. P. and J. V. Z. Brower. 1964. Birds. butterflies, and plant poisons: a study in ecological chemistry. Zoologica (New York) 49:137–159.

Brown, W. L. and E. O. Wilson. 1956. Character displacement. Syst. Zool. 5:49–64.

Bruce, E. J. and F. J. Ayala. 1979. Phylogenetic relationships between man and the apes: electrophoretic evidence. Evolution 33:1040–1056.

Brücher, H. 1943. Experimentelle Untersuchungen über den Selektionswert künstlich erzeugter Mutanten von *Antirrhinum majus*. Z. Botan. 39:1–47.

Brues, A. M. 1969. Genetic load and its varieties. Science 164:1130–1136.

Bryant, E. H. 1974. On the adaptive significance of enzyme polymorphisms in relation to environmental variability. Amer. Nat. 108:1–19.

Bryson, R. A., D. A. Barreis, and W. M. Wendland. 1970. The character of late-glacial and post-glacial climatic changes. In: Pleistocene and Recent Environments of the Central Great Plains. ed. W. Dort and J. K. Jones. University of Kansas Press, Lawrence.

Buffon, G. L. L. 1770. Histoire Naturelle des Oiseaux. Paris.

—— 1808. Natural History of Birds, Fish, Insects, and Reptiles. Translation. 6 vols. Symonds, London.

Buri, P. 1956. Gene frequency in small populations of mutant *Drosophila*. Evolution 10:367–402.

Bush, G. L. 1969a. Sympatric host race formation and speciation in frugivorous flies of the genus *Rhagoletis* (Diptera: Tephritidae). Evolution 23:237–251.

—— 1969b. Mating behavior, host specificity, and the ecological significance of sibling species in frugivorous flies of the genus *Rhagoletis* (Diptera-Tephritidae). Amer. Nat. 103:669–672.

—— 1975. Modes of animal speciation. Ann. Rev. Ecol. Syst. 6:339–364.

Buss, L. W. 1983. Evolution, development, and the units of selection. Proc. Nat. Acad. Sci. USA 80:1387–1391.

Cain, A. J. and P. M. Sheppard. 1952. The effects of natural selection on body colour in the land snail *Cepaea nemoralis*. Heredity 6:217–231.

—— 1954. Natural selection in *Cepaea*. Genetics 39:89–116.

Calvin, M. 1956. Chemical evolution and the origin of life. Amer. Sci. 44:248–263.

Campbell, B. 1966, 1974. Human Evolution. 1st and 2d eds. Aldine-Atherton, Chicago.

Campbell, B., ed. 1972. Sexual Selection and the Descent of Man 1871–1971. Aldine-Atherton, Chicago.

Candela, P. B. 1942. The introduction of blood-group B into Europe. Human Biol. 14:413–443.

Caplan, A., L. Herrera-Estrella, D. Inze, E. Van Haute, M. Van Montague, J. Schell, and P. Zambryski. 1983. Introduction of genetic material into plant cells. Science 222:815–821.

Caple, G., R. P. Balda, and W. R. Willis. 1983. The physics of leaping animals and the evolution of pre-flight. Amer. Nat. 121:455–467.

Carlquist, S. 1962. A theory of paedomorphosis in dicotyledonous woods. Phytomorph. 12:30–45.

—— 1965. Island Life: A Natural History of the Islands of the World. Natural History Press, Garden City, N.Y.

—— 1974. Island Biology. Columbia University Press, New York.

—— 1975. Ecological Strategies of Xylem Evolution. University of California Press, Berkeley.

Carson, H. L. 1959. Genetic conditions which promote or retard the formation of species. Cold Spring Harbor Symp. Quant. Biol. 24:87–105.

—— 1970. Chromosome tracers of the origin of species. Science 168:1414–1418.

—— 1971. Speciation and the founder principle. Stadler Genet. Symp. 3:51–70.

—— 1975. The genetics of speciation at the diploid level. Amer. Nat. 109:83–92.

—— 1977. The unit of genetic change in adaptation and speciation. Annals Missouri Bot. Gard. 63:210–223.

——— 1981. Homosequential species of Hawaiian *Drosophila*. Chromosomes Today (London) 7:150–164.

Carson, H. L. D. E. Hardy, H. T. Spieth, and W. S. Stone. 1970. The evolutionary biology of the Hawaiian Drosophilidae. Evolutionary Biology, suppl. vol. 1970, 437–543.

Carson, H. L. and W. B. Heed. 1964. Structural homozygosity in marginal populations of nearctic and neotropical species of *Drosophila* in Florida. Proc. Nat. Acad. Sci. USA 52:427–430.

Carson, H. L. and K. Y. Kaneshiro. 1976. *Drosophila* of Hawaii: systematics and ecological genetics. Ann. Rev. Ecol. Syst. 7:311–345.

Cavalli-Sforza, L. L. and W. F. Bodmer. 1971. The Genetics of Human Populations. Freeman, San Francisco.

Cavalli-Sforza, L. L. and M. W. Feldman. 1981. Cultural Transmission and Evolution: A Quantitative Approach. Princeton University Press, Princeton, N.J.

Chambers, R. 1844. Vestiges of the Natural History of Creation. London.

Chetverikov, S. S. 1926. On certain aspects of the evolutionary process from the standpoint of modern genetics. Translation. Proc. Amer. Philosoph. Soc. 105:167–195, 1961.

Clarke, B. 1973. Neutralists vs. selectionists. Science 180:600–601.

Clausen, J. 1951. Stages in the Evolution of Plant Species. Cornell University Press, Ithaca, N.Y.

——— 1965. Population studies of alpine and subalpine races of conifers and willows in the California high Sierra Nevada. Evolution 19:56–68.

Clausen, J. and W. M. Hiesey. 1958. Experimental studies on the nature of species. IV. Genetic structure of ecological races. Carnegie Institution of Washington, Washington. D.C., Publ. 615.

Clausen, J., D. D. Keck, and W. M. Hiesey. 1940. Experimental studies on the nature of species. I. Effect of varied environments on western North American plants. Carnegie Institution of Washington, Washington, D.C., Publ. 520.

——— 1948. Experimental studies on the nature of species. III. Environmental responses of climatic races of Achillea. Carnegie Institution of Washington, Washington, D.C., Publ. 581.

Clausen, R. E. 1941. Polyploidy in *Nicotiana*. Amer. Nat. 75:291–306.

Cloud, P. 1974. Evolution of ecosystems. Amer. Sci. 62:54–66.

Cloud, P., G. R. Licari, L. A. Wright, and B. W. Troxel. 1969. Proterozoic eukaryotes from eastern California. Proc. Natl. Acad. Sci. 62:623–630.

Clutton-Brock, T., F. E. Guinness, and S. Albon, 1982. Red Deer, Behavior and Ecology of Two Sexes. University of Chicago Press, Chicago.

Coe, E. H. and M. G. Neuffer. 1977. The genetics of corn. In: Corn and

Corn Improvement. ed. G. F. Sprague. Amer. Soc. Agronomy, Madison, Wis.

Colbert, E. H. 1955, 1969, 1980. Evolution of the Vertebrates. 1st, 2d, and 3d eds. Wiley, New York.

—— 1961. Dinosaurs: Their Discovery and Their World. Dutton, New York.

Colwell, R. N. 1951. The use of radioactive isotopes in determining spore distribution patterns. Amer. Jour. Bot. 38:511–523.

Connell, J. H. 1961. The influence of interspecific competition and other factors on the distribution of the barnacle *Chthamalus stellatus*. Ecology 42:710–723.

—— 1983. On the prevalance and relative importance of interspecific competition: evidence from field experiments. Amer. Nat. 122:661–696.

Connell, J. H. and E. Orias. 1964. The ecological regulation of species diversity. Amer. Nat. 98:399–414.

Connell, J. H. and W. P. Sousa. 1983. On the evidence needed to judge ecological stability or persistence. Amer. Nat. 121:789–824.

Coon, C. S. 1955. Some problems of human variability and natural selection in climate and culture. Amer. Nat. 89:257–279.

Coope, G. R. 1979. Late Cenozoic fossil Coleoptera; evolution, biogeography, and ecology. Ann. Rev. Ecol. Syst. 10:247–267.

Cope, E. D. 1896. The Primary Factors of Organic Evolution. Open Court, Chicago.

Coyne, J. A., I. A. Boussy, T. Prout, S. H. Bryant, J. S. Jones, and J. A. Moore. 1982. Long-distance migration of *Drosophila*. Amer. Nat. 119:589–595.

Cronquist, A. 1968. The Evolution and Classification of Flowering Plants. Houghton Mifflin, Boston.

—— 1981. An Integrated System of Classification of Flowering Plants. Columbia University Press, New York.

Crosby, J. L. 1963. The evolution and nature of dominance. Jour. Theor. Biol. 5:35–51.

Crow, J. F. 1970. Genetic loads and the cost of natural selection. In: Mathematical Topics in Population Genetics. ed. K. Kojima. Springer, New York.

Crow, J. F. and M. Kimura. 1970. An Introduction to Population Genetics Theory. Harper & Row, New York.

Crumpacker, D. W. and J. S. Williams. 1973. Density, dispersion, and population structure in *Drosophila pseudoobscura*. Ecol. Monogr. 43:499–538.

Cunha, A. B. da, H. Burla, and T. Dobzhansky. 1950. Adaptive chromosomal polymorphism in *Drosophila willistoni*. Evolution 4:212–235.

Cunha, A. B. da and T. Dobzhansky. 1954. A further study of chromosomal polymorphism in *Drosophila willistoni* in its relation to the environment. Evolution 8:119–134.

Cunha, A. B. da, T. Dobzhansky, O. Pavlovsky, and B. Spassky. 1959. Genetics of natural populations. XXVIII. Supplementary data on the chromosomal polymorphism in *Drosophila willistoni* in its relation to the environment. Evolution 13:389–404.

Daniels, S. and L. Ehrman, 1974. Embryonic pole cells and mycoplasma-like symbionts in *Drosophila paulistorum*. Jour. Invertebrate Pathology 24:14–19.

Darlington, C. D. 1939, 1958. The Evolution of Genetic Systems. 1st and 2d eds. 1st ed. Cambridge University Press, Cambridge. 2d ed. Basic Books, New York.

—— 1969. The Evolution of Man and Society. Simon and Schuster, New York.

Darlington, C. D. and K. Mather. 1949. The Elements of Genetics. Allen and Unwin, London.

Darlington, P. J. 1980. Evolution for Naturalists: The Simple Principles and Complex Reality. Wiley, New York.

Darwin, C. 1859, 1872. On the Origin of Species by Means of Natural Selection. 1st and 6th eds. Murray, London.

—— 1871, 1874. The Descent of Man and Selection in Relation to Sex. 1st and 2d eds. Murray, London.

—— 1875. The Variation of Animals and Plants Under Domestication. 2d ed. 2 vols. Murray, London.

Darwin, F., ed. 1958. The Autobiography of Charles Darwin and Selected Letters. Appleton, N.Y., 1892. Reprint. Dover, New York.

Dawkins, R. 1976. The Selfish Gene. Oxford University Press, Oxford.

Dayhoff, M. O., ed. 1968. Atlas of Protein Sequence and Structure. Vol. 3, 1967–1968. National Biomedical Research Foundation. Silver Spring, Md.

Dayhoff, M. O., ed. 1969. Atlas of Protein Sequence and Structure. Vol. 4, 1969. National Biomedical Research Foundation. Silver Spring, Md.

—— 1972. Atlas of Protein Sequence and Structure. Vol. 5, 1972. National Biomedical Research Foundation. Silver Spring, Md.

—— 1978. Atlas of Protein Sequence and Structure. Vol. 5, Supplement. National Biomedical Research Foundation. Washington, D.C.

de Beer, G. R. 1951. Embryos and Ancestors. 2d ed. Oxford University Press, Oxford.

—— 1964. Charles Darwin. Doubleday, Garden City, N.Y.

DeFries, J. C. and G. E. McClearn. 1972. Behavioral genetics and the fine structure of mouse populations: a study in microevolution. Evol. Biol. 5:279–291.

DeVore, I., ed. 1965. Primate Behavior: Field Studies of Monkeys and Apes. Holt, Rinehart & Winston, New York.

De Wet, J. M. J. and J. R. Harlan. 1972. Origin of maize: the tripartite hypothesis. Euphytica 21:271–279.

De Wet, J. M. J., J. R. Harlan, and C. A. Grant. 1971. Origin and evolution of teosinte [*Zea mexicana* (Schrad.) Kuntze]. Euphytica 20:255–265.

Dickerson, R. E. 1978. Chemical evolution and the orgin of life. In: Evolution. ed. Scientific American Editorial Board. Freeman, San Francisco.

Dillon, L. S. 1978. The Genetic Mechanism and the Origin of Life. Plenum Press, New York.

—— 1981. Ultrastructure, Macromolecules, and Evolution. Plenum Press, New York.

—— 1983. The Inconstant Gene. Plenum Press, New York.

Dobzhansky, Th. 1937a, 1941, 1951a. Genetics and the Origin of Species. 1st, 2d, and 3d eds. Columbia University Press, New York.

—— 1937b. Genetic nature of species differences. Amer. Nat. 71:404–420.

—— 1940. Speciation as a stage in evolutionary divergence. Amer. Nat. 74:312–321.

—— 1943. Genetics of natural populations. IX. Temporal changes in the composition of populations of *Drosophila pseudoobscura*. Genetics 28:162–186.

—— 1947a. Genetics of natural populations. XIV. A response of certain gene arrangements in the third chromosome of *Drosophila pseudoobscura* to natural selection. Genetics 32:142–160.

—— 1947b. A directional change in the genetic constitution of a natural population of *Drosophila pseudoobscura*. Heredity 1:53–64.

—— 1948. Genetics of natural populations. XVI. Altitudinal and seasonal changes produced by natural selection in certain populations of *Drosophila pseudoobscura* and *Drosophila persimilis*. Genetics 33:158–176.

—— 1950a. Mendelian populations and their evolution. Amer. Nat. 84:401–418.

—— 1950b. Evolution in the tropics. Amer. Scientist 38:209–221.

—— 1951a. (See 1937, 1951a, above.)

—— 1951b. Experiments on sexual isolation in *Drosophila*. X. Reproductive isolation between *Drosophila pseudoobscura* and *Drosophila persimilis* under natural and under laboratory conditions. Proc. Nat. Acad. Sci. USA 37:792–796.

—— 1955. The genetic basis of systematic categories. In: Biological Systematics. Biology Colloquium, Oregon State College, Corvallis.

—— 1956. Genetics of natural populations. XXV. Genetic changes in populations of *Drosophila pseudoobscura* and *Drosophila persimilis* in some localities in California. Evolution 10:82–92.

—— 1958. Genetics of natural populations. XXVII. The genetic changes in populations of *Drosophila pseudoobscura* in the American southwest. Evolution 12:385–401.

—— 1962. Mankind Evolving. Yale University Press, New Haven, Conn.

—— 1965. Mendelism, Darwinism, and evolutionism. Proc. Amer. Philosoph. Soc. 109:205–215.

—— 1970. Genetics of the Evolutionary Process. Columbia University Press, New York.

—— 1971. Evolutionary oscillations in *Drosophila pseudoobscura*. In: Ecological Genetics and Evolution. ed. R. Creed. Blackwell Scientific Publications, Oxford.

Dobzhansky, T., F. J. Ayala, G. L. Stebbins, and J. W. Valentine. 1977. Evolution. Freeman, San Francisco.

Dobzhansky, T., L. Ehrman, O. Pavlovsky, and B. Spassky. 1964. The superspecies *Drosophila paulistorum*. Proc. Nat. Acad. Sci. USA 51:3–9.

Dobzhansky, T. and C. Epling. 1944. Contributions to the genetics, taxonomy, and ecology of *Drosophila pseudoobscura* and its relatives. Carnegie Institution of Washington, Washington, D.C., Publ. 554.

Dobzhansky, T. and H. Levene. 1951. Development of heterosis through natural selection in experimental populations of *Drosophila pseudoobscura*. Amer. Nat. 85:247–264.

Dobzhansky, T. and O. Pavlovsky. 1957. An experimental study of interaction between genetic drift and natural selection. Evolution 11:311–319.

Dobzhansky, T. and J. R. Powell. 1974. Rates of dispersal of *Drosophila pseudoobscura* and its relatives. Proc. Roy. Soc. London, B, 187:281–298.

Dobzhansky, T. and B. Spassky. 1947. Evolutionary changes in laboratory cultures of *Drosophila pseudoobscura*. Evolution 1:191–216.

Dobzhansky, T., B. Spassky, and N. Spassky. 1952. A comparative study of mutation rates in two ecologically diverse species of *Drosophila*. Genetics 37:650–664.

Dobzhansky, T. and N. P. Spassky. 1962. Genetic drift and natural selection in experimental populations of *Drosophila pseudoobscura*. Proc. Nat. Acad. Sci. USA 48:148–156.

Dobzhansky, T. and S. Wright. 1943. Genetics of natural populations. X. Dispersion rates in *Drosophila pseudoobscura*. Genetics 28:304–340.

—— 1947. Genetics of natural populations. XV. Rate of diffusion of a mutant gene through a population of *Drosophila pseudoobscura*. Genetics 32:303–324.

Doebley, J. F. and H. H. Iltis. 1980. Taxonomy of *Zea* (Gramineae). I. A subgeneric classification with key to taxa. Amer. Jour. Bot. 67:982–993.

Dorst, J. 1974. The Life of Birds. Translation. 2 vols. Columbia University Press, New York.

Downs, T. 1961. A study of variation and evolution in Miocene *Merychippus*. Contributions in Science, Los Angeles County Museum, Los Angeles, No. 45.

Dubinin, N. P. 1948. Experimental investigation of the integration of hereditary systems in the processes of evolution of populations. (In Russian). Zh. Obshch. Biol. 9:203–244. Translation. University Library, University of California, Los Angeles.

Dunbar, M. J. 1960. The evolution of stability in marine environments. Natural selection at the level of the ecosystem. Amer. Nat. 94:129–136.

Dunn, L. C. and S. P. Dunn. 1957. The Jewish community of Rome. Sci. Amer. 196(3), pp. 118–128.

Durrant, A. 1962a. The environmental induction of heritable change in *Linum*. Heredity 17:27–61.

—— 1962b. Induction, reversion and epitrophism of flax genotrophs. Nature 196:1302–1304.

Ehrlich, P. R. and A. Ehrlich. 1981. Extinction: The Causes and Consequences of the Disappearance of Species. Random House, New York.

Ehrlich, P. R. and P. H. Raven. 1964. Butterflies and plants: a study in coevolution. Evolution 18:586–608.

—— 1969. Differentiation of populations. Science 165:1228–1232.

Ehrlich, P. R. and R. R. White. 1980. Colorado checkerspot butterflies: isolation, neutrality, and the biospecies. Amer. Nat. 115:328–341.

Ehrman, L. 1965. Direct observation of sexual isolation between allopatric and between sympatric strains of the different *Drosophila paulistorum* races. Evolution 19:459–464.

Ehrman, L. and R. P. Kernaghan. 1971. Microorganismal basis of infectious hybrid male sterility in *Drosophila paulistorum*. J. Hered. 62:66–71.

Ehrman, L. and P. A. Parsons. 1976. The Genetics of Behavior. Sinauer Associates, Sunderland, Mass.

Ehrman, L. and E. B. Spiess. 1969. Rare-type mating advantage in *Drosophila*. Amer. Nat. 103:675–680.

Ehrman, L. and D. L. Williamson. 1965. Transmission by injection of hybrid sterility to nonhybrid males in *Drosophila paulistorum*: preliminary report. Proc. Nat. Acad. Sci. USA 54:481–483.

Eiseley, L. 1958. Darwin's Century. Doubleday, Garden City, N.Y.

Eisenberg, J. F. 1981. The Mammalian Radiations: An Analysis of Trends in Evolution, Adaptation, and Behavior. University of Chicago Press, Chicago.

Eisenberg, J. F., N. A. Muckenhirn, and R. Rudran. 1972. The relation between ecology and social structure in primates. Science 176:863–874.

Eklund, M. W., F. T. Poysky, J. A. Meyers, and G. A. Pelroy. 1974. Interspecies conversion of *Clostridium botulinum* type C to *Clostridium novyi* type A by bacteriophage. Science 186:456–458.

Eldredge, N. and S. J. Gould. 1972. Punctuated equilibria: an alternative to phyletic gradualism. In: Models in Paleobiology. ed. T. J. M. Schopf. Freeman, Cooper, San Francisco.

Emlen, J. M. 1973. Ecology: An Evolutionary Approach. Addison-Wesley, Reading, Mass.

Epling, C., D. F. Mitchell, and R. H. T. Mattoni. 1953. On the role of inversions in wild populations of *Drosophila pseudoobscura*. Evolution 7:342–365.

Evans, G. M., A. Durrant, and H. Rees. 1966. Associated nuclear changes in the induction of flax genotrophs. Nature 212:697–699.

Ewens, W. J. 1965. Further notes on the evolution of dominance. Heredity 20:443–450.

Ewens, W. J. 1972. The substitutional load in a finite population. Amer. Nat. 106:273–282.

Fahraeus, L. E. 1982. Allopatric speciation and lineage zonation exemplified by the *Pygodus serrus—P. anserinus* transition (Conodontophoridae, Ordovician). Newsl. Stratigr. 11 (1):1–7.

Falconer, D. S. 1960, 1981. Introduction to Quantitative Genetics. 1st and 2d eds. 1st ed., Ronald Press, New York; 2d ed., Longman, London and New York.

Farb, P. 1968. Man's Rise to Civilization as Shown by the Indians of North America from Primeval Times to the Coming of the Industrial State. Dutton, New York.

Feduccia, A. 1980. The Age of Birds. Harvard University Press, Cambridge, Mass.

Felsenstein, J. 1971. On the biological significance of the cost of gene substitution. Amer. Nat. 105:1–11.

Fisher, R. A. 1930, 1958. The Genetical Theory of Natural Selection. 1st and 2d eds. 1st ed., Clarendon Press, Oxford; 2d ed., Dover Publications, New York.

Fisher, R. A. and E. B. Ford. 1947. The spread of a gene in natural conditions in a colony of the moth *Panaxia dominula* L. Heredity 1:143–174.

Flake, R. H. and V. Grant. 1974. An analysis of the cost-of-selection concept. Proc. Nat. Acad. Sci. USA 71:3716–3720.

Ford, E. B. 1955. Moths. Collins, London.

—— 1964, 1971. Ecological Genetics. 1st and 3d eds. 1st ed., Methuen, London; 3d ed., Chapman & Hall, London.

—— 1965. Genetic Polymorphism. Faber & Faber, London.

Fothergill, P. G. 1952. Historical Aspects of Organic Evolution. Hollis and Carter, London.

Fox, A. S., W. F. Duggleby, W. M. Gelbart, and S. B. Yoon. 1970. DNA-induced transformation in *Drosophila*: evidence for transmission without integration. Proc. Nat. Acad. Sci. USA 67:1834–1838.

Fox, A. S. and S. B. Yoon. 1966. Specific genetic effects of DNA in *Drosophila melanogaster*. Genetics 53:897–911.

——1970. DNA-induced transformation in *Drosophila*: locus-specificity and the establishment of transformed stocks. Proc. Nat. Acad. Sci. USA 67:1608–1615.

Fox, A. S., S. B. Yoon, and W. M. Gelbart. 1971. DNA-induced transformation in *Drosophila*: genetic analysis of transformed stocks. Proc. Nat. Acad. Sci. USA 68:342–346.

Fry, W. and J. R. White. 1938. Big Trees. 2d ed. Stanford University Press, Stanford, Calif.

Futuyma, D. J. and M. Slatkin, eds. 1983. Coevolution. Sinauer Associates, Sunderland, Mass.

Gadgil, M. and O. T. Solbrig. 1972. The concept of r- and K-selection: evidence from wild flowers and some theoretical considerations. Amer. Nat. 106:14–31.

Galinat, W. C. 1970. The cupule and its role in the origin and evolution of maize. Agricultural Experiment Station, University of Massachusetts, Amherst, Bull. 585.

—— 1971a. The origin of maize. Ann. Rev. Genet. 5:447–478.

—— 1971b. The evolution of sweet corn. Agricultural Experiment Station, University of Massachusetts, Amherst, Bull. 591.

—— 1977. The origin of corn. In: Corn and Corn Improvement. ed. G. F. Sprague. Amer. Soc. Agronomy, Madison, Wisc.

Garn, S. M. 1961. Human Races. Thomas, Springfield, Ill.

Garstang, W. 1922. The theory of recapitulation: a critical re-statement of the biogenetic law. Jour. Linnean Soc. London, Zool., 35:81–101.

Gause, G. F. 1934. The Struggle for Existence. Williams and Wilkins, Baltimore.

—— 1935. Experimental demonstration of Volterra's periodic oscillations in the numbers of animals. Jour. Exp. Biol. 12:44–48.

Geist, V. 1971. Mountain Sheep: A Study in Behavior and Evolution. University of Chicago Press, Chicago.

Geoffroy Saint-Hilaire, E. 1830. Principes de Philosophie Zoologique. Paris.

Ghiselin, M. T. 1974. The Economy of Nature and the Evolution of Sex. University of California Press, Berkeley.

Gilbert, L. E. and P. H. Raven, eds. 1975. Coevolution of Animals and Plants. University of Texas Press, Austin.

Giles, B. E. 1984. A comparison between quantitative and biochemical variation in the wild barley Hordeum murinum. Evolution 38:34–41.

Gillespie, R., D. R. Horton, P. Ladd, P. G. Macumber, T. H. Rich, R. Thorne, and R. V. S. Wright. 1978. Lancefield swamp and the extinction of the Australian megafauna. Science 200:1044–1048.

Gingerich, P. D. 1983. Rates of evolution: effects of time and temporal scaling. Science 222:159–161.

Glaessner, M. F. 1961. Pre-Cambrian animals. Sci. Amer. 204(2):72–78.

Glass, B., M. S. Sacks, E. F. Jahn, and C. Hess. 1952. Genetic drift in a religious isolate: an analysis of the causes of variation in blood groups and other gene frequencies in a small population. Amer. Nat. 86:145–159.

Glick, T. F., ed. 1974. The Comparative Reception of Darwinism. University of Texas Press, Austin.

Goldschmidt, R. B. 1938. Physiological Genetics. McGraw-Hill, New York.

—— 1940. The Material Basis of Evolution. Yale University Press, New Haven, Conn.

—— 1952. Homoeotic mutants and evolution. Acta Biotheor. 10:87–104.

—— 1953. Experiments with a homoeotic mutant, bearing on evolution. Jour. Exp. Zool. 123:79–114.

—— 1955. Theoretical Genetics. University of California Press, Berkeley.

Goodall, J. 1965. Chimpanzees of the Gombe Stream Reserve. In: Primate Behavior. ed. I. DeVore. Holt, Rinehart and Winston, New York.

Goodman, M., ed. 1982. Macromolecular Sequences in Systematics and Evolutionary Biology. Plenum Press, New York.

Gordon, J. W., G. A. Scangos, D. J. Plotkin, J. A. Barbosa, and F. H. Ruddle. 1980. Genetic transformation of mouse embryos by microinjection of purified DNA. Proc. Nat. Acad. Sci. USA 77:7380–7384.

Gottlieb, L. D. 1971. Gel electrophoresis: new approach to the study of evolution. Bioscience 21:939–943.

—— 1974. Genetic confirmation of the origin of Clarkia lingulata. Evolution 28:244–250.

—— 1977. Electrophoretic evidence and plant systematics. Ann. Missouri Bot. Gard. 64:161–180.

Gottschalk, W. 1971. Die Bedeutung der Genmutation für die Evolution der Pflanzen. Fischer, Stuttgart.

Gould, S. J. 1973. The misnamed, mistreated, and misunderstood Irish elk. Nat. Hist. 82(3):10–19.

—— 1974. The origin and function of "bizarre" structures: antler size and skull size in the "Irish elk," *Megaloceros giganteus*. Evolution 28:191–220.

—— 1977. Ontogeny and Phylogeny. Harvard University Press, Cambridge, Mass.

—— 1977. The return of hopeful monsters. Nat. Hist. 86 (June–July):22–30.

—— 1980. Is a new and general theory of evolution emerging? Paleobiol. 6:119–130.

—— 1982. Darwinism and the expansion of evolutionary theory. Science 216:380–387.

Gould, S. J. and N. Eldredge. 1977. Punctuated equilibria: the tempo and mode of evolution reconsidered. Paleobiol. 3:115–151.

Gould, S. J. and R. C. Lewontin. 1979. The spandrels of San Marco and the Panglossian paradigm: a critique of the adaptationist programme. Proc. Roy. Soc. London, ser. B, 205:581–598.

Grant, A. and V. Grant. 1956. Genetic and taxonomic studies in *Gilia*. VIII. The Cobwebby gilias. Aliso 3:203–287.

Grant, B. R. and P. R. Grant. 1982. Niche shifts and competition in Darwin's finches: Geospiza conirostris and congeners. Evolution 36:637–657.

Grant, K. A. and V. Grant. 1964. Mechanical isolation of *Salvia apiana* and *Salvia mellifera* (Labiatae). Evolution 18:196–212.

Grant, V. 1954. Genetic and taxonomic studies in *Gilia*. IV. *Gilia achilleaefolia*. Aliso 3:1–18.

—— 1956. Chromosome repatterning and adaptation. Adv. Gen. 8:89–107.

—— 1958. The regulation of recombination in plants. Cold Spring Harbor Symp. Quant. Biol. 23:337–363.

—— 1959. Natural History of the Phlox Family. Systematic Botany. M. Nijhoff, The Hague.

—— 1960. Genetic and taxonomic studies in *Gilia*. XI. Fertility relationships of the diploid Cobwebby Gilias. Aliso 4:435–481.

—— 1963. The Origin of Adaptations. Columbia University Press, New York.

—— 1964. The Architecture of the Germplasm. Wiley, New York.

—— 1965. Evidence for the selective origin of incompatibility barriers in the leafy-stemmed Gilias. Proc. Nat. Acad. Sci. USA 54:1567–1571.

—— 1966a. Selection for vigor and fertility in the progeny of a highly sterile species hybrid in *Gilia*. Genetics 53:757–775.

—— 1966b. The selective origin or incompatibility barriers in the plant genus *Gilia*. Amer. Nat. 100:99–118.

—— 1971. Plant Speciation. 1st ed. Columbia University Press, New York.

—— 1975. Genetics of Flowering Plants. Columbia University Press, New York.

—— 1977a. Organismic Evolution. Freeman, San Francisco.

—— 1977b. Population structure in relation to macroevolution. Biol. Zentralbl. 96:129–139.

—— 1978. Kin selection: a critique. Biol. Zentralbl. 97:385–392.

—— 1980. Gene flow and the homogeneity of species populations. Biol. Zentralbl. 99:157–169.

—— 1981a. Plant Speciation. 2d ed. Columbia University Press, New York.

—— 1981b. The genetic goal of speciation. Biol. Zentralblatt 100:473–482.

—— 1982. Punctuated equilibria: a critique. Biol. Zentralbl. 101:175–184.

—— 1983. The synthetic theory strikes back. Biol. Zentralblatt 102:149–158.

Grant, V. and R. H. Flake. 1974a. Population structure in relation to cost of selection. Proc. Nat. Acad. Sci. USA 71:1670–1671.

—— 1974b. Solutions to the cost-of-selection dilemma. Proc. Nat. Acad. Sci. USA 71:3863–3865.

Grant, V. and K. A. Grant. 1965. Flower Pollination in the Phlox Family. Columbia University Press, New York.

Grasse, P. P. 1977. Evolution of Living Organisms. Translation. Academic Press, New York.

Grayson, D. K. 1977. Pleistocene avifaunas and the overkill hypothesis. Science 195:691–693.

Greenwood, J. J. D. 1974. Effective population numbers in the snail *Cepaea nemoralis*. Evolution 28:513–526.

—— 1976. Effective population number in *Cepaea*: a modification. Evolution 30:186.

Griffin, J. R. and W. B. Critchfield. 1976. The distribution of forest trees in California. U.S.D.A. Forest Service Res. Paper, PSW-82.

Grun, P. 1976. Cytoplasmic Genetics and Evolution. Columbia University Press, New York.

Gustafsson, A. 1951. Mutations, environment and evolution. Cold Spring Harbor Symp. Quant. Biol. 16:263–281.

Haeckel, E. 1866. Generelle Morphologie der Organismen. 2 vols. Reimer, Berlin.

Hairston, N. G. 1983. Alpha selection in competing salamanders: experimental verification of an a priori hypothesis. Amer. Nat. 122:105–113.

Hairston, N. G., F. E. Smith, and L. B. Slobodkin. 1960. Community structure, population control, and competition. Amer. Nat. 94:421–425.

Haldane, J. B. S. 1932. The Causes of Evolution. Longmans, Green. London.

—— 1933. Science and Human Life. Harper, New York.

—— 1949. Suggestions as to quantitative measurement of rates of evolution. Evolution 3:51–56.

—— 1957. The cost of natural selection. Jour. Genet. 55:511–524.

—— 1960. More precise expressions for the cost of natural selection. Jour. Genet. 57:351–360.

Hamilton. W. D. 1964. The genetical evolution of social behavior. I and II. Jour. Theor. Biol. 7:1–16, 17–52.

—— 1972. Altruism and related topics, mainly in social insects. Ann. Rev. Ecol. Syst. 3:193–232.

Hamrick, J. L. 1979. Genetic variation and longevity. In: Topics in Plant Population Biology. ed. O. Solbrig. Columbia University Press, New York.

Hardin, G. 1960. The competitive exclusion principle. Science 131:1292–1297.

Harding, J,. R. W. Allard, and D. G. Smeltzer. 1966. Population studies in predominantly self-pollinated species. IX. Frequency-dependent selection in *Phaseolus lunatus*. Proc. Nat. Acad. Sci. USA 56:99–104.

Hardy, A. C. 1954. Escape from specialization. In: Evolution as a Process. ed. J. Huxley, A. C. Hardy, and E. B. Ford. Allen and Unwin, London.

Harlan, J. 1975. Crops and Man. Amer. Soc. Agronomy, Madison, Wis.

Harland, W. B., A. V. Cox, P. G. Llewellyn, C. A. G. Pickton, A. G. Smith, and R. Walters. 1982. A Geologic Time Scale. Cambridge University Press, Cambridge.

Harris, C. L. 1981. Evolution; Genesis and Revelations. State University of New York Press, Albany.

Hayes, W. 1968. The Genetics of Bacteria and Their Viruses. 2d ed. Wiley, New York.

Hedrick, P. W., M. E. Ginevan, and E. P. Ewing. 1976. Genetic polymorphism in heterogeneous environments. Ann. Rev. Ecol. Syst. 7:1–32.

Heiser, C. B. 1937a. Seed to Civilization. Freeman, San Francisco.

—— 1937b. Introgression re-examined. Bot. Rev. 39:347–366.

Henry, S. M., ed. 1966–1967. Symbiosis. 2 vols. Academic Press, New York.

Hess, D. 1977. Cell modification by DNA uptake. In: Plant Cell, Tissue, and Organ Culture. ed. J. Reinert and Y. P. S. Bajaj. Springer, Berlin and New York.

Hill, J. 1967. The environmental induction of heritable changes in *Nicotiana rustica* parental and selection lines. Genetics 55:735–754.

Hinton, T., P. T. Ives, and A. T. Evans. 1952. Changing the gene order and number in natural populations. Evolution 6:19–28.

Hiraizumi, Y., L. Sandler, and J. F. Crow. 1960. Meiotic drive in natural populations of *Drosophila melanogaster*. III. Populational implications of the segregation-distorter locus. Evolution 14:433–444.

Hornaday, W. T. 1889. The extermination of the American bison, with a sketch of its discovery and life history. Ann. Report Smithsonian Inst. (Washington, D.C.), 1887, pt. II, 367–548.

Horodyski, R. J. and B. Bloeser. 1978. 1400-million-year-old shale-facies microbiota from the Lower Belt Supergroup, Montana. Science 199:682–684.

Hovanitz, W. 1953. Polymorphism and evolution. Symp. Soc. Exp. Biol. 7:240–253.

Hsü, K. J. et al. (20 authors). 1982. Mass mortality and its environmental and evolutionary consequences. Science 216:249–256.

Hubby, J. L. and R. C. Lewontin. 1966. A molecular approach to the study of genic heterozygosity in natural populations. I. The number of alleles at different loci in Drosophila pseudoobscura. Genetics 54:577–594.

Hubby, J. L. and L. H. Throckmorton. 1965. Protein differences in Drosophila. II. Comparative species genetics and evolutionary problems. Genetics 52:203–215.

Huettel, M. D. and G. L. Bush. 1972. The genetics of host selection and its bearing on sympatric speciation in Proceidochares (Diptera: Tephritidae). Entomol. Exp. Appl. 15:465–480.

Hull, D. L. 1983. Karl Popper and Plato's metaphor. In: Advances in Cladistics. 2 vols. ed. N. Platnick and V. Funk. Columbia University Press, New York.

Hunt, W. G. and R. K. Selander. 1973. Biochemical genetics of hybridisation in European house mice. Heredity 31:11–33.

Hutchinson, G. E. 1957. Concluding remarks. Cold Spring Harbor Symposia Quant. Biol. 22:415–427.

—— 1959. Homage to Santa Rosalia or why are there so many kinds of animals? Amer. Nat. 93:145–159.

—— 1965. The Ecological Theater and the Evolutionary Play. Yale University Press, New Haven, Conn.

—— 1978. An Introduction to Population Ecology. Yale University Press, New Haven, Conn.

Huxley, J. S. 1932. Problems of Relative Growth. Methuen, London.

—— 1938. The present standing of the theory of sexual selection. In: Evolution. ed. G. R. deBeer. Oxford University Press, Oxford.

—— 1942. Evolution: The Modern Synthesis. Allen and Unwin, London.

—— 1954. The evolutionary process. In: Evolution as a Process. ed. J. Huxley, A. C. Hardy, and E. B. Ford. Allen and Unwin, London.

—— 1958. Evolutionary processes and taxonomy with special reference to grades. Uppsala Univ. Årskr., 1958, 21–39.

Iltis, H. H. 1981. The catastrophic sexual transmutation theory (CSTT): the epigenesis of the teosinte tassel spike to the ear of corn. (Abstract.) Bot. Soc. Amer. Misc. Publ. 160:70.

—— 1983a. The catastrophic sexual transmutation theory (CSTT): from the teosinte tassel spike to the ear of corn. Maize Genetics Coop. Newsletter 58:81–92.

—— 1983b. From teosinte to maize: the catastrophic sexual transmutation. Science 222:886–894.

Iltis, H. H., J. F. Doebley, R. Guzman, and B. Pazy. 1979. *Zea diploperennis* (Gramineae): a new teosinte from Mexico. Science 203:186–188.

Ingram, V. M. 1963. The Hemoglobins in Genetics and Evolution. Columbia University Press, New York.

Ives, P. T. 1950. The importance of mutation rate genes in evolution. Evolution 4:236–252.

Jaenicke, J. 1981. Criteria for ascertaining the existence of host races. Amer. Nat. 117:830–834.

Jain, S. K. 1978. Inheritance of phenotypic plasticity in soft chess, *Bromus mollis* L. (Gramineae). Experientia 34:835–836.

Jain, S. K. and A. D. Bradshaw. 1966. Evolutionary divergence among adjacent plant populations. I. The evidence and its theoretical analysis. Heredity 21:407–441.

Jain, S. K. and R. S. Singh. 1979. Population biology of *Avena*. VII. Allozyme variation in relation to the genome analysis. Bot. Gaz. 140:356–362.

Janis, C. 1976. The evolutionary strategy of the Equidae and the origins of rumen and cecal digestion. Evolution 30:757–774.

Janzen, D. H. 1970. Herbivores and the number of tree species in tropical forests. Amer. Nat. 104:501–528.

Jepson, W. L. 1909. A flora of California. Vol. 1, Part 1. Associated Students Store, University of California, Berkeley.

Johannsen, W. 1911. The genotype conception of heredity. Amer. Nat. 45:129–159.

Johnsgard, P. A. 1983. The Hummingbirds of North America. Smithsonian Institution Press, Washington, D.C.

Johnson, G. B. 1974. Enzyme polymorphism and metabolism. Science 184:28–37.

Johnston, J. S. and W. B. Heed. 1976. Dispersal of desert-adapted *Drosophila*: the saguaro-breeding *D. nigrospiracula*. Amer. Nat. 110:629–651.

Jones, G. N. 1941. How many species of plants are there? Science 94:234.

Jones, H. A., J. C. Walker, T. M. Little, and R. M. Larson. 1946. Relation of color-inhibiting factor to smudge resistance in onion. Jour. Agric. Rex. 72:259–264.

Jones, J. S. 1973. Ecological genetics and natural selection in molluscs. Science 182:546–552.

Jukes, T. H. 1966. Molecules and Evolution. Columbia University Press, New York.

—— 1972. Comparison of polypeptide sequences. In: Darwinian, Neo-Darwinian, and Non-Darwinian Evolution. ed. L. M. LeCam, J. Neyman, and E. L. Scott. Berkeley Symposia on Mathematical Statistics and Probability. University of California Press, Berkeley.

Katz, A. J. and S. S. Y. Young. 1975. Selection for high adult body weight in *Drosophila* populations with different structures. Genetics 81:163–175.

Keith, A. 1948. A New Theory of Human Evolution. Watts, London.

Keosian, J. 1964. The Origin of Life. Reinhold, New York.

Kerr, R. A. 1984. Periodic impacts and extinctions reported. Science 223:1277–1279.

Kerr, W. E. and S. Wright. 1954. Experimental studies of the distribution of gene frequencies in very small populations of *Drosophila melanogaster*. I. Forked. Evolution 8:172–177.

Kerster, H. W. 1964. Neighborhood size in the Rusty lizard, *Sceloporus olivaceus*. Evolution 18:445–457.

Kerster, H. W. and D. A. Levin. 1968. Neighborhood size in *Lithospermum caroliniense*. Genetics 60:577–587.

Kettlewell, H. B. D. 1956. Further selection experiments on industrial melanism in the Lepidoptera. Heredity 10:287–301.

—— 1973. The Evolution of Melanism. Oxford University Press, New York.

Kilias, G. and S. N. Alahiotis. 1982. Genetic studies on sexual isolation and hybrid sterility in long-term cage populations of *Drosophila melanogaster*. Evolution 36:121–131.

Kimura, M. 1968. Genetic variability maintained in a finite population due to mutational production of neutral and nearly neutral isoalleles. Genet. Res. 11:247–269.

—— 1979. The neutral theory of molecular evolution. Sci. Amer. 241:98–126.

—— 1981. Possibility of extensive neutral evolution under stabilizing selection with special reference to nonrandom usage of synonymous codons. Proc. Nat. Acad. Sci. USA 78:5773–5777.

—— 1983. The Neutral Theory of Molecular Evolution. Cambridge University Press, Cambridge.

Kimura, M. and T. Ohta. 1971a. On the rate of molecular evolution. Jour. Molec. Evol. 1:1–17.

—— 1971b. Theoretical Aspects of Population Genetics. Princeton University Press, Princeton, N.J.

—— 1972. Population genetics, molecular biometry, and evolution. In: Darwinian, Neo-Darwinian, and Non-Darwinian Evolution. ed. L. M. LeCam, J. Neyman, and E. L. Scott. Berkeley Symposia on Mathematical Statistics and Probability. University of California Press, Berkeley.

King, J. L. 1966. The gene interaction component of the genetic load. Genetics 53:403–413.

King, J. L. and T. H. Jukes. 1969. Non-Darwinian evolution. Science 164:788–798.

King, M. C. and A. C. Wilson. 1975. Evolution at two levels in humans and chimpanzees. Science 188:107–116.

Kircher, H. W. and W. B. Heed. 1970. Phytochemistry and host plant specificity in Drosophila. Recent Adv. Phytochem. 3:191–209.

Klein, R. G. 1977. The ecology of early man in southern Africa. Science 197:115–126.

Kleinhofs, A. and R. Behki. 1977. Prospects for plant genome modification by nonconventional methods. Ann. Rev. Genetics 11:79–101.

Koehn, R. K., A. J. Zera, and J. G. Hall. 1983. Enzyme polymorphism and natural selection. In: Evolution of Genes and Proteins. ed. M. Nei and R. K. Koehn. Sinauer Associates, Sunderland, Mass.

Koestler, A. 1972. The Case of the Midwife Toad. Random House, New York.

Koestler, A. and J. R. Smythies, eds. 1969. Beyond Reductionism: New Perspectives in the Life Sciences. Macmillan, New York.

Kojima, K. and K. N. Yarbrough. 1967. Frequency dependent selection at the esterase-6 locus in Drosophila melanogaster. Proc. Nat. Acad. Sci. USA 57:645–649.

Kolata, G. B. 1974. !Kung hunter-gatherers: feminism, diet, and birth control. Science 185:932–934.

Koopman, K. F. 1950. Natural selection for reproductive isolation between Drosophila pseudoobscura and Drosophila persimilis. Evolution 4:135–148.

Kottler, M. J. 1980. Darwin, Wallace, and the origin of sexual dimorphism. Proc. Amer. Philosoph. Soc. 124:203–226.

Kraus, G. 1973. Homo sapiens in Decline. New Diffusionist Press, Bedfordshire, England.

Krebs, C. J. 1972. Ecology: The Experimental Analysis of Distribution and Abundance. Harper and Row, New York.

—— 1973. Ecology: The Experimental Analysis of Distribution and Abundance. Harper and Row, New York.

Krebs, C. J., M. S. Gaines, B. L. Keller, J. H. Meyers, and R. H. Tamarin. 1973. Population cycles in small rodents. Science 179:35–41.

Kurten, B. and E. Anderson. 1980. Pleistocene Mammals of North America. Columbia University Press, New York.

Lack, D. 1944. Ecological aspects of species formation in passerine birds. Ibis 86:260–286.

—— 1947. Darwin's Finches. Cambridge University Press, Cambridge.

Laird, C. D. and B. J. McCarthy. 1968. Magnitude of interspecific nucleotide sequence variability in *Drosophila*. Genetics 60:303–322.

Lamarck, J. B. P. de. 1809. Philosophie Zoologique. Paris.

—— 1815–1822. Histoire Naturelle des Animaux sans Vertèbres. Paris.

Lamotte, M. 1951. Recherches sur la structure génétique des populations naturelles de *Cepaea nemoralis* (L.). Bull. biol. France Belg. (suppl.) 35:1–238.

—— 1959. Polymorphism of natural populations of *Cepaea nemoralis*. Cold Spring Harbor Symp. Quant. Biol. 24:65–86.

Lancaster, J. B. 1975. Primate Behavior and the Emergence of Human Culture. Holt, Rinehart and Winston, New York.

Lande, R. 1976. Natural selection and random genetic drift in phenotypic evolution. Evolution 30:314–334.

Laporte, L. F. 1983. Simpson's Tempo and Mode in Evolution revisited. Proc. Amer. Philosoph. Soc. 127:365–417.

Laughlin, W. S. 1950. Blood groups, morphology and population size of the Eskimos. Cold Spring Harbor Symp. Quant. Biol. 15:165–173.

Lavie, B. and E. Nevo. 1981. Genetic diversity in marine molluscs: a test of the niche-width variation hypothesis. Marine Ecology (Naples), 2:335–342.

—— 1982. Heavy metal selection of phosphoglucose isomerase allozymes in marine gastropods. Marine Biol. 71:17–22.

Laycock, G. 1974. Dilemma in the desert: bighorns or burros? Audubon Magazine, September 1974.

Ledig, F. T. and M. T. Conkle. 1983. Gene diversity and genetic structure in a narrow endemic, Torrey pine (*Pinus torreyana* Parry ex Carr.). Evolution 37:79–85.

Lee, R. D. and I. Devore, eds. 1976. Kalahari Hunter-Gatherers. Harvard University Press, Cambridge, Mass.

Lemen, C. and P. W. Freeman. 1983. Quantification of competition among coexisting heteromyids in the southwest. Southwestern Nat. 28:41–46.

Leng, E. R. 1960. Long-term selection of corn for oil and protein content. Mimeographed annual report, Illinois Agricultural Experiment Station, Urbana.

Lerner, I. M. 1954. Genetic Homeostasis. Oliver and Boyd, Edinburgh and London.

—— 1958. The Genetic Basis of Selection. Wiley, New York.

Lerner, I. M. and F. K. Ho. 1961. Genotype and competitive ability of *Tribolium* species. Amer. Nat. 95:329–343.

Levene, H. 1953. Genetic equilibrium where more than one ecological niche is available. Amer. Nat. 87:331–333.

Levin, D. A. 1970. Reinforcement of reproductive isolation: plants versus animals. Amer. Nat. 104:571–581.

—— 1971. Plant phenolics: an ecological perspective. Amer. Nat. 105:157–181.

—— 1972. Low frequency disadvantage in the exploitation of pollinators by corolla variants in *Phlox*. Amer. Nat. 106:453–460.

—— 1978. The origin of isolating mechanisms in flowering plants. Evol. Biol. 11:185–317.

—— 1979. The nature of plant species. Science 204:381–384.

—— 1981. Dispersal versus gene flow in plants. Ann. Missouri Bot. Gard. 68:233–253.

Levin, D. A. and W. L. Crepet. 1973. Genetic variation in *Lycopodium lucidulum*: a phylogenetic relic. Evolution 27:622–632.

Levin, D. A. and H. W. Kerster. 1967. Natural selection for reproductive isolation in *Phlox*. Evolution 21:679–687.

—— 1968. Local gene dispersal in *Phlox*. Evolution 22:130–139.

—— 1969. Density-dependent gene dispersal in *Liatris*. Amer. Nat. 103:61–74.

Levin, D. A. and B. A. Schaal. 1970. Corolla color as an inhibitor of interspecific hybridization in *Phlox*. Amer. Nat. 104:273–283.

Levins, R. 1968. Evolution in Changing Environments. Princeton University Press, Princeton, N.J.

Lewin, R. A., ed. 1962. Physiology and Biochemistry of Algae. Academic Press, New York.

Lewin, R. 1981. Evolutionary history written in globin genes. Science 214:426–429.

—— 1983. Do ape-size legs mean ape-like gait? Science 221:537–538.

Lewis, H. 1962. Catastrophic selection as a factor in speciation. Evolution 16:257–271.

Lewis, H. and C. Epling. 1959. *Delphinium gypsophilum*, a diploid species of hybrid origin. Evolution 13:511–525.

Lewis, H. and M. E. Lewis. 1955. The genus *Clarkia*. Univ. Calif. Publ. Bot. 20:241–392.

Lewis, H. and P. H. Raven. 1958. Rapid evolution in *Clarkia*. Evolution 12:319–336.

Lewis, H. and M. R. Roberts. 1956. The origin of *Clarkia lingulata*. Evolution 10:126–138.

Lewontin, R. C. 1955. The effects of population density and composition on viability in *Drosophila melanogaster*. Evolution 9:27–41.

—— 1970. The units of selection. Ann. Rev. Ecol. Syst. 1:1–18.

—— 1973. Population genetics. Ann. Rev. Genet. 7:1–17.

—— 1974. The Genetic Basis of Evolutionary Change. Columbia University Press, New York.

Lewontin, R. C. and J. L. Hubby. 1966. A molecular approach to the study of genic heterozygosity in natural populations. II. Amount of variation and degree of heterozygosity in natural populations of Drosophila pseudoobscura. Genetics 54:595–609.

Licari, G. R. and P. Cloud. 1972. Prokaryotic algae associated with Australian proterozoic stromatolites. Proc. Nat. Acad. Sci. USA 69:2500–2504.

Lillie, F. R. 1921. Studies of fertilization. VIII. On the measure of specificity in fertilization between two associated species of the sea-urchin genus Strongilocentrotus. Biol. Bull. 40:1–22.

Lindsay, D. W. and R. K. Vickery. 1967. Comparative evolution in Mimulus guttatus of the Bonneville basin. Evolution 21:439–456.

Littlejohn, M. J. 1965. Premating isolation in the Hyla ewingi complex (Anura: Hylidae). Evolution 19:234–243.

Loehlin, J. C., G. Lindzey, and J. N. Spuhler. 1975. Race Difference in Intelligence. Freeman, San Francisco.

Loomis, W. F. 1967. Skin-pigment regulation of vitamin-D biosynthesis in man. Science 157:501–506.

Löve, A. 1949. Mutations at the crater of Hekla in eruption. Hereditas, suppl. vol. 1949, 621–622.

Løvtrup, S. 1974. Epigenetics; A Treatise on Theoretical Biology. Wiley, New York.

—— 1978. On von Baerian and Haeckelian recapitulation. Syst. Zool. 27:348–352.

Ludwig. W. 1950. Zur Theorie der Konkurrenz. Die Annidation (Einnischung) als fünfter Evolutionsfaktor. Neue Ergeb. Probleme Zool., Klatt-Festschrift 1950, 516–537. (Not seen)

Lumsden, C. J. and E. O. Wilson. 1981. Genes, Mind, and Culture; The Coevolutionary Process. Harvard University Press, Cambridge, Mass.

Lysenko, T. 1948. The Science of Biology Today. Translation. International Publishers, New York.

MacArthur, R. 1955. Fluctuations of animal populations, and a measure of community stability. Ecology 36:533–536.

—— 1958. Population ecology of some warblers of northeastern coniferous forests. Ecology 39:599–619.

Mangelsdorf, P. C. 1952. Hybridization in the evolution of maize. In: Heterosis, ed. J. Gowen. Iowa State University Press, Ames.

—— 1958. Reconstructing the ancestor of corn. Proc. Amer. Phil. Soc. 102:454–463.

—— 1974. Corn: Its Origin, Evolution, and Improvement. Harvard University Press, Cambridge, Mass.

—— 1983. The mystery of corn: new perspectives. Proc. Amer. Philosoph. Soc. 127:215–247.

Mangelsdorf, P. C., ed. 1959. The origin of corn. Bot. Mus. Leafl. Harvard Univ. 18:329–440.

Mangelsdorf, P. C., E. S. Barghoorn, and U. C. Banerjee. 1978. Fossil pollen and the origin of corn. Bot. Mus. Leafl. Harvard University 26:237–255.

Mangelsdorf, P. C., H. W. Dick, and J. Camara-Hernandez. 1967. Bat Cave revisited. Bot. Mus. Leafl. Harvard University 22;1–31.

Mangelsdorf, P. C. and D. F. Jones. 1926. The expression of Mendelian factors in the gametophyte of maize. Genetics 11:423–455.

Mangelsdorf, P. C., R. S. MacNeish, and W. C. Galinat. 1964. Domestication of corn. Science 143:538–545.

Mangelsdorf, P. C. and R. G. Reeves. 1959. The origin of corn. III. Modern races, the product of teosinte introgression. Bot. Mus. Leafl. Harvard University 18:389–411.

Mangelsdorf, P. C. and E. C. Smith. 1949. A discovery of remains of primitive maize in New Mexico. Jour. Heredity 40:39–43.

Margulis, L. 1970. Origin of Eukaryotic Cells. Yale University Press, New Haven, Conn.

—— 1981. Symbiosis in Cell Evolution: Life and Its Environment on the Early Earth. Freeman, San Francisco.

Martin, M. M. and J. Harding. 1982. Estimates of fitness in *Erodium* populations with intra- and interspecific competition. Evolution 36:1290–1298.

Martin, P. S. 1983. The discovery of America. Science 179:969–974.

Martin, P. S. and R. G. Klein, eds. 1984. Quaternary Extinctions. University of Arizona Press, Tucson.

Martin, P. S. and H. E. Wright, Jr., eds. 1967. Pleistocene Extinctions: The Search for a Cause. Yale University Press, New Haven, Conn.

Marx, J. 1981. More progress on gene transfer. Science 213:996–997.

—— 1982a. Still more about gene transfer. Science 218:459–460.

—— 1982b. Building bigger mice through gene transfer. Science 218:1298.

Maynard Smith, J. 1968. "Haldane's dilemma" and the rate of evolution. Nature 219:1114–1116.

—— 1976. Group selection. Quart. Rev. Biol. 51:277–283.

—— 1978. The Evolution of Sex. Cambridge University Press, Cambridge.

Mayo, O. 1966. On the evolution of dominance. Heredity 21:499–511.

Mayr, E. 1942. Systematics and the Origin of Species. Columbia University Press, New York.

—— 1954. Change of genetic environment and evolution. In: Evolution as a Process. ed. J. Huxley, A. C. Hardy, and E. B. Ford. Allen and Unwin, London.

—— 1957a. Species concepts and definitions. In: The Species Problem. ed. E. Mayr. American Association for the Advancement of Science, Washington, D.C.

—— 1957b. Difficulties and importance of the biological species concept. In: The Species Problem. ed. E. Mayr. American Association for the Advancement of Science, Washington, D.C.

—— 1963. Animal Species and Evolution. Harvard University Press, Cambridge, Mass.

—— 1969. Principles of Systematic Zoology. McGraw-Hill, New York.

—— 1970. Populations, Species, and Evolution. Harvard University Press, Cambridge, Mass.

—— 1972a. The nature of the Darwinian revolution. Science 176:981–989.

—— 1972b. Sexual selection and natural selection. In: Sexual Selection and the Descent of Man 1871–1971. ed. B. Campbell. Aldine-Atherton, Chicago.

—— 1982a. The Growth of Biological Thought; Diversity, Evolution, and Inheritance. Harvard University Press, Cambridge, Mass.

—— 1982b. Adaptation and selection. Biol. Zentralblatt 101:161–174.

Mayr, E. and W. B. Provine, eds. 1980. The Evolutionary Synthesis. Harvard University Press, Cambridge, Mass.

McClure, M. S. and P. W. Price. 1975. Competition among sympatric Erythroneura leafhoppers (Homoptera: Cicadellidae) on American sycamores. Ecology 56:1388–1397.

McVeigh, I. and C. J. Hobdy. 1952. Development of resistance by Micrococcus pyogenes var. aureus to antibiotics: morphological and physiological changes. Amer. Jour. Bot. 39:352–359.

Merrell, D. J. 1953. Gene frequency changes in small laboratory populations of Drosophila melanogaster. Evolution 7:95–101.

—— 1975. An Introduction to Genetics. Norton, New York.

Mettler, L. E. and T. G. Gregg. 1969. Population Genetics and Evolution. Prentice-Hall, Englewood Cliffs, N.J.

Michener, C. D. 1947. A revision of the American species of Hoplitis (Hymenoptera, Megachilidae). Bull. Amer. Mus. Nat. Hist. 89:257–318.

—— 1975. The Brazilian bee problem. Ann. Rev. Entomol. 20:399–416.

Milkman, R. D. 1960a. The genetic basis of natural variation. I. Crossveins in Drosophila melanogaster. Genetics 45:35–48.

—— 1960b. The genetic basis of natural variation. II. Analysis of a polygenic system in Drosophila melanogaster. Genetics 45:377–391.

—— 1961. The genetic basis of natural variation. III. Developmental lability and evolutionary potential. Genetics 46:25–38.

Miller, A. H. 1955. A hybrid woodpecker and its significance in speciation in the genus *Dendrocopos*. Evolution 9:317–321.

Miller, S. L. 1953. A production of amino acids under possible primitive earth conditions. Science 117:528.

Millicent, E. and J. M. Thoday. 1961. Effects of disruptive selection. IV. Gene flow and divergence. Heredity 16:199–217.

Monson, G. and L. Sumner, eds. 1980. The Desert Bighorn: Its Life History, Ecology, and Management. University of Arizona Press, Tucson.

Moos, J. R. 1955. Comparative physiology of some chromosomal types in *Drosophila pseudoobscura*. Evolution 9:141–151.

Mourant, A. E. 1954. The Distribution of the Human Blood Groups. Blackwell Scientific Publications, Oxford.

Mourant, A. E., A. C. Kopec, and K. Domaniewska-Sobczak. 1976. The Distribution of the Human Blood Groups and Other Polymorphisms. 2d ed. Oxford University Press, London.

Muller, C. H. 1966. The role of chemical inhibition (allelopathy) in vegetational composition. Bull. Torrey Bot. Club 93:332–351.

—— 1970. The role of allelopathy in the evolution of vegetation. In: Biochemical Coevolution. Oregon State University Press, Corvallis.

Muller, H. J. 1932. Some genetic aspects of sex. Amer. Nat. 66:118–138.

Müntzing, A. 1930. Über Chromosomenvermehrung in *Galeopsis*-Kreuzungen und ihre phylogenetische Bedeutung. Hereditas 14:153–172.

—— 1932. Cyto-genetic investigations on synthetic *Galeopsis tetrahit*. Hereditas 16:105–154.

Murray, J. and B. Clarke. 1980. The genus *Partula* on Moorea: speciation in progress. Proc. Roy Soc. London, B, 211:83–117.

Nassar, R., H. J. Muhs, and R. D. Cook. 1973. Frequency-dependent selection at the Payne inversion in *Drosophila melanogaster*. Evolution 27:558–564.

Nei, M. 1972. Genetic distance between populations. Amer. Nat. 106:283–292.

Nei, M. and Y. Imaizumi. 1966. Genetic structure of human populations. I and II. Heredity 21:9–35, 183–190.

Nei, M. and R. K. Koehn, eds. 1983. Evolution of Genes and Proteins. Sinauer Associates, Sunderland, Mass.

Nevo, E. 1978. Genetic variation in natural populations: patterns and theory. Theoret. Pop. Biol. 13:121–177.

Nevo, E., R. Ben-Shlomo, and B. Lavie. 1984. Mercury selection of allozymes in marine organisms: prediction and verification in nature. Proc. Nat. Acad. Sci. USA 81:1258–1259.

Newell, N. D. 1949. Phyletic size increase-an important trend illustrated by fossil invertebrates. Evolution 3:103–124.

—— 1967. Revolutions in the history of life. Geol Soc. Amer., Spec. Pap. 89:63–91.

NRC Board on Agriculture. 1984. Genetic Engineering of Plants. National Academy Press, Washington, D.C.

Numbers, R. L. 1982. Creationism in 20th century America. Science 218:538–544.

Nyberg, D. 1982. Sex, recombination, and reproductive fitness: an experimental study using *Paramecium*. Amer. Nat. 120:198–217.

Odum, E. P. 1969. The strategy of ecosystem development. Science 164:262–270.

—— 1971. Fundamentals of Ecology. 3d ed. Saunders, Philadelphia.

Officer, C. B. and C. L. Drake. 1983. The Cretaceous-Tertiary transition. Science 219:1383–1390.

Ohta, T. and M. Kimura. 1971. On the constancy of the evolutionary rate of cistrons. Jour. Molec. Evol. 1:18–25.

Oparin, A. I. 1938. The Origin of Life on Earth. Translation. Macmillan, New York.

—— 1964. Life: Its Nature, Origin, and Development. Academic Press, New York.

Osborn, H. F. 1910. The Age of Mammals in Europe, Asia, and North America. Macmillan, New York.

—— 1929. The titanotheres of ancient Wyoming, Dakota, and Nebraska. U.S. Geological Survey, Washington, D.C., Monograph 55.

—— 1934. Aristogenesis, the creative principle in the origin of species. Amer. Nat. 68:193–235.

Ostrom, J. H. 1974. *Archaeopteryx* and the origin of flight. Quart. Rev. Biol. 49:27–47.

—— 1976. *Archaeopteryx* and the origin of birds. Biol. Jour. Linnean Soc. 8:91–182.

Overton, W. R. 1982a. Creationism in schools: the decision in McLean vs. Arkansas Board of Education. Science 215:934–943.

—— 1982b. Opinion. Rev. Bill McLean vs. Arkansas Board of Education. Academe 68(2):27–36.

Owen, D. F. 1961. Industrial melanism in North American moths. Amer. Nat. 95:227–233.

Palmiter, R. D., R. L. Brinster, R. E. Hammer, M. E. Trumbauer, M. G. Rosenfeld, N. C. Birnberg, and R. M. Evans. 1982. Dramatic growth of mice that develop from eggs microinjected with metallothionein-growth hormone fusion genes. Nature 300:611–615.

Pandey, K. K. 1976. Genetic transformation and "graft-hybridization" in flowering plants. Theoretical and Applied Genetics 47:299–302.

—— 1978. Gametic gene transfer in *Nicotiana* by means of irradiated pollen. Genetica 49:53–69.

—— 1980. Further evidence for egg transformation in *Nicotiana*. Heredity 45:15–29.

—— 1981. Tissue culture, genetic transformation and plant improvement. Symposium on Tissue Culture of Economically Important Plants. ed. A. N. Rao. COSTED, Singapore.

Pandey, K. K. and M. R. Patchell. 1982. Genetic transformation in chicken by the use of irradiated male gametes. Mol. Gen. Genet. 186:305–308.

Parthier, B. 1982. Transfer RNA and the phylogenetic origin of cell organelles. Biol. Zentralblatt 101:577–596.

Passingham, R. 1982. The Human Primate. Freeman, San Francisco.

Paterniani, E. 1969. Selection for reproductive isolation between two populations of maize, *Zea mays* L. Evolution 23:534–547.

Patterson, B. 1949. Rates of evolution in taeniodonts. In: Genetics, Paleontology, and Evolution. ed. G. L. Jepsen, E. Mayr, and G. G. Simpson. Princeton University Press, Princeton, N.J.

Pianka, E. R. 1978. Evolutionary Ecology. 2d ed. Harper and Row, New York.

Pilbeam, D. 1984. The descent of hominoids and hominids. Sci. Amer., 250(3):84–96.

Pimentel, D., G. J. C. Smith, and J. Soans. 1967. A population model of sympatric speciation. Amer. Nat. 101:493–504.

Poole, A. L. and D. Cairns. 1940. Botanical aspects of ragwort (*Senecio jacobaea* L.) control. Bull. Dept. Sci. Industr. Res., New Zealand, 82:1–66.

Powell, J. R., T. Dobzhansky, J. E. Hook, and H. E. Wistrand. 1976. Genetics of natural populations. XLIII. Further studies on rates of disperal of *Drosophila pseudoobscurra* and its relatives. Genetics 82:483–506.

Prazmo, W. 1965. Cytogenetic studies on the genus *Aquilegia*. III. Inheritance of the traits distinguishing different complexes in the genus *Aquilegia*. Acta Soc. Bot. Polon. 34:403–437.

Protsch, R. and R. Berger. 1973. Earliest radiocarbon dates for domesticated animals. Science 179:235–239.

Prout, T. 1962. The effects of stabilizing selection on the time of development in *Drosophila melanogaster*. Genet. Res. 3:364–382.

Race, R. R. and R. Sanger. 1962. Blood Groups in Man. 3d ed. Blackwell Scientific Publications, Oxford.

Radinsky, L. 1976. Oldest horse brains: more advanced than previously realized. Science 194:626–627.

Raff, R. A. and T. C. Kaufman. 1983. Embryos, Genes and Evolution. The Developmental-genetic Basis of Evolutionary Change. Macmillan, New York.

Ralls, K. 1977. Sexual dimorphism in mammals: avian models and unanswered questions. Amer. Nat. 111:917–938.

Raup, D. M. and J. J. Sepkowski. 1982. Mass extinctions in the marine fossil record. Science 215:1501–1503.

—— 1984. Periodicity of extinctions in the geologic past. Proc. Nat. Acad. Sci. USA 81:801–805.

Raven, P. H. 1970. A multiple origin for plastids and mitochondria. Science 1969:641–646.

Rejmanek, M. and J. Jenik. 1975. Niche, habitat, and related ecological concepts. Acta Biotheor. 24:100–107.

Remington, C. L. 1954. The genetics of Colias (Lepidoptera). Adv. Genet. 6:403–450.

Rensch, B. 1947. Neuere Probleme der Abstammungslehre. 1st ed. Ferdinand Enke Verlag, Stuttgart.

—— 1960a. The laws of evolution. In: Evolution After Darwin. 1st vol. ed S. Tax. University of Chicago Press, Chicago.

—— 1960b. Evolution Above the Species Level. Translation. Columbia University Press, New York.

Richards, P. W. 1952. The Tropical Rain Forest; An Ecological Study. Cambridge University Press, Cambridge.

Richerson, P., R. Armstrong, and C. R. Goldman. 1970. Contemporaneous disequilibrium, a new hypothesis to explain the "paradox of the plankton." Proc. Nat. Acad. Sci. USA 67:1710–1714.

Ridley, H. N. 1930. The Dispersal of Plants Throughout the World. Reeve, Ashford, Kent.

Rieger, R., A. Michaelis, and M. M. Green. 1976. A Glossary of Genetics and Cytogenetics: Classical and Molecular. 4th ed. Springer, New York.

Ris, H. and W. Plaut. 1962. Ultrastructure of DNA-containing areas in the chloroplast of Chlamydomonas. Jour. Cell Biol. 13:383–391.

Romanes, G. J. 1890. Mr. A. R. Wallace on physiological selection. The Monist 1:1–20.

Romer, A. S. 1966. Vertebrate Paleontology. 3d ed. University of Chicago Press, Chicago.

—— 1967. Major steps in vertebrate evolution. Science 158:1629–1637.

Roose, M. L. and L. D. Gottlieb. 1976. Genetic and biochemical consequences of polyploidy in Tragopogon. Evolution 30:818–830.

Rosenthal, G. A. and D. H. Janzen, eds. 1979. Herbivores: Their Interaction with Secondary Plant Metabolites. Academic Press, New York

Rosin, S., J. K. Moor-Jankowski, and M. Schneeberger. 1958. Die Fertilität im Bluterstamm von Tenna (Hämophilie B). Acta Genet. 8:1–24.

Ross, H. H. 1957. Principles of natural coexistence indicated by leafhopper populations. Evolution 11:113–129.

—— 1958. Evidence suggesting a hybrid origin for certain leafhopper species. Evolution 12:337–346.

Rothschild, M., T. Reichstein, J. von Euw, R. Aplin, and R. R. M. Harman. 1970. Toxic Lepidoptera. Toxicon 8:293–299.

Roughgarden, J. 1979. Theory of Population Genetics and Evolutionary Ecology: An Introduction. Macmillan, New York.

Rubin, G. M. and A. C. Spradling. 1982. Genetic transformation of Drosophilia with transposable element vectors. Science 218:348–353.

Rundel, P. W. 1972a. An annotated check list of the groves of Sequoiadendron giganteum in the Sierra Nevada, California. Madrono 21:319–328.

—— 1972b. Habitat restriction in giant sequoia: the environmental control of grove boundaries. Amer. Midl. Nat. 87:81–99.

Ruse, M. 1979. The Darwinian Revolution. University of Chicago Press, Chicago.

—— 1982. Darwinism Defended. A Guide to the Evolution Controversies. Addison-Wesley, Reading, Mass.

Sakai, K. and K. Gotoh. 1955. Studies on competition in plants. IV. Competitive ability of F_1 hybrids in barley. Jour. Hered. 46:139–143.

Salisbury, E. J. 1942. The Reproductive Capacity of Plants. Bell, London.

Sandler, L. and E. Novitski. 1957. Meiotic drive as an evolutionary force. Amer. Nat. 91:105–110.

Savile, D. B. O. 1959. Limited penetration of barriers as a factor in evolution. Evolution 13:333–343.

—— 1972. Arctic adaptations in plants. Monograph no. 6, Canada Dept. Agriculture, Ottawa.

—— 1975. Evolution and biogeography of Saxifragaceae with guidance from their rust parasites. Ann. Missouri Bot. Gard. 62:354–361.

Schaal, B. A. 1980. Measurement of gene flow in Lupinus texensis. Nature 284:450–451.

Schaeffer, B. 1948. The origin of a mammalian ordinal character. Evolution 2:164–175.

Schiemann J. 1982. Gentransfer bei Eukaryoten. Biol. Zentralblatt 101:5–25.

Schindewolf, O. H. 1950. Grundfragen der Paläeontologie. Schweizerbart, Stuttgart.

Schmalhausen, I. I. 1949. Factors of Evolution: The Theory of Stabilizing Selection. Translation. Blakiston, Philadelphia.

Schoener, T. W. 1982. The controversy over interspecific competition. Amer. Sci. 70:586–595.

—— 1983. Field experiments on interspecific competition. Amer. Nat. 122:240–285.

Schopf, J. W. 1974. Paleobiology of the Precambrian: the age of blue-green algae. Evol. Biol. 7:1–43.

ogy, and Evolution. ed. G. L. Jepsen, E. Mayr, and G. G. Simpson. Princeton University Press, Princeton, N.J.

—— 1950. Variation and Evolution in Plants. Columbia University Press, New York.

—— 1969. The Basis of Progressive Evolution. University of North Carolina Press, Chapel Hill.

—— 1974. Flowering Plants: Evolution Above the Species Level. Harvard University Press, Cambridge, Mass.

—— 1982. Darwin to DNA, Molecules to Humanity. Freeman, San Francisco.

Stebbins, R. C. 1949. Speciation in salamanders of the plethodontid genus *Ensatina*. University Calif. Publ. Zool. 48:377–526.

—— 1957. Intraspecific sympatry in the lungless salamander *Ensatina eschscholtzi*. Evolution 11:265–270.

Stern, C. 1958. Selection for subthreshold differences and the origin of pseudo-exogenous adaptations. Amer. Nat. 92:313–316.

—— 1959. Variation and hereditary transmission. Proc. Amer. Phil. Soc. 103:183–189.

—— 1960, 1973. Principles of Human Genetics. 2d and 3d eds. Freeman, San Francisco.

Stewart, W. N. 1983. Paleobotany and the Evolution of Plants. Cambridge University Press, Cambridge.

Stirton, R. A. 1947. Observations on evolutionary rates in hypsodonty. Evolution 1:32–41.

Stommel, H. and E. Stommel. 1979. The year without a summer. Sci. Amer., 240(6):176–186.

Stothers, R. B. 1984. The great Tambora eruption in 1815 and its aftermath. Science 224:1191–1198.

Streams, F. A. and D. Pimentel. 1961. Effects of immigration on the evolution of populations. Amer. Nat. 95:201–210.

Strid. A. 1970. Studies in the Aegean flora. XVI. Biosystematics of the *Nigella arvensis* complex with special reference to the problem of non-adaptive radiation. Opera Bot., No. 28, pp. 1–169.

Struhsaker, T. T. 1967. Auditory communication among vervet monkeys (*Cercopithecus aethiops*). In: Social Communication Among Primates. ed. S. A. Altmann. University of Chicago Press, Chicago.

Stubbe, H. 1960. Mutanten der Wildtomate *Lycopersicon pimpinellifolium* (Jusl.) Mill. Kulturpflanze 8:110–137.

Sudworth, G. B. 1908. Forest Trees of the Pacific Slope. U.S. Department of Agriculture, Washington, D.C.

Sukatchev, W. 1928. Einige experimentelle Untersuchungen über den Kampf ums Dasein zwischen Biotypen derselben Art. Z. indukt. Abstammungs-Vererbungsl. 47:54–74.

Susman, R. L. and J. T. Stern. 1982. Functional morphology of *Homo habilis.* Science 217:931–934.

Suzuki, D. T., A. J. F. Griffiths, and R. C. Lewontin. 1981. An Introduction to Genetic Analysis. 2d ed. Freeman, San Francisco.

Symposium. Diversity and Stability in Ecological Systems. Brookhaven Symp. Biol., no. 22.

Takhtajan, A. 1959a. Die Evolution der Angiospermen. Gustav Fischer, Jena.

—— 1959b. Essays on the Evolutionary Morphology of Plants. Translation. American Institute of Biological Sciences, Washington, D.C.

—— 1969. Flowering Plants: Origin and Dispersal. Translation. Oliver and Boyd, Edinburgh.

—— 1976. Neoteny and the origin of flowering plants. In: Origin and Early Evolution of Angiosperms. ed. C. B. Beck. Columbia University Press, New York.

—— 1983. Macroevolutionary processes in the history of plant world. (In Russian with English summary.) Botanicheskii Zhurnal 68:1593–1603.

Teilhard de Chardin, P. 1959. The Phenomenon of Man. Translation. Harper and Row, New York.

Terrace, H. S., L. A. Petitto, R. J. Sanders, and T. G. Bever. 1979. Can an ape create a sentence? Science 206:891–902.

Thoday, J. M. 1958. Natural selection and biological progress. In: A Century of Darwin. ed. S. A. Barnett. Allen and Unwin, London.

—— 1972. Disruptive selection. Proc. Roy. Soc. London, B, 182:109–143.

Thoday, J. M. and T. B. Boam. 1959. Effects of disruptive selection. II. Polymorphism and divergence without isolation. Heredity 13:205–218.

Thoday, J. M. and J. B. Gibson. 1962. Isolation by disruptive selection. Nature 193:1164–1166.

—— 1970. The probability of isolation by disruptive selection. Amer. Nat. 104:219–230.

Thompson, D. W. 1917, 1942, 1961. On Growth and Form. 1st, 2d, and abridged eds. Cambridge University Press, London.

Tilman, D. 1982. Resource Competition and Community Structure. Princeton University Press, Princeton, N.J.

Timmis, J. N. and J. Ingle. 1974. The nature of the variable DNA associated with environmental induction in flax. Heredity 33:339–346.

Timofeeff-Ressovsky, N. W. 1934a. The experimental production of mutations. Biol. Rev. 9:411–457.

—— 1934b. Über den Einfluss des genotypischen Milieus und der Aussenbedingungen auf die Realisation des Genotyps. Nachr. Ges. Wissensch. Göttingen, N.F., 1:53–106.

—— 1940. Mutations and geographical variation. In: The New Systematics. ed. J. Huxley. Clarendon Press, Oxford.

Timofeeff-Ressovsky, N. W., A. V. Jablokov, and N. V. Glotov. 1977. Grundriss der Populationslehre. Translation. Fischer, Jena.

Trivers, R. L. 1972. Parental investment and sexual selection. In: Sexual Selection and the Descent of Man. ed. B. G. Campbell. Aldine, Chicago.

Trivers, R. L. and H. Hare. 1976. Haplodiploidy and the evolution of the social insects. Science 191:249–263.

Turesson, G. 1922. The genotypical response of the plant species to its habitat. Hereditas 3:211–350.

—— 1925. The plant species in relation to habitat and climate: contributions to the knowledge of genecological units. Hereditas 6:147–236.

Turner, B. J. 1974. Genetic divergence of Death Valley pupfish species. Biochemical vs. morphological evidence. Evolution 28:281–294.

Turner, M. E., J. C. Stephens, and W. W. Anderson. 1982. Homozygosity and patch structure in plant populations as a result of nearest-neighbor pollination. Proc. Nat. Acad. Sci. USA 79:203–207.

Urey, H. C. 1952. The Planets: Their Origin and Development. Yale University Press, New Haven, Conn.

Uzzell, T. and C. Spolsky. 1974. Mitochondria and plastids as endosymbionts: a revival of special creation? Amer. Sci. 62:334–343.

Van Valen, L. 1971. Adaptive zones and the orders of mammals. Evolution 25:420–428.

—— 1973. A new evolutionary law. Evol. Theory 1:1–30.

Van Valen, L. and R. E. Sloan. 1977. Ecology and the extinction of the dinosaurs. Evol. Theory 2:37–64.

Vries, H. de. 1901–1903. Die Mutationstheorie. 2 vols. Veit, Leipzig.

Wade, M. J. 1976. Group selection among laboratory populations of *Tribolium*. Proc. Nat. Acad. Sci. USA 73:4604–4607.

—— 1977. An experimental study of group selection. Evolution 31:134–153.

—— 1978. A critical review of the models of group selection. Quart. Rev. Biol. 53:101–114.

—— 1982. Group selection: migration and the differentiation of small populations. Evolution 36:949–961.

Waddington, C. H. 1953. Genetic assimilation of an acquired character. Evolution 7:118–126.

—— 1956. Genetic assimilation of the bithorax phenotype. Evolution 10:1–13.

—— 1957. The Strategy of the Genes. Allen and Unwin, London.

Wagner, G. P. 1980. Empirical information about the mechanism of typogenetic evolution. Naturwissenschaften 67:258–259.

—— 1981. Feedback selection and the evolution of modifiers. Acta Biotheoretica 30:79–102.

Wald, G. 1955. The origin of life. In: The Physics and Chemistry of Life. ed. Scientific American. Simon and Schuster, New York.

Walker, E. P. 1975. Mammals of the World. 2 vols. 3d ed. Johns Hopkins University Press, Baltimore.

Walker, J. C. and M. A. Stahmann. 1955. Chemical nature of disease resistance in plants. Ann. Rev. Plant Physiol. 6:351–366.

Wallace, A. R. 1889. Darwinism; An Exposition of the Theory of Natural Selection. Macmillan, London.

Wallace, B. 1968. Topics in Population Genetics. Norton, New York.

—— 1970. Genetic Load: Its Biological and Conceptual Aspects. Prentice-Hall, Englewood Cliffs, N.J.

—— 1981. Basic Population Genetics. Columbia University Press, New York.

Washburn, S. L. 1960. Tools and human evolution. Sci. Amer. 203(3):62–75.

Washburn, S. L. and D. A. Hamburg. 1965. The study of primate behavior. In: Primate Behavior. ed. I. DeVore. Holt, Rinehart and Winston, New York.

Watt, W. B. 1969. Adaptive significance of pigment polymorphisms in *Colias* butterflies. II. Thermoregulation and photoperiodically controlled melanin variation in *Colias eurytheme*. Proc. Nat. Acad. Sci. USA 63:767–774.

Weismann, A. 1889–1892. Essays upon Heredity and Kindred Biological Problems. 2 vols. Clarendon Press, Oxford.

—— 1892. The Germ-plasm: A Theory of Heredity. Translation. Walter Scott, London.

Wells, P. V. 1969. The relation between mode of reproduction and extent of speciation in woody genera of the California chaparral. Evolution 23:264–267.

Werth, E. 1956. Bau und Leben der Blumen: Die blütenbiologischen Bautypen in Entwicklung und Anpassung. Enke, Stuttgart.

Westoll, T. S. 1949. On the evolution of the Dipnoi. In: Genetics, Paleontology and Evolution. ed. G. L. Jepsen, E. Mayr, and G. G. Simpson. Princeton University Press, Princeton, N.J.

Whalen, M. D. 1978. Reproductive character displacement and floral diversity in *Solanum* section *Androceras*. Systematic Bot. 3:77–86.

White, M. J. D. 1945. Animal Cytology and Evolution. 1st ed. Cambridge University Press, Cambridge.

—— 1973. Animal Cytology and Evolution. 3d ed. Cambridge University Press, Cambridge and London.

—— 1978. Modes of Speciation. W. H. Freeman, San Francisco.

Whittaker, R. H. and P. P. Feeny. 1971. Allelochemics: chemical interactions between species. Science 171:757–770.

Whittaker, R. H., S. A. Levin, and R. B. Root. 1973. Niche, habitat, and ecotype. Amer. Nat. 107:321–338.

Whitten, W. M. 1981. Pollination ecology of *Monarda didyma*, *M. clinopodia*, and hybrids (Lamiaceae) in the southern Appalachian mountains. Amer. Jour. Bot. 68:435–442.

Whyte, L. L. 1965. Internal Factors in Evolution. Braziller, New York.

Wiener, A. S. and J. Moor-Jankowski. 1971. Blood groups of non-human primates and their relationship to the blood groups of man. In: Comparative Genetics in Monkeys, Apes and Man. ed. A. B. Chiarelli. Academic Press, London and New York.

Wiklund, C. and T. Järvi. 1982. Survival of distasteful insects after being attacked by naive birds: a reappraisal of the theory of aposematic coloration evolving through individual selection. Evolution 36:998–1002.

Wilkes, H. G. 1967. Teosinte: The Closest Relative of Maize. Bussey Institution, Harvard University, Cambridge, Mass.

—— 1972. Maize and its wild relatives. Science 177:1071–1077.

—— 1982. The origin of maize—is teosinte the answer? (Abstract) Bot. Soc. Amer. Misc. Publ. 162:113.

Williams, C. B. 1958. Insect Migration. Macmillan, New York.

Williams, G. C. 1966. Adaptation and Natural Selection. Princeton University Press, Princeton, N.J.

Williams, G. C., ed. 1971. Group Selection. Aldine-Atherton, Chicago.

Williams, G. C. 1975. Sex and Evolution. Princeton University Press, Princeton, N.J.

Williamson, D. L. and L. Ehrman. 1967. Induction of hybrid sterility in nonhybrid males of *Drosophila paulistorum*. Genetics 55:131–140.

Wilson, A. C. 1975. Evolutionary importance of gene regulation. Stadler Genetics Symposia (Univ. Missouri) 7:117–134.

Wilson, D. S. 1980. The Natural Selection of Populations and Communities. Benjamin Cummings, Menlo Park, Calif.

Wilson, E. O. 1971. Competitive and aggressive behavior. In: Man and Beast: Comparative Social Behavior. ed. J. F. Eisenberg and W. S. Dillon. Smithsonian Institution, Washington, D.C.

—— 1975. Sociobiology: The New Synthesis. Harvard University Press, Cambridge, Mass.

—— 1978. On Human Nature. Harvard University Press, Cambridge, Mass.

Winter, F. L. 1929. The mean and variability as affected by continuous selection for composition in corn. Jour. Agric. Res. 39:451–476.

Wolf, C. W. 1948. Taxonomic and distributional studies of the New World cypresses. Aliso 1:1–250.

Wolpoff, M. H. 1980. Paleo-Anthropology. Knopf, New York.

Wood, C. E. 1975. The Balsaminaceae in the southeastern United States. Jour. Arnold Arboretum 56:413–426.

Woodworth, C. M., E. R. Leng, and R. W. Jugenheimer. 1952. Fifty generations of selection for protein and oil in corn. Agron. Jour. 44:60–65.

Wright, S. 1931. Evolution in Mendelian populations. Genetics 16:97–159.

—— 1932. The roles of mutation, inbreeding, crossbreeding, and selection in evolution. Proc. 6th Internat. Congr. Genetics 1:356–366.

—— 1943. Isolation by distance. Genetics 28:114–138.

—— 1946. Isolation by distance under diverse systems of mating. Genetics 31:39–59.

—— 1949. Adaptation and selection. In: Genetics, Paleontology, and Evolution. ed. G. L. Jepsen, E. Mayr, and G. G. Simpson. Princeton University Press, Princeton, N.J.

—— 1956. Modes of selection. Amer. Nat. 90:5–24.

—— 1960. Physiological genetics, ecology of populations, and natural selection. In: Evolution After Darwin. 1 vol. ed. S. Tax. University of Chicago Press, Chicago.

—— 1977. Evolution and The Genetics of Populations. Vol. 3. Experimental Results and Evolutionary Deductions. University of Chicago Press, Chicago.

—— 1978. Evolution and the Genetics of Populations. Vol. 4. Variability Within and Among Natural Populations. University of Chicago Press, Chicago.

Wright, S. and W. E. Kerr. 1954. Experimental studies of the distribution of gene frequencies in very small populations of Drosophila melanogaster. II. Bar. Evolution 8:225–240.

Wynne-Edwards, V. C. 1962. Animal Dispersion in Relation to Social Behaviour. Oliver and Boyd, Edinburgh and London.

Yablokov, A. V., A. S. Baranov, and A. S. Rozanov. 1980. Population structure, geographic variation, and microphylogenesis of the sand lizard (Lacerta agilis). Evolutionary Biology 12:91–127.

Yamakake, T. A. K. 1975. Cytological studies of maize (Zea mays L.) and teosinte (Zea mexicana Schrader Kuntze) in relation to their origin and evolution. Massachusetts Agric. Expt. Station Bull. 635.

Zeuner, F. E. 1963. A History of Domesticated Animals. Hutchinson, London.

Zeveloff, S. I. and M. S. Boyce. 1980. Parental investment and mating systems in mammals. Evolution 34:973–982.

Zirkle, C. 1946. The early history of the idea of the inheritance of acquired characters and of pangenesis. Trans. Amer. Philosoph. Soc. 35:91–151.

Organism Index

Prepared by Glorieux Dougherty

Phlox, 269; *P. drummondii*, 30, 124; *P. glaberrima*, 269; *P. pilosa*, 60, 63, 81, 82, 269

Phoenicopteriformes, 323

Phytoplankton, 223

Piciformes, 323

Pig, 351

Pigeons, 111, 323, 352, 387

Pines, 130, 241; arboreal form, 130; elfinwood form, 130; hard-seeded, 310; white, 202; yellow, 202

Pinus, 228 *P. albicaulis*, 130, 201; *P. attenuata*, 211; *P. murrayana*, 201; *P. radiata*, 211

Pipevine swallowtail (*Battus philenor*), 229

Pipilo erythrophthalmus, 30

Placental mammals, 398

Placentals, 332; archaic meat-eating (Creodonta), 315

Placoderms, 281

Plankton, 308

Plant flagellates (Euglenoids), 5

Plants, 98, 100n1, 174, 314, 339, 349, 359; adaptation, 6; allopolyploid speciation, 252-53; angiospermous, 213; animal-disseminated, 310; aquatic, 281; autogamous, 207; balance polymorphism, 131; biotic sympatry, 267; bird-pollinated, 59; competition, 121; defense mechanisms, 382; desert annual, 377; diploid, 46; disruptive selection, 129-30; domestication, 111; drift, 156; ecological niche, 227; enzyme polymorphism, 29; evolutionary trends, 284-99; evolution of, 16; experimental transformation, 182-83; frequency-dependent selection, 123-24; hybrid breakdown, 216-17; hybrid speciation, 253-55; hybrid sterility, 214-16; insect-pollinated, 59, 332; interracial character differences, 51; interspecific competition, 220; introgression, 236; isolation, 211; large-bodied, 228, 279, 280; length of generations, 311; living fossils, 304; marine, 332; natural hybridization, 245; neighborhood size, 79-81; origin of, 9; paedomorphic character, 371; paedomorphosis, 363; phenodeviants, 248; phenotypic plasticity, 175-76; progress, 376; quantum speciation, 250-52; racial variation, 233-34; reproductive isolation, 238; reticulate evolution, 333; seed, 56, 376; selection, 92-93; self-fertilizing,

75; sessile, 59-60, 63, 65, 67; speciational shifts, 305; speciational trends, 342-43; sun leaves/shade leaves, 176, 177; sympatric speciation, 256-57; terrestrial, 281, 335, 384-86, 397-98, 399-400; vascular, 377, 378-79; wind-dispersed, 310; woody, 201; *see also* Flowering plants; Higher plants; Leguminous plants; Lower plants

Platyrrhini, 408

Plesiosaurs, 399

Plethodon glutinosus, 220; *P. jordani*, 220

Plethodont salamanders, 219

Pliohippus, 287, 288, 289, 290, 293, 302

Plovers, 323

Pneumococcus, 178, 179

Podicipitiformes, 323, 324

Polemonium (Polemoniaceae), 228-29, 317, 343; *P. carneum*, 343; *P. caerulem*, 343; *P. californicum-delicatum* group, 343; *P. eximium* group, 343; *P. pulcherrimum*, 343

Polygala monticola brizoides (Polygalaceae), 212; *P. vauthieri*, 212

Polyprion, 366

Pomacanthus, 366

Pongidae, 408, 409; sensu lato, 409; sensu stricto, 409

Pongids, 412

Pongo (orangutan), 408; *P. pygmaeus*, 420

Porpoises, 11

Potentilla glandulosa (Rosaceae), 167

Predators, 376, 380, 382, 383, 417, 421

Presbytis, 418; *P. obscurus*, 420

Primates, 30, 281, 299, 363, 426; altruism, 146; biological aspects, 407-15; duration of life stages, 420; encephalization quotients, 420; intelligence, 411-12, 413, 419-21; language, 425, 431; learning ability, 433; non-human, 416-21, 422, 434; social aspects, 416-27; social group, 430-32

Primroses, 248

Primula, 337

Proboscidians, 285

Procecidochares, 201, 257

Procellariiformes, 323

Proconsul, 409

Prokaryotes, 184, 185, 187, 277, 278

Prosimians, 407-9, 416, 420, 422, 433

Protistans, 5, 403

Protitanotherium emarginatum, 367

Protohippus, 287, 293

Author Index

Prepared by Glorieux Dougherty

Subject Index